KEY ISSUES IN
CROSS-CULTURAL PSYCHOLOGY

KEY ISSUES IN CROSS-CULTURAL PSYCHOLOGY

*Selected Papers from the
Twelfth International Congress of the
International Association for Cross-Cultural Psychology
held in Pamplona-Iruña, Navarra, Spain*

Edited by
Hector Grad
Amalio Blanco
James Georgas

Published for the International Association
for Cross-Cultural Psychology

SWETS & ZEITLINGER PUBLISHERS

| LISSE | ABINGDON | EXTON (PA) | TOKYO |

Library of Congress Cataloging-in-Publication Data

International Association for Cross-Cultural Psychology.
  International Congress (12th : 1994 : Pamplona, Spain)
    Key issues in cross-cultural psychology : selected papers from the
  Twelfth International Congress of the International Association for
  Cross-Cultural Psychology held in Pamplona-Iruña, Navarra, Spain.
  July 24-27, 1994 / edited by Hector Grad, Amalio Blanco, James Georgas.
      p.   cm
  Published for the International Association for Cross-Cultural Psychology
    Includes bibliographical references.
    ISBN 9026514417
    1. Ethnopsychology--Congresses.   I. Grad, Hector, 1958, II. Blanco
  Abarca, Amalio. III. Georgas, James, 1934-
  IV. Title.
  GN502.I56   1996
  155.8--dc20   96-12891
                                                                    CIP

Copyright © 1996 Swets & Zeitlinger B.V., Lisse

All rights reserved. No part of this publication may be reproduced, stored in a retrieval system, or transmitted in any form or by any means, electronic, mechanical, photocopying, recording, or otherwise, without the prior written permission of the publisher

Printed in the Netherlands on acid free paper.

ISBN 90 265 1441 7

Contents

Prologue. H. Grad, A. Blanco, & J. Georgas	1
Presidential Address. R. S. Malpass Face Recognition at the Interface of Psychology, Law, and Culture	7

Part I. Conceptual and Methodological Issues

P. Boski Cross-Cultural Psychology at the Crossroads or: Lake Victoria is not Lake Mwanza, While Cross-Cultural Psychology is not Cultural (Enough)	25
A. C. Paranjpe The Interpretive Turn: Implications for Cross-Cultural Psychology	42
H. Helfrich Cross-Cultural Psychology in Germany	52
S. C. Carr Social Psychology and Culture: Reminders from Africa and Asia	68

Part II. Consequences of Acculturation

R. C. Mishra, D. Sinha, & J. W. Berry Cognitive Functioning in Relation to some Eco-Cultural and Acculturational Features of Tribal Groups of Bihar	89
A. Ostrowska & D. Bochenska Ethnic Stereotypes among Polish and German Silesians	102
J. Georgas & D. Papastylianou Acculturation and Ethnic Identity: The remigration of Ethnic Greeks to Greece	114
D. Hocoy Empirical Distinctiveness between Cognitive and Affective Elements of Ethnic Identity and Scales for their Measurement ..	128
C. Ward & A. Kennedy Before and After Cross-Cultural Transition: A Study of New Zealand Volunteers on Field Assignments	138

Part III. Cognitive Processes

J. R. J. Fontaine, Y. H. Poortinga, B. Setiadi, & S. S. Markam The Cognitive Structure of Emotions in Indonesia and the Netherlands: A preliminary report	159
A. Akande The Perception of Ability Scale for Students in Africa and New Zealand	172
M. Juan-Espinosa & A. Palacios Urban and Rural People's Conceptions of Intelligence in Equatorial Guinea	182

Part IV. Values

G. Hofstede, L. Kolman, O. Nicolescu, & I. Pajumaa Characteristics of the Ideal Job among Students in Eight Countries	199

G. Sagie & S. H. Schwartz National Differences in Value Consensus .. 217
K. Boehnke, M. P. Regmi, B. O. Richmond, S. Chandra, & C. Stromberg Worries, Values and Well-being: A Comparison of East and West German, Nepalese and Fijian Undergraduates. ... 227
R. H. van den Berg & N. Bleichrodt Personal Values and Acculturation 240
M. Varma Values in Indian Villages. 251

Part V. Social Psychology

J. Adamopoulos & A. Stogiannidou The Perception of Interpersonal Action: Culture-General and Culture-Specific Components 263
D. M. Keats & F. Fu-Xi The Development of Concepts of Fairness in Rewards in Chinese and Australian Children 276
M. Yuki & S. Yamaguchi Long-Term Equity within a Group: An Application of the Seniority Norm in Japan 288
W. G. Stephan, C. W. Stephan, M. Abalakina, V. Ageyev, A. Blanco, M. Bond, I. Saito, P. Turcinovic, & B. Wenzel Distinctiveness Effects in Intergroup Perceptions: An International Study 298
H. Y. Tsai Concept of "Mien Tzu" (Face) in East-Asian Societies: The Case of Taiwanese and Japanese 309

Part VI. Personality, Developmental Psychology, Health Psychology

E. Diener Subjective Well-Being in Cross-Cultural Perspective 319
B. Ataca, D. Sunar, & Ç. Kagitçibasi Variance in Fertility due to Sex-Related Differentiation in Child-Rearing Practices 331
G. Musitu Ochoa, J. Herrero Olaizola, & E. Garcia Fuster The Psychosocial Ecology of Child Maltreatment: A Cross-Cultural and Discriminant Analysis 344
P. H. Cook The Application of an Ecocultural Framework to the Study of Disability Programs in a Rural and Urban Indian Community . 355
E. Mc Auliffe AIDS: Barriers to Behavior Change in Malawi 371

Prologue

This book, *Key Issues in Cross-Cultural Psychology*, comprises a selection of papers presented at the Twelfth International Congress of the International Association for Cross-Cultural Psychology, July 24-27, 1994, in Pamplona-Iruña, Navarra, Spain. The Congress was held in the School of Health Sciences of the Universidad Publica de Navarra.

The setting of the Congress was, indeed, consistent with the aims of IACCP. Spanish culture springs from a rich diversity of ethnic, religious, and national sources, and this cultural richness was appropriately reflected by the site of the Congress. Pamplona is the capital of a region marked by the confluence of two ancient cultures with two languages that define an intercultural space propitious for social research.

Nevertheless, cross-cultural teaching and research are in fact underrepresented in Spanish academic reality. Cross-cultural psychology is seldom included as a specific field in the curriculum of psychological departments in Spain. On the one hand, many Spanish researchers are involved in international networks, and there is considerable research in areas we would define as cross-cultural, e.g., national identity, intergroup relations, language. On the other hand, this research is seldom identified and classified as "cross-cultural", but is perceived as an extension of the subject area, i.e., social psychology, developmental psychology, educational psychology. In this sense, the success of the Pamplona Congress signified a turning point in the status of cross-cultural psychology as an academic field in Spain.

The Congress was a stimulating meeting place for the presentation and discussion of approximately 300 contributions: 26 symposia, 1 round table discussion, 129 papers, 48 posters, and 4 workshops. Both the scientific program and lively discussions during the Congress suitably reflected the current issues and research trends in cross-cultural psychology.

Roy Malpass of the University of Texas at El Paso, USA, gave the Presidential Address, "Face Recognition in the Interface of Psychology, Law, and Culture". A Keynote Lecture on, "Subjective Well-Being in Cross-Cultural Perspective" was given by Ed Diener of the University of Illinois, USA. Janak Pandey of the University of Allahabad, Allahabad, India, the incoming President of IACCP, spoke on "Socio-Cultural Dimensions of Poverty" at the closing ceremony of the Congress.

The main goal in editing this volume was to offer an appropriate representation of this rich vitality in cross-cultural psychology. Ninety-four manuscripts were submitted, a very large number. The quality of the manuscripts was exceptionally high, and the comments of the external reviewers, two for each manuscript, were crucial to the editor's selection of contributions to the book. We would like to thank each of these anonymous reviewers for their careful attention to the manuscripts, and their thoughtful suggestions to us and to the authors. The result is evident in this volume, in which a platform has been given to cross-cultural researchers from all the continents. This outcome is now rendered for your consideration.

The selected papers are organized into six main Parts: I. Conceptual and Methodological Issues; II. Consequences of Acculturation, III; Cognitive Processes, IV; Values, V; Social Psychology; and VI. Personality, Developmental Psychology, Health Psychology.

Some of the critical issues and controversies of current cross-cultural psychology, which were discussed during the Congress, are undercurrents in the different chapters of this book. One is the current controversy regarding "cross-cultural" vs. "cultural" or "indigenous" psychology. The issue is directly addressed in the chapter by Boski. In addition, its methodology has been directly employed in many of the chapters of this book, characterized by the mono-cultural method, its emphasis on "indigenous" psychology, and on the detailed description of the historical and context variables of the culture. Some examples are the chapters by: Mishra, Sinha, & Berry; Juan-Espinosa & Palacios; Ostrowska & Bochenska; and Cook;

Part II reflects an area, acculturation, that is currently receiving considerable attention in cross-cultural psychology. We titled the section, "Consequences of Acculturation", in that different authors studied different aspects of acculturation processes. For example, Mishra, Singa, and Berry focused on differential cognitive functioning as a consequence of acculturation; Ostrowska and Bochenska; Hocoy; and Georgas and Papastylianou studied ethnic identity as part of the process of acculturation of migrants or ethnic groups; Ward and Kennedy studied changes in personality and behavior with sojourners.

The interest in values, Part IV, remains very strong in cross-cultural psychology. The chapter by Hofstede et al., focuses on Masculinity-Femininity, another aspect of his seminal work on Culture's Consequences which has not received the same attention from researchers as individualism-collectivism. Sagie & Schwartz is a further report on Schwartz's theory of values, which has created so much interest during the past few years. In the chapters by Boehnke et al.; Varma; and van den Berg and Bleichrodt, aspects of Schwartz's theory are explored.

This book was possible due to the support granted by the Colegio Oficial de Psicologos de Navarra to the organization of the Congress, and particularly by the enthusiastic and successful work of the Organizing Committee chaired by Camino Urrutia in Pamplona. We wish to acknowledge the coordination of the Scientific Committee by Amalio Blanco and Hector Grad, both from the Universidad Autonoma de Madrid.

The Congress was held under the auspices of the Executive Council of IACCP: President Roy S. Malpass; President-Elect Janak Pandey; Past-President Cigdem Kagitcibasi; Secretary-General Colleen Ward; Deputy Secretary-General Paul G. Schmitz; and Treasurer Deborah L. Best, whom we would like to sincerely thank for choosing Pamplona for the site of the Congress. We would particularly like to thank Paul Schmitz for his continual encouragement in the organization of the Congress.

We would also like to warmly thank the members of the Publications Committee: Peter B. Smith, John E. Williams, Fons J.R. van de Vijver, Ype Poortinga, John Adamopoulos, Walter J. Lonner, and Yoshihisa Kashima, for their helpful advice and continual support in the preparation of these Proceedings.

The editorial assistance for the preparation of the final manuscript was due to the dedication, expertise and hard work of the following graduate students of the Department of Psychology of the University of Athens. We would particularly like to thank: Vassilis Pavlopoulos, Angelika Mouyer, Tsabika Bafiti, Penny Panagiotopoulou, Alexandros Pantzopoulos, and Litsa Papadimou, for their help. Thanks also to Christina Mylonas. The critical comments and suggestions of Elias Besevegis and Nikos Giannitsas were, as always, extremely helpful.

Hector Grad and Amalio Blanco, Madrid
James Georgas, Athens

Presidential Address

Face Recognition at the Interface of Psychology, Law, and Culture

Roy S. Malpass
University of Texas at El Paso, El Paso, Texas, U.S.A.

A talk such as this one provides an opportunity to speak more broadly than in other places, and so I'll take advantage of that today. I want to talk about a problem I have been working on for many years, and on which I hope to work for many more. This problem has taken me from community contexts, to the experimental laboratory, to the legal system, and into cross-cultural research. I'll trace some of these threads, and take the time to comment more extensively on some of them. The focus is on a phenomenon, rather than a particular theory, and the many approaches, methods and perspectives encountered while attempting to understand it.

The phenomenon on which I have been working for 25 years has become known as the "cross-race" effect. This is simply the finding - reliable across cultures and decades - that observers are better able to recognize faces of their own group as compared with other, visibly distinct, groups. I will briefly summarize some things we know about the phenomenon, and emphasize some recent findings. Then I will discuss briefly the idea of Social Framework evidence in the legal process and how the cross-race effect might figure in this area. Finally, I will discuss some cross-cultural aspects of the phenomenon and the important theoretical questions it poses for psychology.

From Activism into the Laboratory

My first research experience was a project in the field in which a sociologist colleague and I were interested in structural and motivational correlates of economic dependency. We spent two field seasons with the MicMac, one of Maritime Canada's original peoples. I thought of myself as a cross-cultural psychologist. I thought that was the kind of research I would do for a lifetime. But as we learn from the introductory remarks in standard treatments of cross-cultural research methods, this is a research strategy, one to be applied when the problem at hand demands it.

After an interlude in graduate school, I found myself in the late 1960's, surrounded by the issues of the time. Another psychologist and I participated in a group of professionals working to do what we could for racial justice in our community, Champaign-Urbana, Illinois. At that time there were no African-Americans in either the Police or Fire Departments. The most commonly offered explanation was that they could not pass the state qualification examinations. So we helped to develop training materials for both the fire and police examinations. When we went to the Police Headquarters to view the police examination we got a surprise. The second item on the examination displayed 8 faces on a two-page spread. Accompanying each face was some written information, such as the person's name, home town, age, favorite offense, and so on. Persons taking the test were given 8 minutes to study this information, after which they were to turn to the next page where four of the eight faces were displayed. Test-takers were asked to fill in information about these four individuals. We were delighted to see that the faces were all white! For African-American applicants, the exam was discriminatory on the face of it (sorry)! I volunteered to go to the library and find the research literature to document this point. I could find no such literature. After graduate and undergraduate students I sent to find the literature was sure I had missed also came up empty handed, I could only conclude that no such literature existed. We did the first empirical study of the cross-race effect, beginning in the summer of 1968 (Malpass & Kravitz, 1969).

After 25 years and three or four dozen studies we know that the cross-race effect is stable across time and space. Considering black and white faces and subjects in the USA, there is about a 10%-14% lesser recognition for other-race faces, with black subjects showing a slightly smaller decrease for whites than whites do for blacks. Perhaps of equal interest, in a standard face recognition experiment there is about a 15% false identification rate for other-race faces.

But the cross-race effect has not been studied only for American blacks and whites, nor has it been restricted to standard recognition studies. It has been found in naturalistic studies of eyewitness identification, with blacks, white and Hispanics (Brigham, Maass, Snyder and Spaulding, 1982; Platz and Hosch, 1988). The effect has been found for Asian subjects and faces in Asia and North America, and for African and European faces in the UK and in Africa. So "they" look more alike for number of different kinds of "theys".

Comment on the Idea of "Race"

At this early point let me digress for a moment on the topic of "race". Race is a term of deceptive concreteness. It is an idea that can apparently be taken at face value (sorry). It is interesting that in the study of face perception and recognition we confront this peculiar concept. Now if our responsibilities for truth and to accurately reflect scientific knowledge could be set aside, and we could live as everyday individuals with little knowledge of the world outside our residential communities, ideas like race would cause little trouble. But what happens when a researcher says, in the method section of his paper, that she used 60 faces, 20 black, 20 Asian and 20 Caucasian? What a mixed metaphor! One set is identified with reference to a unidimensional category of physical appearance; another is identified with reference to global geography; and the third is identified by reference to a 19th century concept of race, based primarily on linguistic data - not physical appearance. None of these bases for categorization are appropriate for research on face recognition, or for describing contemporary human populations. Within any of these categories there are so many important variations on fundamental psychological and cultural dimensions that these categories are fundamentally useless. Someone brought up in England, a native speaker of English from one or another area of London, may have parents who came from Jamaica, Kenya, India, one or another Chinese community, Mediterranean Africa, or even Canada or the United States! And any of the latter may have had parents who came from Mexico, Guatemala, Poland, Italy, Ireland or Africa!

We can no longer use nation of origin, first language, physical appearance, or other cues as a code for culture. We have to be much more differentiated about things. In my studies, for example, how should I classify an American student whose great-grandfather was the son of a French man and a Native American woman, who married an Irish woman, and whose maternal grandparents were a female English missionary to China, married to a Chinese civil engineer? To my eyes this "subject" is very European in appearance, except for a hint of something different. I don't know whether it is his Huron great grandmother or his Chinese grandfather. But this really doesn't matter. What do I call him? Caucasian? If I want to assess his cross-race experience, do I have to ask about how many times he met his grandfather, and whether he still looks at the family photo album! Well, I don't have to ramble on about this, as I'm sure you see the point. So long as our questions are not very refined we won't have to worry. But the rate of increase of intercultural families is far larger than conventional thinking allows, and is growing. Soon we will no longer know what "culture" a person grew up in when we know they were born and spent all their life in the Netherlands! If we were more anthropological in our orientation there would be two decades of really interesting research that could follow this problem.

So, I will continue to refer to this as the "Cross-Race effect", and you will hopefully understand that when I do so I mean it with quotation marks around it, and you may even perceive a smile on my face at the futility of finding a better term. I use the term as a folk concept, not as a technical term, and if pressed I will not defend it.

We know that the Cross-Race effect is stable over time and populations, and has been found on four continents. But why does the cross-race effect occur? Over the years we have followed a strategy of systematically eliminating plausible rival hypotheses. Here are some "explanations" and the results of research into them.

Explanations of the Cross-Race Effect
   1. Cross-race differences in intragroup facial similarity. This ethnocentric hypothesis can be dismissed because across studies, blacks and whites have

approximately equal difficulty with each others faces, and approximately equal facility with faces of their own group.

2. Racial attitudes. It seems obvious that cross-racial attitudes could follow a similar and symmetric pattern across groups, and might be responsible for decreasing recognition. There are two problems with this: first, there is no satisfactory theory of how cross-race attitude would effect learning about other race faces, or how it would effect performance in a recognition experiment. Second, from the studies that have investigated this relationship, there is only occasional and marginal evidence for the relationship.

3. Differential linguistic coding. Malpass, Lavigueur & Weldon, (1973), proposed that there is a deficiency in the linguistic categories required to encode other-race faces in memory. Attempts to train subjects to recognize faces on the basis of verbal descriptions found that such verbal recognition increased with training, but that there was no corresponding effect on face recognition. Subsequently, students of eyewitness identification have had forced upon them the idea that visual recognition of faces contains hardly any verbal components, and that verbal descriptions of faces have little to do with the faces being described. All verbal descriptions of faces appear to be inadequate, regardless of the race of the face, and it appears that face recognition does not depend upon verbal encoding anyway.

4. Differential encoding - the modern version. Chance & Goldstein (1982) suggested that other-race faces might be encoded in a more superficial (less inferential) manner than own-race faces. Devine and Malpass (1985) found that whether subjects were given inferential instructions (judge the face's friendliness) or no instructions, recognition was at its maximum level, and the cross-race effect was found. When subjects were given superficial instructions (judge the face's race) recognition performance was reduced, but it was reduced approximately equally for both own and other race faces. Again differential encoding cannot serve as an explanation of the cross-race effect.

5. Extension to other methodologies. With the exception of the naturalistic studies of Brigham et al. (1982) and Platz & Hosch (1988) laboratory research on the cross-race effect had been narrowly identified with a classic recognition paradigm in which subjects are shown a series of faces for one or two seconds each, and then asked to view a series containing these and an additional set of new faces, responding by saying which they saw previously. Malpass, Erskine and Vaughn (1988) investigated a related cognitive task. They placed a matrix of face photographs, 50% black, 50% white, on a table. At the end of the table, facing 90 degrees away, was a lectern on which was placed a stack of face photographs containing half of those on the table and an equal number of new faces. Subjects were asked to turn up each face in turn and decide whether its' duplicate was among those on the table. If it was they were to pick up the photo from the lectern and place it on its match. If it was not, they were to pick up the photo and place it in a box. The time it took to reach a decision (from turning up the face to picking it up) was recorded, as was the number of times the subject looked at the face on the lectern. Subjects took more time to make decisions for faces that were not in the matrix on the table and also for other-race faces. More interestingly, however, they looked more often at other-race faces on the lectern than own race faces. Subjects had more difficulty retaining information about other-race faces from the lectern to the table, only a turn of the head away.

On the basis of the evidence reviewed so far, it is not clear what antecedent, or developmental processes are responsible for the effect. Appropriate data are needed to point our way.

6. Experience is implicated. Social experience is everyone's favorite explanation for the cross-race effect, but the evidence is very sparse. The first attempt to show an experiential base for the cross-race effect was in the first empirical study of this phenomenon (Malpass and Kravitz, 1969). There was little theoretical development of the experience notion at that time. Here's what our theory was like: The more other-race individuals a person is aware of, the more the person will learn to differentiate among them. This theory lacks depth. But it certainly has conviction, and it has survived to the present with hardly any change (Ng & Lindsay, 1994). Like our findings from the 1960's these investigators found no relationship between their measures of social experience and cross-race facial recognition - even though the cross-race effect was strongly replicated. Kravitz and I asked subjects many questions about their cross-race experience, including the number in their school at various times of their lives, the number who were friends, whose home they went into and who came to the subject's home. We asked about how many other-persons filled various roles in the subject's community, how many were known by name, and whether this was reciprocal. We apparently were far more desperate to find such a relationship than Ng & Lindsay (1994), because we used canonical correlation methods to attempt to identify any linear combination of experience variables with any linear combination of face recognition variables. In this we failed, as have subsequent studies. Chance & Goldstein (in press) make a case for the experience hypothesis based on presumptions of differential experience. In a series of studies they developed at least circumstantial evidence for the experience hypothesis. For example, Chance, Goldstein, and Andersen (1986) studied college aged subjects' recognition for faces of adults and of infants. While they found a difference for these two kinds of faces, they also found a correlation between reported experience with infants and recognition for infant faces. The supposition is, of course, that this is an analog of the cross-race effect. Surely the social context is different.

This is where we stood until recently on the explanation of the cross-race effect. It's not because of inherent properties of the stimulus faces, such as certain groups actually being more homogeneous. It is *NOT* because of racial attitudes, at least not in any simple way. It has little if anything to do with the verbal codes we use, partly because there is little relationship between verbal processes and face recognition in the first place, and partly because people appear to encode faces in an enriched, meaningful manner regardless of their race. And the evidence that social experience itself is involved is very weak - nearly nonexistent.

7. Face Recognition and racially ambiguous features.

The IdentiKit-2 is a collection of acetate transparencies, about 4 x 6 inches, each one containing the image of a component of human faces (eyes, nose, mouth, chin, hair, etc.). Using the IdentiKit 2 it is possible to categorize the kit's eyes, noses and mouths as uniquely European-American, uniquely African-American, or acceptable as either of these. The latter we termed racially ambiguous features. We asked judges from both groups to help us select features for these categories. Using a computer based image processing program we constructed faces from each set. Those faces constructed from racially ambiguous features were made "black" by including an African-American hair image and facial texture. They were made "white" by using European-American hair and no facial texture. We did three studies: the first used our standard face photographs and again found the cross-race effect in the subject population we would use for the remaining experiments. The second study found the cross-race effect using black and white faces constructed from the IdentiKit-2 features. The third used "black" and "white" faces to study the cross-race effect. Student judges of both groups agreed that the resulting faces (in their appropriate "white" or "black" guises) were acceptable as members of their respective groups. The study was constructed so that across subjects each face appeared equally often in its "white" and "black" guise. Two possible outcomes are of interest.

If the cross-race effect were found, the most obvious interpretation would be that, since the facial features were actually identical between the "white" and "black" faces, the identification of a face as an "own" or "other" race face has an impact on the cognitive processing of the remaining facial information. The presence of racial markers (skin tone, hair) could invoke some out-group processing strategy, causing the inter-individual differentiation of outgroupers to be less than ingroupers. On the other hand, were the cross-race effect not found, experience with facial features would be implicated, since the effect would disappear when the feature differences are taken away.

We found that the cross-race effect did not appear when the "black" and "white" faces shared the same features. Our interpretation is that face recognition actually does depend upon processing of facial features, and that when the features in two groups of facial stimuli are equal the recognition differential disappears.

Now we have more data to point the way to future studies. And the data point in the direction of the social experience that observers bring with them concerning the somewhat different feature sets which appear in faces of different "racial" groups.

Re-Entering the Legal Context

Research on the cross-race effect began with practical concerns of social justice, and at least the anticipation of its figuring in legal action. It has turned out quite differently. The initial issue on which we were willing to go to court never got there, as the question was settled by the time the research was done. But the cross-race effect has had many days in court. The resistance of this finding to explanation has not hindered the interest of defense attorneys in using testimony about deficits in cross-race face recognition to increase doubt in the minds of jurors about the accuracy of a cross-race identification. A small number of us who have done the primary research on this problem have given testimony in court, in a handful of cases for each of us. Others have also offered testimony on this topic, and I am told that there are some who frequently testify on the cross-race effect.

Social Framework Evidence

The cross-race effect fits well within an emerging category of evidence offered to the court: Social Framework evidence. The term was introduced by Monahan & Walker (1988). It is in part a response to the discrepancy between the understanding of the finders of fact - judges or juries - and scientific knowledge. Usually the focus is on facts which are contested and about which the judge or jury will have to decide. There are many areas of testimony that could be listed under this heading, but I'll draw three examples from a very interesting paper by Vidmar & Schuller (1989).

The Battered-Woman Syndrome

The question here is whether a woman bears co-responsibility for her battering by not having left the relationship. Another important component is whether after having tolerated a battering relationship over many episodes she forfeits the right to claim self defense if she does attack the batterer, especially if her attack does not come in direct response to an attack on her. Testimony can be developed to inform the jury of several pertinent facts: that there is a complex of actions that are frequently seen in such cases; that "battered women" are an identifiable class of persons; that such women do not feel free to leave the relationship; and that a reasonable perception of imminent threat of harm can occur in the absence of an overt attack. Thus jurors can be provided with a framework for understanding the complexities of this syndrome.

The Rape Trauma syndrome

If a defendant concedes that sexual intercourse occurred, but argues that it was consensual, the jury is left with the problem of deciding whether intercourse was forced. If the victim has displayed components of the rape-trauma syndrome, testimony on this point may assist the jury in their evaluation. Again, jurors can be provided with a framework for their considerations.

Whether the Testimony of an Eyewitness Ought to be Doubted

Social framework evidence in the area of eyewitness identification is somewhat different, as it is often aimed at assisting jurors to evaluate the performance of eyewitnesses from a human factors point of view. Here are a few examples. Each is associated with a moderately large research literature.

(a) It is commonly believed that the confidence with which an eyewitness identification is made can be used as an indication of the probable accuracy of the identification. The research literature finds there is no practical relationship between confidence and accuracy.

(b) It is commonly believed that memory for the face of an offender is enhanced by such dramatic events as the brandishing of a weapon. The research literature, however, supports the opposite view - that witnesses are more likely to focus their attention on the weapon, to the detriment of their ability to describe and to identify the offenders face.

Cross-Cultural Applications of Social Framework Testimony

These examples represent social science contributions to our understanding of social and personal processes that form a context for the decision making of judges and juries. There are other examples that are more clearly cultural in nature.

South African Murder Trials: SARHWU, and the Queenstown Six

The South African courts employed the "doctrine of common purpose" to expand prosecution of persons associated with murders associated with crowd behavior. I'll describe two examples, drawn from Colman (1991a, 1991b).

The case of *Sibisi and others* (1989) involved the murder of 4 black men who had refused to join others in a strike. The strike was a particularly bitter one by members of the South African Railway and Harbor Workers Unions (SARHWU) against the railway over the unjustified firing of one of their members. Large numbers of people gathered at the Union headquarters in subsequent days, and finally the police shot dead six of the strikers. On the following payday strikers found that they had also been fired, and subsequently five nonstrikers were kidnapped from their work sites by an angry group of fired strikers. Many people were involved. One of the five escaped, but the others were killed by a mob. Eight persons were charged with the murder, although only 3 or 4 had participated in the actual killings. The others were convicted on grounds of the common purpose doctrine. Four were sentenced to death, although this sentence was later lifted on the basis of psychological evidence given in extenuation.

The case of *Gqeba and Others* (1990) - the Queenstown Six - followed a conflict between Xhosa-speaking ANC supporters who had sponsored a boycott of merchants in Queenstown, and police units loyal to the Zulu Inkatha leader Chief Buthelezi. During the conflict Inkatha police killed 11 people among a large number who were gathered at a church to discuss the boycott. On the day following the funeral a group of young activists sought to punish a young woman who was accused of being an informant, and of sleeping with an Inkatha policeman. After a flogging, the mob grew to more than 200 people. She was taken to a central area of the township where a car tire was placed around her neck, doused with gasoline, and set alight. The mob sang and danced while she burned to death. Six members of the mob were convicted of the murder, through the doctrine of common purpose. Apparently none of them had taken part in the actual killing. Five of these men were sentenced to death, but due to technical issues the trial was set aside. These same persons were later retried and sentenced to 60 months in prison, with 40 months suspended. The difference was the introduction of social psychological evidence in the second trial.

Psychological evidence - social framework evidence - was offered in the extenuation phase of these trials. It was based on an interrelated set of social psychological principles, including deindividuation and related processes (Diener, 1980), conformity and obedience (Asch, 1956; Milgram, 1974; Tanford and Penrod, 1984), and bystander apathy, among other social psychological concepts and findings. Colman (1991a, 1991b) argued that the conditions of crowding, and the continuous singing and dancing contributed to deindividuation, and enhanced obedience and conformity pressures. The court was warned to not underestimate the importance of external, situational pressures and overestimate the importance of internal motives - the fundamental attribution error (Ross, 1977).

The Cross-Race Effect as Social Framework Evidence

The cross-race effect is another instance of social framework evidence. Over 70% of experts surveyed by Kassin, Ellsworth & Smith (1989) agreed that a cross-race bias exists, for white subjects. At one level, it is simply a human performance question, similar to many other aspects of eyewitness behavior. But as soon as one asks whether the witness's social experience has created a condition where recognition memory for one of more cultural groups is impaired, the problem becomes social framework evidence in a different way: it involves the social history of the witness, and becomes a question of the witness's culture.

The demands of the legal context are different from scientific psychology. While a well implemented scientific account might involve describing a comprehensive set of interrelationships, the law can (and does) ask much more. It can ask us to go beyond our generalized account and address the specifics of individual cases, or individual assessments, such as my clinical colleagues provide. So our consideration of the experience hypothesis will proceed from this perspective. Differences between research done for the usual scientific purposes and research done for the purposes of testimony are quite apparent. For example, in the area of cross-race face recognition, I know of no paper in the literature where it is reported what percentage of subjects showed own race recognition that was greater than other race recognition, whether a "raw" report, or using some confidence interval. And the inferential statistics we use to detect influences of one variable on another are *FAR* away from the concrete level of questions that might be asked of us in the specifics of a particular case.

There are some who might say that to deal in specifics is not the job of those offering social framework evidence, and that generalities are all that are called for. My own view is that generalities are what one settles for when one cannot responsibly be specific. In a different legal tradition, in Germany, for example, scientific psychologists are called upon regularly to give testimony, and to address the individual case to the extent that they are able, based on the knowledge available.

So, what are the demands of social framework testimony in the area of the cross-race effect? What does one need to know? I'd like now to begin slipping into the future, to what I see as some possible trajectories for work on this interesting and slippery problem.

1. Is the effect stable?

Yes, it seems to be. Chance and Goldstein (1996) cite it as one of the most easily replicated effects in all of psychology.

2. Is the effect preserved in realistic settings, beyond the laboratory?

Yes, it seems to be. Brigham et al. (1982) and Platz and Hosch (1988) both did studies in convenience store settings and found the effect. Platz and Hosch found that it extended to Mexican-Americans and Mexican faces as well.

3. Does it lead to errors of identification in every identification setting?

To answer this I need to identify some different expectations for different settings. Recall the matching study I reported earlier, where subjects had difficulty retaining information for making facial image matches. Try to imagine that each face represents not a specific object but instead represents a location in a complex space containing all possible faces and transformations of those faces. Since a face may be different from moment to moment, or day to day, the same face in its various transformations will occupy a region in this space. Similar faces may overlap with this region, and we may confuse these faces. We can think of a sphere around each location in this system, and within this sphere we cannot differentiate faces from each other in memory - they're too similar. I like to call this the "sphere of confusion". It may be useful to think of the memory images of cross-race faces as involving a larger sphere of confusion than own-race faces. Now let's consider various identification settings.

Mugshot searches. From the sphere of confusion notion, this ought to be the most dangerous setting. The witness has to compare an internally held image with a large sphere of confusion to individual images. It seems likely that for other-race faces, more in the mug-book will seem similar to the internal image than will be the case with own-race faces, where the sphere of confusion is smaller.

Simultaneous lineups. The task here is to differentiate one person from a set of people presented for identification. Any single face will stand out less. With a large sphere of confusion, it is more likely that any given pair of faces can *not* be easily differentiated, and so there would likely be fewer identifications, and fewer false identifications than with smaller spheres of confusion, as in own-race identifications.

Sequential lineups. The task here is more like a mugbook, and from this perspective seems to be a task in which cross-race identifications may be more likely to be made in error. How the absolute rates would compare with simultaneous lineups is a different, and interesting question.

Show-ups. The reasoning here is similar to mugbooks - although recent research suggests a different result.

For some of these there are one or two studies, but not a basis for a stable idea of how the cross-race comparison will come out.

Who will be the better cross-race face recognizers? It would be useful to know which individuals will be subject to a cross-race deficit and which will not. At the present time, one cannot make a case based on good data that there is *ANY* person attribute that can be used to make such a judgment. Moreover if you look at any cross-race recognition data-set you will find subjects who are actually better at recognizing other race faces than own race faces. We have not yet devised a way to differentiate these individuals from those with the opposite pattern.

Diagnosing Experience - a Matter of Culture

For the final exercise of this presentation, I'd like to revisit the experience question. This really is the heart of the cultural background of the cross-race effect. It is underdeveloped both theoretically and empirically, and there are many interesting question that can be asked. There is more to do here than a platoon of investigators could do in their entire careers.

A discussion of cross-cultural social experience and its relationship to differentiating among own- and other- "race" individuals could be interesting enough simply as a problem in social-cognitive psychology. It is also of importance in the law at more than one level. First, it is of importance to defendants who are the subject of a cross-race identification, and second it is of interest as a matter of justice in a complex multi-racial society.

I want to begin by examining some possible models of social experience as it might be related to cross-race face recognition.

1. Proximity model: Proximity is weighted by frequency. The more other-race individuals are near you, and the nearer they are, the better they should be recognized. This model is just a foil, as it comments not at all on the motivational base, or incentive system that might lie behind people making or failing to make distinctions among individuals.

Our initial notions about social experience embodied the simple assumption that the more other-race persons who passed before ones eyes the better able one would be to differentiate among them. There are both theoretical and empirical reasons to move away from this simple theory. Empirically, it is because all attempts to use this approach in studying an experience - recognition relationship have failed- and more than half a dozen have tried. Theoretically, we know that there is a tendency to categorize out-group persons into fewer categories than own-group persons. If they are more alike socially - that is if for purposes of social interaction there is less demand to differentiate among them - then why should we differentiate among them perceptually? We need another model that takes account of social interaction and its consequences.

2. Utility model: This model is more complex. Its claim is that greater other-race recognition will occur the greater the variance of the distribution of the value of other-group social interaction outcomes (Malpass, 1990). Perhaps you did not swallow that whole the first time, so let me explain. I assume that every social interaction we have with persons has a social value associated with it. We can therefore think of a distribution of values associated with each person with whom we interact, and with each category of persons with whom we interact. We can conceptualize a distribution of social interaction outcomes characterizing the history of one's experience with a particular group.

For people in complex urban social systems, the "others" in their social environment will rank from neutral, or unimportant, to highly positive and highly negative. Probably those who are unimportant or neutral are those we see but with whom we do not interact. Among those with whom we do interact, some are bearers of positive resources in our social exchange with them, and others are bearers of negative resources. Possibly the most salient persons are those who have the capacity to be either positive or negative in their interactions. Recently in South Africa I was waiting with a friend for an elevator in the corporate offices of a major South Africa corporation. Next to us was young white man, obviously as new to his job as he was to his blue suit. The elevator door opened and displayed nine White and seven Black men in suits. My companion leaned towards me and said, "This kid has got a real problem. He's got to figure out which one of these people he works for!" In that environment the incentives for the young man to learn how to differentiate among people from the other group ought to be substantial.

Now if there were markers that would help to identify those who are unimportant they could be safely ignored. Those who are at least potentially important are another matter, however, and there is real value in being able to differentiate among them. The reason is that unless there are reliable cues to assist you in knowing who are the good guys and the bad guys, you just have to do it on the basis of recognizing them as individuals, and recalling the history of your interactions with them. This was solved in the Saturday matinee "Westerns" watched by children of my generation: the bad guys wore black hats, and the good guys rode white horses! But like many other things about those "movies", real life isn't much like that.

By this reasoning, diversity in the quality of social outcomes associated with interaction with group members should cause us to differentiate among them as individuals. This is why we are so very good at it for persons within our own social ingroups. And this also is why we are so bad at it for certain other groups. But that is too broad a statement. It is not the groups that is the issue, but the special circumstances in which we interact with their members. If they are all dangerous, there is little need to differentiate among them, except for a small number of "friendlies". If they are all inconsequential, there is likewise little gained by learning to differentiate among them.

Now we are faced with the problem of how to investigate cross-race social experience. There are, it seems to me, two major possibilities. The first is to construct a questionnaire to measure the number and quality of the respondent's cross-race social experience. There are at least three difficulties with this approach. First, the history of such attempts is not encouraging. Second, such a questionnaire is likely to be detailed, and thus long and complex. Third, people may not have clear memories for the aspects of experience in which we have an interest. An alternative would be to take a more culturally based approach. This would involve identifying interaction settings that are diverse with respect to other-race interaction, identifying long- and short- term participants and measuring their recognition.

This is a demanding direction for investigation because it requires one to develop a theory of the kinds of social experience that should have an impact on cross-race recognition and to develop means of identifying the social interaction settings that fulfill the conditions of the theory. I find this an attractive direction for continued research on the cross-race effect because it takes me closer to where I started, and closer to the kinds of cultural studies that I had in mind 30 years ago when I found a real affinity for the culturally sensitive aspects of human behavior.

social interaction settings that fulfill the conditions of the theory. I find this an attractive direction for continued research on the cross-race effect because it takes me closer to where I started, and closer to the kinds of cultural studies that I had in mind 30 years ago when I found a real affinity for the culturally sensitive aspects of human behavior.

References

Asch, S. (1956). Studies of independence and conformity: I. Minority of one against a unanimous majority. *Psychological Monographs, 70* (Whole No. 416).
Brigham, J.C., Maass, A., Snyder, D., & Spaulding, K. (1982). The accuracy of eyewitnesses in a field setting. *The Journal of Personality and Social Psychology, 42,* 673-678.
Chance, J.E., & Goldstein, A.G. (1982). *Depth of processing and recognition of own- and other-faces.* Paper presented at the meetings of the Midwestern Psychological Association, Minneapolis, Minn.
Chance, J.E., Goldstein, A.G., & Andersen, B. (1986). Recognition memory for infant faces: An analog of the other-race effect. *Bulletin of the Psychonomic Society, 24,* 257-260.
Chance, J.E., & Goldstein, A.G. (1996). The Other-Race Effect and Eyewitness Identification. In S. Sporer, R.S. Malpass, & G. Koehnken (Eds.), *Psychological issues in eyewitness identification.* Hillsdale, NJ: Lawrence Erlbaum.
Colman, A. (1991a). Crowd psychology in South African murder trials. *American Psychologist, 46,* 1071-1079.
Colman, A. (1991b). Psychological evidence in South African murder trials. *The Psychologist, 4,* 482-486
Devine, P.G., & Malpass, R.S. (1985). Orienting strategies in differential face recognition. *Personality and Social Psychology Bulletin, 11,* 33-40.
Diener, E. (1980). Deindividuation: The absence of self awareness and self regulation in group members. In P. Paulus (Ed.), *The Psychology of Group Influence.* Hillsdale, New Jersey: Lawrence Erlbaum.
Kassin, S.M., Ellsworth, P.C., & Smith, V. (1989). The "general acceptance" of psychological research on eyewitness testimony. *American Psychologist, 44,* 1089-1098.
Malpass, R.S. (1990). An excursion into utilitarian analysis. *Behavior Science Research, 24,* 1-15.
Malpass, R.S, Erskine, D.W., & Vaughn, L.L. (1988). *Matching own- and other-race faces.* Paper presented at the meetings of the Eastern Psychological Association, Buffalo, N. Y.
Malpass, R.S., & Kravitz, J. (1969). Recognition for faces of own and other "race". *Journal of Personality and Social Psychology, 13,* 333-334.
Malpass, R.S., Lavigueur, H., & Weldon, D. E. (1973). Verbal and visual training in face recognition. *Perception and Psychophysics, 14,* 330-334.
Milgram, S. (1974). *Obedience to authority.* New York: Harper & Row.
Monahan, J., & Walker, L. (1988). Social science research in law: A new paradigm. *American Psychologist, 43,* 465-474.

Ng, W., & Lindsay, R.C.L. (1994). Cross-race facial recognition: Failure of the contact hypothesis. *Journal of Cross-Cultural Psychology, 25,* 217-232.

Platz, S. J., & Hosch, H. M. (1988). Cross-racial/ethnic eyewitness identification: A field study. *Journal of Applied Social Psychology, 18,* 972-984.

Ross, L. (1977). The intuitive psychologist and his shortcomings: Distortions in the attribution process. In L. Berkowitz (Ed.), *Advances in Experimental Social Psychology* (Vol. 10, pp. 173-220). New York: Academic Press.

S. v. Sibisi and Others, 1989, Case No. 187/87, Supreme Court of South Africa, Witwatersrand Local Division.

S. v. Gqeba and Others, 1990, Case No. 53/89, Supreme Court of South Africa, Eastern Cape Local Division.

Tanford, S., & Penrod. S. (1984). Social influence model: A formal integration of research on majority and minority influence processes. *Psychological Bulletin, 95,* 189-225.

Vidmar, N. J., & Schuller, R.A. (1989). Juries and expert evidence: Social framework testimony. *Law and Contemporary Problems, 52,* 133-176.

Part I

Conceptual and Methodological Issues

Cross-Cultural Psychology at the Crossroads or: Lake Victoria is not Lake Mwanza, while Cross-Cultural Psychology is not Cultural (Enough)

Pawel Boski
Institute of Psychology, Polish Academy of Sciences, Warsaw, Poland

For the past two years an unusually interesting and fruitful discussion has been going on in the pages of the *Cross-Cultural Psychology Bulletin* about the name, content and purpose of our discipline: "Who are we and what makes (or should make) our work distinct in relation to the broader family of psychological sciences?" Inspired by this discussion and my own thoughts on this problem of professional identity, I convened a symposium at the 12th IACCP Congress in Pamplona, *Cross-cultural psychology at the cross-roads: Comparative international studies or psychological analyses of the cultures*[1] to further discuss these issues. Similar symposia were simultaneously organized at the 23rd IAAP Congress in Madrid and a year later for the IVth European Congress of Psychology in Athens. Judging from all these recent and ongoing events, the renewed concern about the basics reflects the state of our collective professional consciousness rather than idiosyncrasies of few individual minds.

The paper has two aims: 1) to provide a critical appraisal of current cross-cultural psychology, which - in my view - is insufficiently cultural; and 2) to offer a positive conceptual framework for a postulated model of cultural psychology, combined with empirical examples which illustrate my ideas.

There is Not Enough Culture in Cross-Cultural Psychology

In this section I will attempt to show conceptual and empirical evidence supporting the above thesis; arguments will also be presented that the status of the discipline suffers because of this deficit.

Back to the Semantic Basics: Why CROSS-Cultural Psychology?

Let me start with a naive question: "Why the hyphenated *Cross-* and not simply *Cultural* Psychology?" Historically speaking, the question should be addressed to *JCCP* and *IACCP*'s founding fathers: "Did you consider, around

---

[1] The speakers at the symposium were (in alphabetical order): P. Boski, L. Eckensberger, G. Hofstede, C. Kagitcibasi, U. Kim, W. Lonner, Y. H. Poortinga, and H. Triandis. The present paper elaborates my personal ideas, but its scope has been broadened by the contributions from my distinguished colleagues and CCPB discussants. I also wish to thank John Berry, Geert Hofstede and Harry Triandis, as well for anonymous reviewers for their comments to the draft versions of the paper.

25 years ago, any terminological option other than *CROSS-cultural*, or was this one consciously agreed upon as the most adequate name to the goals which were then outlined? At this point, there is no answer to this question in Lonner's (1994) *JCCP* anniversary reminiscences.

Let's look around our neighboring disciplines. I can not see any *cross-cultural* sociology or anthropology, although both of them are well established academic disciplines without the hyphen. Or consider such semantic stimulus: *cross-social psychology*. To me it simply does not sound good, neither in English nor in Polish.

Theoretically, there is a persuasive argument that one should first establish the more general (basic) level of analysis, e.g., *cultural*, before going to the more specific (subordinate) levels. Thus, we have a (a) general *PSYCHOLOGY*, next (b) *developmental, social or clinical psychology,* and only later (c) can they become, say, *cognitive-developmental, cognitive-social, or clinical-developmental psychologies*.

Let's have another set of examples. We have a growing number of international associations, some of them obviously familiar to our members: *Political, Economic, etc. Psychology*. International comparisons are not rarities in their workings; if they are not a rule yet, such studies will definitely become the rule in the future. And yet, these disciplines do not bear anything like *Cross-, Trans-*, hyphenated name tags (e.g., Cross-Political Psychology). Moreover, these new emerging disciplines do not aspire to be cross-cultural; international will suffice, both for membership and for the type of work being done. Thus, our founding fathers took two steps at one time (which is a well understood youthful reaction), believing that it was feasible to put culture(s) and international comparisons on the same plate.

My intuition is that the initial emphasis on defining our discipline was on its comparative/methodological character, but since the term *comparative* had already been reserved for the inter-species analyses and *international* appeared in the names of associations, *CROSS-* became the agreeable option of the day. The position that our discipline bore mainly methodological character at birth is shared by other authors (see, Triandis, 1990; Eckensberger, 1994). But the hyphenated double meaning contributed, in practice, to a semantic confusion which has been more and more with us: To compare psychological phenomena across (above!) various social (national, ethnic, cultural) groups or to study in-depth the cultural context and determinants of psychological processes? - That is the question.

Vague or Peripheral Image of Cross-Cultural Psychology in Psychology

There hardly is any situation more embarrassing for a scientific discipline (as well as for an individual scientist) than being considered marginal or peripheral. And yet, evidence exists that things have been like this with cross-cultural psychology. Lonner (1989) found few and poorly representative references to cross-cultural research in American introductory textbooks. Recently, Zimbardo observed in the forword to Triandis's Culture and Social Behavior (1994): "Yet, the work of cross-cultural psychologists has remained on the periphery of general psychology, given short shrift in introductory texts and even in many social psychology texts. [...] The Zeitgeist is finally right for the emergence of cross-cultural psychology into the mainstream of social and general psychology" (pp. xii-xiii). The first part of this statement reflects the status quo; the other expresses a hope which is socially desirable in this type of text.

Rather than externalizing the blame for our relative absence in psychological literature, it may be more productive to turn the critical-attributional attention inward. I would like to share my editorial experience with the preceding volume of the IACCP proceedings, Journeys into Cross-Cultural Psychology (Bouvy, van de Vijver, Boski, & Schmitz, 1994). In reviewing the manuscripts, we found among many submitting authors, two types of conceptual ambiguities about the criteria that a standard text in an IACCP publication should meet.

The first type concerns studies considered cross-cultural, if it derives from a single non-Western culture, and assumes that its results should be automatically compared with the already existing mainstream findings. A number of such papers by psychologists from the *majority world* (i.e., developing countries) were submitted for consideration to the Liege conference volume. The problem with such studies is not that they are *monocultural*, but that do not make any convincing claim for the operation of identifiable cultural variables.

The second type consists of using the term *cross-cultural* as equivalent to *comparative/international*. Thus a personality-temperament researcher may want to broaden the scope of his work on basic factorial dimensions and decides to administer translated versions of the original questionnaire for obtaining data from other national samples. The only cultural concern of such research is to ascertain an equivalent back-translation of the research instrument.

In both cases, either an off-mainstream single culture study without in depth cultural analysis or a comparative international replication, the outcome is similar - peripheral - because it lacks creative discovery of new knowledge.

Let's focus on JCCP and examine, more systematically, the content of works published in our flag journal.

The 25th Anniversary Review of JCCP Confirms the Thesis: There is Little of Culture in Comparative Cross-Cultural Research.

The silver anniversary issue of JCCP reveals some empirical data about the type of research that has been published under the name of cross-cultural psychology. I would like to draw the reader's attention to the following facts reported in Ongel & Smith (1994) paper: (a) 593 (93%) papers in their sample were classified as imposed *etic* (i.e., fundamentally acultural); (b) 97% of the published papers reported studies where culture was treated as a causal variable rather than a context "within which specified variables interact with each other"; (c) "88% studied samples of individuals with no reference to the cultural context by which they are influenced" (1994, p. 46). Lonner concurs with the latter by saying that "there have been too many occasions where the (sample) description has been incomplete" (1994, p. 16).

Coupled with other evidence on North-American (or Anglophone) dominance in authors' nationality, education, and current academic address, the above meta-analysis points to one clear conclusion: Cross-cultural psychology has become an academic discipline conducted by WASP researchers and their followers, who are interested in testing the limits of the mainstream theories (imposed etic) around the world by using various national/ethnic samples, erroneously believing that cultural causation can be obtained this way. The mistake of a standard *JCCP* text consists of taking "empty labels" of national samples for cultural variables or of treating culture in a naive, amateurish perspective, added as a post-hoc speculative ornament at the end of Discussion section. The above conclusion coincides with Schweder's (1990) criticism of cross-cultural psychology as another version of Platonic universalist ideas of the mind; but similar are some of our *ingroup* judgments: take Hofstede's (1990) text in which he claims that there has been no culture in a standard cross-cultural text, other than culture = country or Ss' nationality. More recent arguments by Betancourt and Lopez (1993) follow the same lines.

In this context, Lonner's (1994) and Ongel & Smith's (1994) calls for more detailed sample description, and for more emic or derived etic theoretical approach will remain cries in the wilderness until a deeper process in the paradigmatic thinking takes place in our minds. So far, a standard JCCP-style practice of calling *culture* "independent variable" or "causal factor" is a misnomer, since neither is this "variable" conceptually defined nor measured and manipulated. It is exactly the same type of erroneous or shorthand expression, at best, which we would attribute to someone treating *gender* or *age* as causal agents.

Some Examples of No-Cultural Cross-Cultural Studies

Even the fact that the statistical evidence provided by Lonner (1994) and Ongel & Smith (1994) is so abundant, it still makes sense to select some illustrations to document my conclusions on the "acultural nature of a standard JCCP text.

The special JCCP issue with Amir & Sharon's (1987) replication studies in the area of person perception, attribution, attitudes, interpersonal attraction and group dynamics, offers an example often referred to in the literature. Their main conclusion was that the original main effects were easier to demonstrate in Israel, than were the interactions. Pepitone & Triandis (1987) wrote a critical commentary arguing that interaction effects are by necessity less robust and more context dependent than main effects. Yet, what is of most importance here is the fact, that culture was completely absent in Amir & Sharon's work; neither did it appear at the conceptualization stage nor in the interpretation of their results. Interactions (or lack of them) were not seen through the "filter" provided by cultural contexts: American (mainstream) vs Israeli.

In the 1993(4) issue, Domino and Regmi (1993), *Attitudes toward cancer - A cross-cultural comparison of Nepalese and U.S. students*, culture(s) enter the picture in the discussion section after the authors have reported the finding that, with their Cancer Metaphors Test, "Nepalese students see cancer both more pessimistically and more optimistically than do United States students". They offer a post-hoc free association type of thinking that because of Nepalese being Hindu or Buddhist, and reincarnation beliefs held in these religions, strong bipolar emotions can be generated with them, which is not the case with the U.S. subjects whose background must have been predominantly Judeo-Christian (p. 395). The result was neither predicted from any cultural analysis nor is the post hoc interpretation convincing, i.e., reincarnation beliefs > bipolar affects.

In the silver anniversary issue of JCCP, (Schaufeli & Janczur 1994), the cultural references which might justify a comparison between the two countries are limited to a list of conditions making life in the Netherlands (for nurses and everyone) easier than it is in Poland (pp. 98-99). Should the editor accept this as a legitimate cultural difference justifying publication in *JCCP*? The results confirmed this rather common-sense hypothesis: all the indices of burnout were higher among Polish nurses; the interpretation remained the same as was the introduction: because the life in Poland is tougher than it is in the Netherlands. We have thus a thesis documenting that economic and technological-organizational development are related to stress and burnout but it tells us nothing about cultural variables or context.

JCCP publishes more studies conducted entirely in America on student samples of different ethnic background. In Yan & Gaier's (1994) paper, *Causal attributions for college success and failure (An Asian-American comparison)*, little is known about the seven Asian samples except that as foreign students they were fluent in English and given test materials in that language. The results showed differences in terms of importance of ability attribution, considered as more important by American than by Asian students; which was later interpreted along the Individualism-Collectivism lines (p. 154). And although the importance of this cultural dimension can not be overestimated, I believe it becomes all too easy to treat it as a good-for-all explanatory device.

Most of the above examples are selected papers from recent issues of JCCP. The reader may be more surprised if a similar critique of insufficiently cultural content is raised against such a leading research theme as studies of acculturation (Berry, 1994). If we examine Berry, Kim, Power, Young, and

Bujaki's (1989) original scales of acculturative tactics or the more recent formulations in Schmitz's (1994) work, we will not find any culture dimensions there. Here are two examples of Schmitz's Integration and Assimilation items: "This country would be a better place if we immigrants would keep our own way of life alive "(I); "Immigrants should forget their cultural background and adapt"(A).

The studies were carried out in Germany but otherwise we do not know *to which country, to which immigrants* and, what is the most important, *to which way of life* these questions addressed. How can acculturation be studied if we wipe out culture from our research tools?

As the above mentioned review by Ongel and Smith shows, almost all of our published papers are conceptualized within the framework of one mainstream psychological theory or another and without any a priori reference to cultural variables. With continuation of this still dominant approach, which Berry calls *transport and test* (Berry, Poortinga, Segall & Dasen, 1992, p. 3), and I would name *pseudo-cultural*, we are doomed to the marginality status of replications.

Fortunately, positive alternatives exist. One of them is a *two-stage universalist model* in planning and executing a research project. It will be exemplified below with studies on human values. Another alternative is the author's *three-stage culture specific model*, which will outlined later in this paper.

The Two Stage Model: Studies on Universal Dimensions and Structure of Human Values

Stage 1 in this model consists of exploring variables of objective or subjective cultures, with the attempt to arriving at a world-map of cultures: *What type(s) of C?* only after this is done, can stage two be initiated, which consists of theorizing and asking empirical questions about psychological consequences of these cultural types or dimensions: $C > P$.

When answering what objective culture (C_o) is, anthropologists list ecology, family/kinship structure, subsistence economy, etc. HRAF represent the source of this kind of archive information. Berry's eco-cultural model (1976, 1979) is perhaps the best example of studying $C_o > P$ relationships: How do eco-cultural factors affect social conformity and abilities adaptive to a specific ecological niche? Linking the subsistence of hunting-gathering vs. sedentary agriculturalists with their dominant cognitive styles, i.e., psychological differentiation, has been a remarkable achievement in our discipline.

More recently, subjective cultures (C_s) have been investigated by psychologists themselves. These works are largely equated with the studies of values and associated with the names of Hofstede (1980, 1991), Triandis (1990, 1994), Schwartz (1992, 1994b). Here stage 1 consists of the attempt to describe/diagnose cultures in psychological terms, which is by itself a formidable task. So far we have the four dimensions postulated by Hofstede and seven culture-level value types documented by Schwartz.

Stage 2 in this model consists of drawing hypotheses, regarding how these various value differences affect the multiplicity of psychological processes and findings, reported in the mainstream literature. Triandis (1994, pp. 167-172) has an impressive list of differences between collectivist and individualist cultures in the functioning of Self, perception of ingroups and outgroups, the regulatory role of attitudes vs. norms, attributions, and emotional-motivational differences.

Schwartz and his collaborators (Schwartz, 1994a; Huismans, 1994) have proposed a more complex pattern of hypotheses by unfolding the circular value structure and relating the sine curve to other variables. For instance, readiness for social contact with outgroup members was predicted to vary with value types in a sinusoid way from highest (positive) for Universalism through lowest (negative) for Conservation. The pattern of correlations confirmed these predictions. Huismans (1994) successfully employed the same "curve unfolding" method to relate the importance of value types to religiosity.

The logic of the Two Stage process of cultural-psychological research is demonstrated below with hypothetical examples, anchored in the study of values.

Acculturation

It has been shown that cultures B and C differ from culture A (the host country) in the following way: C shows emphasis on conservatism and hierarchy; B stresses mastery; while A is high on autonomy and equality. Thus, cultural distance between C and A is greater than between B and A. Consequently, in their adaptation efforts to culture A, individuals from culture C will experience more of acculturative stress than immigrants from culture B. Also, Assimilation or Integration strategies will be more favored with the B --> A than C --> A change while Separation will be more preferred with the C --> A move.

Rather than proceeding straight to the measurement of acculturative attitudes (Berry et al., 1989) or other measures of acculturation, a potential researcher delays such activities (to step 2) until his hypothesis has been formulated at the level of psychological analysis of cultures involved (step 1).

Burnout among Nurses

Another hypothetical example of a similar reasoning, refers to the earlier mentioned study on burnout among nurses. Since nursing is a profession which epitomizes the values of femininity or benevolence, it is hypothesized that, all other things being equal (e.g., technological-organizational advancement, salaries), nurses should have more work satisfaction and less burnout in countries where cultures are high on these values, i.e., consistent with the role-profile of nursing.

Unlike the Schauffeli and Janczur (1994) study with their Dutch-Polish comparison of convenience, the model approach starts with search for genuine cultural difference which may be related to psychological phenomenon of burnout in a particular profession.

A Three Stage Model: Studies on Humanism-Materialism in Polish Culture and its Consequences

In this section another model of cultural psychology will be presented; its main difference with the previous consists of *emic* rather than universalistic departure point, regarding values or other cultural syndromes.

Culture-Specific Value Dimensions; Measuring Humanism-Materialism

Although they have provided a major breakthrough in cross-cultural psychology, studies of universal value dimensions are not the only and last word in the progress of our discipline. Even if we endeavor to measure Individualism-Collectivism, Masculinity-Femininity, Power and a host of other value-dimensions worldwide, and the result may be useful for some broad comparisons (similar to economic indices, like GNP), the sense and understanding of cultural specificity will still be missed.

Paradoxically, the universalist approach does not assume any prior cultural knowledge of the phenomena studied. This is clearly seen in the study of values. Terms denoting values in the Rokeach-Schwartz survey, such as *pleasure, wealth, healthy, wisdom*, etc, are as free from any cultural context as a person from the Euro-American tradition can only think of.

By contrast, the essence of the first stage in the model of cultural psychology is an extraction of *emic* or indigenous values (syndromes), which can only be done by intrinsic knowledge of one's culture of origin or settlement. Syndromes like: Greek *philotimo* (Triandis, & Vassiliou, 1972; Georgas, 1993); Mexican-Hispanic *simpatia* (Triandis, Marin, Lisansky & Betancourt, 1984; Marin, 1994) or Chinese *Confucian dynamism* (Bond, 1987; Hofstede, 1991) would not be discovered without this kind of expertise. Coming to my own cultural roots, *romanticism-positivism* has been the leading theme in arts and literature, political thought and mentality of Poles, since the early 20th century. Roughly, it corresponds to an idealist, dream-like vs. sober/realistic approach to social life; it can not be reduced though, to any of the currently proposed value dimensions. I have no empirical measure of *romanticism-positivism*, yet in some of my recent works, a related construct of *humanism* has been proposed as a core element of Polish subjective culture (Boski, 1993; 1994b).

Humanism-Materialism was derived from *Emic Culture Value Questionnaire* (Boski, 1994b). ECVQ was used to study individual value preferences as well as the construals of national-cultural prototypes: Polish and Canadian or American. From seven research samples, where Ss were Poles living in Warsaw; Polish immigrants to Canada and the U.S.; and Polish-Americans in these two countries (Boski, 1990, 1992, 1994a), separate and combined factor analyses generated identical solutions, whereby the first value-dimension was called *humanism-materialism*. Materialist items are, e.g., *I am happy with myself and the world when I see my bank account grow; It has been my life ideal to be a business-person, someone associated with big money-making;* and Humanist are e.g., *I often offer people my selfless sympathy and my helpful hand; In my heart I believe and in practice I show that people of all races and nationalities are equal and should be treated equally,*

Figure 1. Humanism-Materialism in Self and Cultural Prototypes

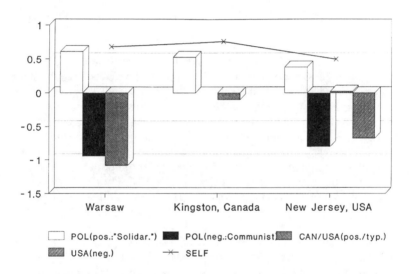

Materialist items are associated with *business* and *money making*, while the *humanist* end embraces a wider range of prosocial concerns about social injustice and the well-being of individuals as well as groups. The internal consistency level scale of the scale is *alpha* = .87.

How distinctive is *humanism* for Polish culture in comparison to some other national cultures, when all are measured at the level of prototype construction? A summary of results is reported in Figure 1.

In all three conditions, Polish prototype (culture-ideal) is rated high in *humanism* (and so is Self). Polish anti-culture prototype, which in the years

1989/90 was still an active representative of the now defunct communist regime, was located at the opposite, *materialist* end. Finally, American/Canadian prototypes were perceived as highly materialist in Warsaw (even more than the negative Pole), with an attenuating tendency among immigrants and Polish-Americans. Yet their contrasts with the humanist values of the Polish ideal remained statistically significant (for all contrasts, $p < .001$). We may conclude then, that *humanism* turned out to be a central value dimension of Polish subjective culture (or mentality); *materialism*, on the other hand appeared as its antithesis.

The objective origin of humanism has been traced to catholicism, neglect of pragmatic-economic concerns, history of freedom fighting against oppression, and ethos of nobility-turned-intelligentsia (see Boski, 1994b for a more detailed analysis).

Integrating the Culture Specific and Universal: Humanism-Materialism vs. Other Value-Dimensions

Questions can be raised in the second stage as to the position of this emic-derived dimension vis-a-vis those value types that have acquired the status of universality. Of all value dimensions currently studied by cross-cultural psychologists, Humanism seems to be closest to Hofstede's Femininity (and Materialism to Masculinity). This conclusion can be drawn from inspection of Hofstede's questionnaire items and from his comparisons of the Feminine and Masculine cultures. In Feminine cultures, dominant values are caring for others and preservation; people and warm relationships are important; sympathy is for the weak; the needy should be helped; welfare society ideal; stress is on equality, solidarity and quality of work life. (Hofstede, 1991, p. 96, table 4.2, and p. 103, table 4.3). Also, females score higher than males on both dimensions.

Next is the question of possible relationships between Humanism and Individualism-Collectivism. From inspection of the HUM-MAT scale and analysis of objective Polish culture (Boski, 1994b) it is unlikely that the two constructs converge. Thus, to integrate them, an orthogonal conceptual framework has been postulated. The theoretical scheme is based on the taxonomy displayed in Figure 2.

The framework partly derives from Greenwald (1982), and particularly from his observations that a well know phenomenon of *deindividuation* can assume a pattern of alienation in anonymous crowd or a format of "melting" within group uniformity. A similar argument can be extended to individuation, where a self-directed person may strive either for predominantly individual interests or be motivated by concerns for other people's (community) well-being.

The original IND/COL distinction represents, in my view, a confounding of these two dimensions, such that Individualism = Individuation *or*

Desociation, while Collectivism = Deindividuation *or* Sociation (see, Schwartz, 1990, for a similar type of critique). By deconfounding the sense of agency (*I* or *not-I*) from value-objects (*Me* or *Others*), we have conceptual space of four separate mentality types. Thus, Humanism shares some elements common with Collectivism and some with Individualism; with the former it is the social embeddedness of value-orientation, with the latter it is self-directedness in pursuing these goals. With the conceptual integration of Humanism with Individualism-Collectivism, the goal of *derived etic* (Berry *et al.*, 1992, p. 234) has additionally been accomplished.

Testing the empirical consequences: From Humanism and other value-types to representation of political democracy

The four value-type model (Figure 2) has been tested within a broader research project on the understanding and evaluation of political democracy. As argued by Reykowski (1994), individualism should be positively related to the principles of political democracy, while collectivism (considered as communist legacy) should be an impediment to democracy in the post-communist transformation. Humanism, which earlier appeared to characterize an anti-communist evaluation in cultural prototypes (Figure 1), should also be a positive predictor of democratic political thinking.

Figure 2. A Taxonomy of Value Orientations: Humanism and its Alternatives

	Humanism (Social involvement)	Collectivism (Ingroup conformity)
Sociation		
Unsociation	Individualism (Self-interest)	Alienation (Marginalization)

Individuation (Agency)   Deindividuation

The project has been a longitudinal study conducted in five waves on a convenience sample of 160 teachers, bank employees, city councilors, and skilled laborers from public and private industrial sectors of Warsaw and Lodz, the two largest cities in Poland[1]. Humanism was measured at culture and individual levels with the items reported earlier. Sample items of *Collectivism* were: *I am at distance with others until we have decided to call each other by our first names; I feel comfortably with others only if I have come to know them as the palm of my hand*. *Individualism* was measured with items like: *I wish I lived in a house full of people: relatives and visiting guests;* (reversed) *I tend to trust written documents (contracts) rather than words given by others.*

Political alienation was measured after Korzeniowski (1993) with the following sample items: *One can never be sure whether the decisions made by political authorities are correct and appropriate or not; It is frequently difficult to determine whether the political situation in Poland is changing for better or for worse; Ordinary people have no influence on decisions made by politicians.*

The concept of *democracy* was measured with a task consisting of sorting its normative markers (e.g., free elections, freedom of expression, equality in law, protection of minority rights, etc) from other, unrelated statements (e.g., market full of goods affordable to everyone, safety on our streets and communities, etc).

The measures of Humanism were found to be correlated with the three other normative dimensions[2]. The best test of discriminant validity is provided by comparing their levels in five social categories of our Ss. As Figure 3 shows, Humanism was consistently high (through three completed waves of the project) among categories *working with people*, i.e., City councilors and teachers, while those employed in private sector of economy, i.e., physical laborers or educated bank employees, had significantly lower scores on measures of Humanism.

Figure 3. Humanism in Five Social Categories and Three Waves

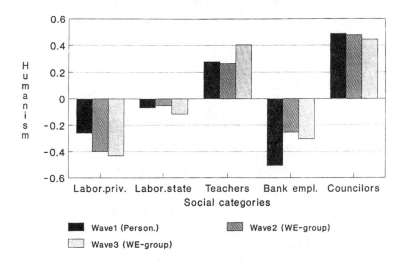

The picture is very different with Collectivism and Alienation measures (Figure 4). Here, more educated City Councilors and Bank employees had the lowest scores, in contrast to less educated laborers, whatever their job sector.

It is possible then, for two categories of people to have equally anti-collectivist, self-directed orientation and at the same time opposite scores in Humanism; Bank employees and city councilors exemplify this situation.

Figure 4. Collectivism and Alienation in Five Social Categories, Wave 2 and 3

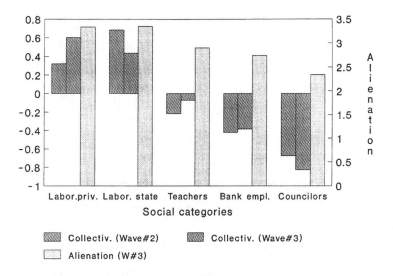

Finally, Figure 5 summarizes results of hierarchical regression analyses where the four value types and social categories (dummies) were predictors of Understanding Democracy. The results are clearly in favor of theoretical expectations. All four value types contributed significantly, and in the predicted direction, to explaining the variance of the dependent variable. Humanism, in particular, is related to better understanding and higher endorsement of political democracy.

General Conclusion

The intent of the above empirical section has been to demonstrate an approach to conducting cultural-psychological research, rather than to report a whole study for its own sake. To do so, the three stage model has been implemented and exemplified: (a) From culture level of analysis and discovery of *Humanism* as an *emic* value dimension in Polish culture; through (b) Integrating this dimension into the broader theoretical framework of value dimensions (*derived etic*); to (c) Testing of its consequences in one selected domain of political cognition and evaluation.

Figure 5. Predictors of Understanding Democracy. Wave 1 & 2 Predictors, Wave 2 Dependent

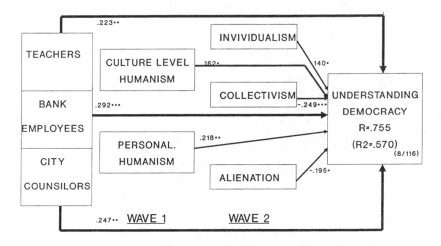

I regard the approach here advocated as an addition to the works by Hofstede, Triandis and Schwartz, all of whom have propagated a universalist, two stage model of cross-cultural psychology.

In closing, I wish to urge the reader not to forget the critical tone with which I have started this paper. I found (not unlike some other authors) that many of the published works make no serious reference to culture at all and are limited to comparing results from any number of national samples, instead. This type of *pseudo*-(cross)cultural psychology will hopefully be on decline as we are entering the second 25th period of *JCCP* and a similar anniversary of the IACCP will be coming soon.

And finally, a comment is needed to justify the subtitle of this paper. It comes as a reaction to Segall's (1993) CCPB article where he discusses the cultural vs. cross-cultural controversy: "Finally, I ask, what's in a name? Is it cultural psychology or cross-cultural? ... Is it Lake Victoria or Lake Mwanza? Does it matter?" (p 4). Sure, it matters, as it was already argued by Davidson (1994). It could be a nice cultural psychological experiment to ask British and Kenyan/Ugandan Ss of their social representations of *Lake Mwanza* and of *Lake Victoria*. My expectation is that the differences between the objects and the interaction with Ss' ethnicity would be significant; and I also hope that most (cross-) cultural psychologists would endorse this version of Sapir-Whorf hypothesis. Similarly, Leningrad is not St.Petersburg (in Russia, not in Florida); and so Gdansk is not Danzing. The list of such examples is endless.

Cultural psychologists object to this approach and insist on taking the indigenous meaning of the lake as well as that of intelligence and many other concepts, central to psychological inquiry.

References

Amir, Y., & Sharon, I. (1987). Are social psychological laws cross-culturally valid? *Journal of Cross-Cultural Psychology, 18*, 383-470.

Berry, J.W. (1976). *Human ecology and cognitive style: Comparative studies in cultural and psychological adaptation.* London: Sage.

Berry, J.W. (1979). A cultural ecology of social behavior. In L. Berkowitz (Ed), *Advances in experimental social psychology* (Vol. 12, pp. 177-206).

Berry, J.W. (1994). Acculturation and psychological adaptation: An overview. In A.M. Bouvy, F.J.R. van de Vijver, P. Boski, & P. Schmitz (Eds.), *Journeys into cross-cultural psychology* (pp. 129-141). Lisse: Swets & Zeitlinger.

Berry, J.W., Kim, U., Power, S., Young, M., & Bujaki, M. (1989). Acculturation attitudes in plural societies. *Applied Psychology, 38*, 185-206.

Berry, J.W., Poortinga, Y.H., Segall, M.H., & Dasen, P.R. (1992). *Cross-cultural psychology.* Cambridge: Cambridge University Press.

Betancourt, H., & Lopez, S.R. (1993). The study of culture, ethnicity and race in American psychology. *American Psychologist, 48*, 629-637.

Bond, M. (1987). Chinese values and the search for culture-free dimensions of culture. *Journal of Cross-Cultural Psychology, 18*, 143-164

Boski, P. (1990). Correlative national self-identity of Polish immigrants in Canada and the United States. In N. Bleichrodt & P.J.D. Drenth (Eds.), *Contemporary issues in cross-cultural psychology* (pp. 207-216). Lisse: Swets and Zeitlinger.

Boski, P. (1992). In the homeland and in the chosen-land: National self-identity and well-being of Poles in Poland and in America. In S. Iwawaki, Y. Kashima, & K. Leung (Eds.) *Innovations in cross-cultural psychology* (pp. 199-213). Lisse: Swets & Zeitlinger.

Boski, P. (1993). Between West and East: Humanistic values and concerns in Polish psychology. In U. Kim & J.W. Berry (Eds.), *Indigenous psychologies.* Newbury Park, CA: Sage.

Boski, P. (1994a). Psychological acculturation via identity dynamics: Consequences for subjective well-being. In A.M. Bouvy, F.J.R. van de Vijver, P. Boski, & P. Schmitz (Eds.), *Journeys into cross-cultural psychology* (pp. 197-215). Lisse: Swets & Zeitlinger.

Boski, P. (1994b). Psychological analysis of a culture: Stability of core values among Poles in the homeland, immigrants and Polish-Americans. *Polish Psychological Bulletin, 25.*

Davidson, G. (1994). Cultural, cross-cultural or intercultural? Comments on Lonner, Dasen and Segall. *Cross-Cultural Psychology Bulletin, 28*, 1-4.

Domino, G., & Regmi, M.P. (1993). Attitudes toward cancer: A cross-cultural comparison of Nepalese and U.S. students. *Journal of Cross-Cultural Psychology*, 24, 389-398.
Eckensberger, L.H. (1994). On the social psychology of cross-cultural research. In A.M. Bouvy, F.J.R. van de Vijver, P. Boski, & P. Schmitz (Eds.), *Journeys into cross-cultural psychology* (pp. 197-215). Lisse: Swets & Zeitlinger.
Georgas, J. (1993). Ecological-social model of Greek psychology. In U. Kim & J.W. Berry (Eds.), *Indigenous psychologies*. Newbury Park: Sage.
Greenwald, A.G. (1982). Is anybody in charge: Personalysis versus the principle of personal unity. In W.J. Suls (Ed.), *Psychological perspectives on the self* (vol. 1, pp. 151-184). Hillsdale, NJ: Erlbaum.
Hofstede, G. (1980). *Culture's consequences: International differences in work-related values*. Beverly Hills: Sage.
Hofstede, G. (1990). Empirical models of cultural differences. In N. Bleichrodt & P.J.D. Drenth (Eds.) *Contemporary issues in cross-cultural psychology* (pp. 4-20). Lisse: Swets and Zeitlinger.
Hofstede, G. (1991). *Cultures and organizations: Software of the mind*. London: McGraw-Hill.
Huismans, S. (1994). The impact of differences in religion on the relation between religiosity and values. In A.M. Bouvy, F.J.R. van de Vijver, P. Boski, & P. Schmitz (Eds.), *Journeys into cross-cultural psychology* (pp. 255-267). Lisse: Swets & Zeitlinger.
Korzeniowski, K. (1993). Is it possible to build a democracy in Poland? A psychological analysis of threats. *Polish Psychological Bulletin*, 24, 109-120.
Lonner, W.J. (1989). The introductory psychology text: Beyond Ekman, Whorf biased IQ tests. In D.M. Keats, D. Munro, & L. Mann (Eds.), *Heterogeneity in cross-cultural psychology*. Amsterdam: Swets and Zeitlinger.
Lonner, W.J. (1994). Reflections on 25 years of JCCP. *Journal of Cross-Cultural Psychology*, 25, 8-24.
Marin, G. (1994). The experience of being a Hispanic in the United States. In W.J. Lonner & R. Malpass (Eds.), *Psychology and culture*. Boston: Allyn & Bacon.
Ongel, U., & Smith P.B.(1994). Who we are and where are we going: JCCP approaches its 100th issue. *Journal of Cross-Cultural Psychology*, 25, 25-53.
Pepitone, A., & Triandis, H.C. (1987). On the universality of social psychological theories. *Journal of Cross-Cultural Psychology*, 18, 471-498.
Reykowski, J. (1994). Collectivism-individualism and the demise of communism. In U. Kim, G. Yoon, & H.C. Triandis (Eds.), *Individualism and collectivism*. Newbury Park, CA: Sage.
Schaufeli, W.B., & Janczur, B. (1994). Burnout among nurses: A Polish-Dutch comparison. *Journal of Cross-Cultural Psychology*, 25, 95-113.
Schmitz, P.G. (1994). Acculturation and adaptation processes among immigrants in Germany. In A.M. Bouvy, F.J.R. van de Vijver, P. Boski, & P.Schmitz (Eds.), *Journeys into cross-cultural psychology* (pp.142-157). Amsterdam: Swets & Zeitlinger.

Segall, M. (1993). Cultural Psychology: Reactions to some claims and assertions of dubious validity. *Cross Cultural Psychology Bulletin*, 27, 2-4.

Schwartz, S.H. (1990). Individualism-Collectivism: Critique and proposed refinements. *Journal of Cross-Cultural Psychology*, 21, 139-157.

Schwartz, S.H. (1992). Universals in the content and structure of values: Theoretical advances and empirical tests in 20 countries. In M. Zanna (Ed.), *Advances in experimental social psychology* (vol. 25, pp. 1-65). Orlando: Academic Press.

Schwartz, S.H. (1994a). Studying human values. In A.M. Bouvy, F.J.R. van de Vijver, P. Boski, & P. Schmitz (Eds.), *Journeys into cross-cultural psychology* (pp. 239-254). Lisse: Swets & Zeitlinger.

Schwartz, S.H. (1994b). Cultural dimensions of values: Toward an understanding of national differences. In U. Kim, G. Yoon, & H.C. Triandis (Eds.), *Individualism and collectivism*. Newbury Park, CA: Sage.

Schweder R. (1990). Cultural psychology: What is it? In J.W. Stigler, R.A. Schweder, & G. Herdt (Eds.), *Cultural psychology: Essays on comparative human development*. Cambridge: Cambridge University press.

Triandis, H.C. (1990). Cross-cultural studies of individualism-collectivism. In J.J. Berman (Ed.), *Nebraska Symposium on Motivation: Cross-cultural perspectives* (pp. 41-133). Lincoln: University of Nebraska Press.

Triandis, H.C. (1994). *Culture and social behavior*. New York: cGraw-Hill.

Triandis, H.C., & Vassiliou, V. (1972). Comparative analysis of subjective culture. In H.C. Triandis (Ed.), *The analysis of subjective culture*. New York: Wiley.

Triandis, H.C., Marin, G., Lisansky, J., & Betancourt, H. (1984). *Simpatia* as a cultural script of Hispanics. *Journal of Personality and Social Psychology*, 47, 1362-1375.

Yan, W., & Gaier, E.L. (1994). Causal attributions for college success and failure: An Asian-American comparison. *Journal of Cross-Cultural Psychology*, 25, 146-158.

Zimbardo, P.G. (1994). *Preface* to H.C. Triandis. *Culture and social behavior*. NY: McGraw-Hill.

Notes

1 J.Reykowski is the principal investigator in this KBN (Committee for Scientific Research) sponsored project. Five waves have been planned, the data here presented come from wave #1 and #2.

2 On the psychometric side, personal (individual) and cultural measures of humanism were negatively correlated with all the three remaining variables. The strongest were the correlations between personal humanmism and individualism, $r = -.237$, $p < .01$ and cultural humanism and collectivism, $r = -.258$, $p < .01$. Individualism and collectivism were orthogonal and they correlated positively with alienation: $r = .254$, $p < .01$ and $r = .317$, $p < .001$, respectively.

The Interpretive Turn: Implications for Cross-Cultural Psychology

Anand C. Paranjpe
Simon Fraser University, Burnaby, B.C., Canada

The Role of Interpretation in Science: Historical background

In my view, the current controversy over the role of interpretation is a new phase of a long, ongoing debate between empiricism and rationalism, the two rival epistemologies that have shaped Western thought since the Enlightenment. Francis Bacon (1620/1905), the founding father of British empiricism, explicitly used the metaphor of the mind as a mirror (p. 264). He suggested that if the mind is rid of all preconceptions - which he called the "Idols"-it would receive an accurate image of the world, which would amount to perfect knowledge. John Locke thought that mind is at birth a *tabula rasa*, and that experience or observation is both the source of all knowledge, as well as the only means to its validation. As noted by Charles Taylor (1980) the main thrust of modern science at its origin during the Enlightenment was to try to know the world as it is, regardless of what meaning or value it might have for an inquiring mind. This goal may be achieved, it was thought, if we stick to what is "given in experience," and avoid meanings, interpretations and values that are "added by the mind." This view sharply divides facts from values and asks us to leave the latter out of scientific inquiry. Values as well as interpretation and understanding were assigned to the field human sciences (Geisteswissenschaften), and the territories of natural and human sciences were sharply divided.

Writing in the nineteen fifties, Gordon Allport (1955, pp. 7-12) placed behaviorist psychology in the tradition of British empiricism. The behaviorists, he noted, see organisms as relatively passive creatures acted upon by environmental forces even as Locke saw the mind as a "passive receptacle receiving external engravings" (pp. 16-17). They viewed human behavior as governed by stimuli according to laws of nature just as Bacon suggested that all events in nature are the results or efficient causes i.e., forces external to objects, as opposed to intentions or Aristotle's "final causes" internal to persons as purposive agents. In his analysis Allport goes on to note that a dialectical response to Locke's vision was articulated by Leibniz. Allport speaks of a Leibnitzian tradition in psychology, a tradition in which he places phenomenology, gestalt, and some - but not all - varieties of cognitive psychology.

The key feature of the Leibnitzian vision is its rejection of both the Lockean notion of the emptiness of the mind, and also of the Baconian notion that only external forces cause all changes in the world. Leibniz insisted that only a mind that is attentive and prepared can have meaningful experience. This idea later evolved into Kant's thesis that scientific knowledge is impossible without "categories of the understanding," such as cause and effect, that are not "given in experience," but are added by a pre-equipped mind. Also, as noted by Allport (1955), the Leibnitzian tradition maintains that "the person is not a collection of acts, nor simply the locus of acts; the person is the *source* of acts" (p. 12, emphasis original). To put in another way, humans are not seen as passive recipients of external sources of information and causal influence, but as active meaning makers and initiators of action.

As is often recognized, neo-positivism of the Vienna Circle may be viewed as the twentieth century culmination of Locke's molecularist and reductionist view of knowledge. The logical empiricists hoped to reduce all knowledge to a vast array of propositions describing direct observations, all tightly connected by rules of Russell an Frege's symbolic logic. Carnap's (1938/1949) scheme of the unity of science aimed at a grand unification of science by reducing, i.e., translating, statements of all branches of knowledge in the language of physics.

As noted by Richard Rorty (1979, pp. 167-171), serious challenges to this vision came from within the British empiricist tradition itself. Willard Quine (1953/1964), a student of Bertrand Russell, pointed out that the statements of science must form a dynamic "field" where changes in any statement demanded by a new observation or argument necessitates changes elsewhere. This view echoes a holistic perspective similar to that of the Gestaltists, much *unlike* the molecularism of Lockeans and the logical positivists. Moreover, as argued by Quine (1953/ 1964), no statement of science should be immune from revision - or else claims to unshakable "foundations" of knowledge would result in dogmatism. Science must be an open system, open to new information and novel *interpretation*.

Cross-Cultural Psychology and the Search for Universals

To help understand the cross-cultural psychologists' search for psychological universals, it is necessary to explore the history of Western thought to a medieval era prior to Enlightenment. As noted by W.T. Jones (1969, Vol. 2, pp. 185-190) in his history of Western thought, the status of universals was the topic of a long and heated controversy during the middle ages. The issue of concern here is ontological, or about the nature and "existence" of the universals, as well as epistemological, i.e., about how they are known. The "realists," following Plato, considered universals such as "chair" or "dog" are *real* in the sense that they are imperishable essences, or generalized and rationally comprehensible "Forms" that are only approximately represented by particular, perishable chairs or dogs accessible to our senses. By sharp contrast, the nominalists held that the so-called universals are mere names, verbal labels used to designate all objects in a class. Conceptualists suggest a view different from both these; for them, universals are categories *constructed*-fabricated or invented - by the mind in trying to cope with an infinite variety of objects. Let us examine the implications for cross-cultural psychology of universals as meant in these three classical meanings of the term universal.

Let us *first* consider the nominalist position. If universals were mere labels, there would not be much of a point in looking for them in every culture. It seems to me that in contemporary psychology, Skinner's view comes closest to the nominalist position. Since words are mere labels, they are to be analyzed as mere emitted responses or "verbal behavior," not in terms of the meanings they may have for the speakers and listeners. Although Skinner (1953, p. 35) sometimes seemed to search for universal causal laws, his ideas of "functional analysis" of behavior follows the physicist Mach's view that bypasses the notion of causes, and treats variables as simply changeable values in an equation specifying quantitative relationships among symbols representing sets of observations. Also, Skinner refused to use statistical averages, and chose instead to predict behaviors of particular pigeons or rats by extrapolating learning curves of specific organisms, simply assuming that past trend will continue in the future. Most cross-cultural psychologists, who are commonly interested in comparing average scores of samples from different cultures, could hardly mean universals in the nominalist sense implied in Skinner's work. Nor would they be interested in predicting and controlling behaviors of particular individuals on the basis of the cumulative records of their past behaviors, for such focus on particulars is irrelevant for the search for cross-cultural universals.

The *second*, "realist" position assumes the reality of essences envisioned by reason, regardless of how, and even whether or not, they are observed. If "intelligence" is defined as a capacity to solve problems, or "field independence" as capacity to perceive accurately regardless of context, in this abstract sense they are real existents. As investigators committed to empirical methodology, most cross-cultural psychologists seem to care little about purely rational abstractions or Platonic essences. Their focus is often on empirical, measurement, and generalization. Take, for instance, Triandis's (1978) oft-quoted definition of a universal as, "a psychological process or relationship which occurs in all cultures" (p. 1), and his view (1993, p. 250) that "[t]he glory of science is generalization." What is implied here is not some rationally comprehensible concept, but a phenomenon or relationship that "occurs" in the sense that it is *observed* in *all* cultures.

Van de Vijver and Poortinga (1982) have pointed out, rightly in my opinion, the limitations of universals as matters of empirical observation and inductive generalization. First, they note, that "according to Triandis's definition, the occurrence of a phenomenon in at least one individual in each culture is sufficient condition for the universality of the phenomenon.

"...[T]his requirement is rather loose, and virtually all phenomena studied by psychologists are universals in this sense" (p. 388). Second, they point out that "[u]niversality is an absolute concept. It is meaningless to say that a phenomenon is more or less universal..." (p. 104). What they are suggesting, I think, are the limits of induction. As David Hume pointed out long time ago, empirical observations must be limited to a finite number of instances; and no matter how large the sample size might be, the generalization from *some* to *all* instances in any category must be a matter of guesswork of what might be true of the uncounted instances. Besides, the limits of culture are not finite; it is often hard to delineate cultures and subcultures from other cultures and subcultures. There can be no finite census of cultures either, for cultures are not static; they change over time, interact, subdivide, and integrate. So it makes little sense to try to find out with certainty something that occurs in *all* cultures, and at all times. Of course this is where statistical methods come into picture; they help us estimate the *degree* to which the observations from a sample may be *generalizable* across the universe of observation. It is perfectly sensible to rely on actuarial methods in cross-cultural psychology as in many other areas of life sciences. But this means abandoning aspirations for universality in an absolutist sense.

Van de Vijver and Poortinga (1982, p. 389) have pointed out, again rightly I believe, that the issue of great concern in the context of measurement is that of the *construct validity* of the measuring instruments. This brings us to the third, *conceptualist* view of universals. According to this view, universals are neither mere labels, nor reals in the sense of Platonic Forms or Kantian *noumena*, but rather concepts or *constructs*. Constructs are cognitive devises of making meaning or *interpreting* the world; they are important to the extent that they may influence behavior. The relevance of constructs follows from the phenomenological assumption, implied in the Leibnitzian tradition, that humans respond to the world in terms of the meaning they attach to objects, rather than their physical properties. Based on this fundamental premise, different varieties of constructionist theories have arisen. Some, like those of Piaget (1954) and Kelly (1955) emphasize the development and change of cognitive constructions within individuals, while the social constructionists like Harre (1987) and Gergen (1985) tend to emphasize the socially shared nature of constructs. Regardless of the many differences among them, consturctionist approaches emphasize the meanings we attach to events and objects around us. The same physical object may mean different things to different people, or to the same person at different times: a piece of chalk to write, or a projectile to throw, or a chemical to test or transform. What one does with it would differ according to the meaning assigned to it. Meanings and constructs may influence behavior of those who hold them, regardless of whether or not they occur universally, i.e., in all cultures around the world.

We may speak of cognitive constructs at two levels, at the level of the subjects studied by the psychologists, and at the level of psychologists. Both psychologists and their subjects employ shared constructs to help make sense of their world and respond to it in the light of the resulting understandings. Many subjects from India, Mexico, and China would understand the world respectively in terms of *karma*, *machismo*, and *filial piety*. The constructs common to specific cultures might have little or no meaning and relevance in other cultures. Likewise, constructs such as "anal retentive character," "formal operations," or "extroversion" have specific meanings in respectively Freudian, Piagetian and Eysenckian theories, but mean little, or may even be considered nonsense, from the viewpoint of proponents of other theories. Whether or not the phenomena described by such constructs "occur" in every culture around the world, they do make sense within the interpretive framework shared by subcultures of psychoanalytical, Piagetian or Eysenckian psychologists. To speak of their "cross-cultural validity" is of little or no value outside the interpretive frameworks. This brings us to the next issue, namely the absolutist, relativist, and universalist orientations in cross-cultural research.

Absolutism, Relativism and Universalism in Cross-Cultural Psychology

As noted by Jahoda (1983, p. 24) and Berry et al. (1992, p. 233), in cross-cultural psychology the focus is primarily on the measurement of variables and the relationships among variables, or on isolated cultural elements rather than cultural and conceptual systems as a whole. Consistent with such molecularist focus, cross-cultural psychologist tend to use the terms *emic* and *etic* to refer primarily, but not exclusively, to elements rather than systems as a whole. By contrast, in linguistics where the terms originated, emic refers to an entire approach or conceptual system that restricts itself to criteria relative to specific domain (a specific language, or by extension, culture), while etic refers to an approach that insists on absolute or universal criteria applicable everywhere. Against this background, although the idea of universality is most commonly viewed as applicable to specific variables, it is sometimes also used to an entire approach to psychological studies. For instance, psychoanalysis may be viewed as a Western emic, and Zen as an Eastern emic.

In his seminal essay published back in the sixties Berry (1969) explicitly recognized that in the early days of cross-cultural research, approaches to psychology such as psychoanalytical were simply *assumed* to be etic or universally applicable, although they are but emics of Western origin. As is well known, in his later work Berry (1989) has suggested an iterative procedure according to which research is begun with an emic of whatever cultural origin, transported to another culture where it is imposed as if it were an etic, and suitably modified in light of observations in that culture so as to remove some of the limits imposed by its cultural origin. Through repeated or iterative cross-cultural applications, an emic may be gradually refined make it applicable for an increasing number of cultures. As suggested by Berry et al. (1992) extension of this research methodology will "ultimately lead to so much evidence that it can be reasonably concluded that a psychological characteristic is universal present" (pp. 233-234).

Indeed, it is ironic that despite the recognition of the importance of the distinction between focus on elements versus system as a whole, the above quote from Berry et al. ends with "a psychological characteristic" - a mere element. Here the influence on cross-cultural psychology of molecularism, typical of the Lockean tradition comes through most clearly. Nevertheless, the long debate over emic and etic psychology gives me the impression that cross-cultural psychologists like Berry are aiming at ultimately producing an etic or universal psychology, no matter how long it takes to do it. Presumably, some time in the future, we will actually have a psychology that is rid of all limitations put on it by its origin in a specific culture.

Here it may be worth asking the following questions: If and when a truly universal, etic psychology emerges, what kind of a psychology will it be? To be universal, would it not have to be one single united body of knowledge where everyone speaks in one voice, so to speak, ending the cacophony of innumerable incommensurable emics? Does that mean that an etic psychology will have to establish itself by dissolving, or making redundant, all competing alternatives or differing systems of psychology within the same culture (behaviorism, psychoanalysis, humanistic and the like) as well as across cultures (indigenous psychologies of Japan, India and so on)?

Opinions are radically divided on these matters. This was illustrated in a recent exchange of ideas in the International Journal of Psychology. In an article in that journal, Misra and Gergen (1993a) argued that the approach that prevails in contemporary cross-cultural psychology is primarily a Euro-American product. It is "basically a monocultural academic enterprise", they said, but "it did pretend and successfully manage to project a universalistic image" (p. 231). They propose an indigenous Indian approach as a welcome addition, and suggest that a pluralistic spirit which encourages more such alternatives will enrich psychology. Poortinga (1993) and Triandis (1993) express their strong disagreement with this pluralist vision of psychology in their comments on the Misra and Gergen article published in the same issue. According to Triandis, "[i]f we take the argument in favor of indigenous psychologies seriously, we will have a myriad of such psychologies" (p. 250). Citing a survey in which respondents named 600 different languages, and assuming that there would be as many different cultures and corresponding ethnopsychologies, Triandis asks: "Does that mean we should have 600 Indian psychologies?" (p. 250). Triandis's words clearly express his fear of relativism; in his view, encouraging pluralism will simply be ruinous. To help remedy the situation, he advises: "follow Western patterns" (p. 249). In response, Misra and Gergen (1993b) ask psychologists to go "beyond scientific colonialism." In my view, following "Western science" would hardly ensure unity for, as Staats (1983) has pointed out, Western psychology is itself suffering from a crisis of disunity. Moreover, in the heyday of Carnap's project for the unity of science, even positivists of the Vienna Circle could not agree on fundamental issues about the nature of science. This issue clearly implies the problem of absolutism and relativism, an issue identified by Berry et al. (1992) as one of the fundamental issues facing cross-cultural psychologists today. Indeed, it is a fundamental epistemological issue, and Segall (1993) touches on the same in his debate with Shweder (1990). Indeed, the deeper implications of this issue are being hotly debated within in the field of the philosophy of social sciences over the past decade. Many prominent authors have important contributions to this debate. Notable among them are Krausz and Meiland (1982), Barnes and Bloor (1982), Richard Bernstein (1983, 1988), Clifford Geertz (1984), Michael Krausz (1989), and Richard Rorty (1991). Notwithstanding the relevance of their work to the issue we face in cross-cultural psychology today, their work is rarely referred to in our field. This is perhaps indicative of the common feeling among psychologists that we are scientists, and as such have little if anything to learn from philosophers. The issue is certainly too complex to review here let alone solve it. However, a few thoughts may be expressed with regard to specific implications for cross-cultural psychology.

1. As noted by McRae (1973) most historical attempts to unify science were based on restricting the scope of knowledge to a single overarching goal, or one exclusive method to the neglect of the others, and these attempts have not been successful. Positivism, for instance, aimed at accurate description as the only goal, and Baconian science aimed only at the control of nature, and behaviorism generally followed suit. As Habermas (1971) has put it, the pursuit of knowledge has been guided by many legitimate interests: practical, critical, and emancipatory. Control over others, emancipation of self from external influence, and personal edification are different goals that demand the pursuit of radically different *types* of knowledge, which in turn demand differing approaches to psychology. Many indigenous psychologies of East are devoted to self-realization and self-control as opposed to control of others as Skinnerian and other Western psychologies are. The cannot either substitute of supplant one another. Their plurality and independent existence is justified by the legitimacy of their differing goals. A strong current trend in cross-cultural psychology is to consider accurate description and discovery of pan-human variable as perhaps the only goal psychology may pursue. This goal is no different from that of positivism, a doctrine imperiled by the interpretive turn in contemporary philosophy of science.

2. During a period of more than a century that has passed since the beginning of modern psychology, numerous alternatives theories have been proposed in all sorts of areas, particularly personality. The current trend

indicates further proliferation of theories, not their diminution. Psychology lacks unity and homogeneity not only within each hemisphere, but within each nation, subculture, and perhaps every department of psychology. Against this background, expecting to unite psychology in a single universal etic psychology seems incongruous.

3. There is no point in bemoaning the diversity of perspectives in psychology as Arthur Staats (1983) has been, nor does it make sense to rationalize the current disunity as indicative of psychology's primitive and proto-scientific state of development. Thomas Kuhn's view of psychology as being in a pre-paradigmatic state implied a comparison with physics, and suggests an attitude that T.H. Leahey (1987, p. 25) has dubbed "physics envy." This was Kuhn's early view prior to his "discovery of hermeneutics" and his realization of the importance of alternative interpretations. Construing alternative interpretations is not weakness, but the strength of science; indeed, science cannot make progress without being able to generate novel hypotheses and diverse alternative interpretations to choose from.

4. As the physicist Ernst Mach (1901/1960, pp. 132-133) put it, it is in principle always possible to construct more than one alternative conceptual framework, each of which satisfies epistemic criteria such as accuracy, comprehensives, parsimony and predictability. This is the principle that the psychologist George Kelly (1955) called "constructive alternativism," and was echoed by the Vedas of ancient India in saying that "truth is one, but wise men construe it differently." (Rig Veda, 1.164.46; 10.114.5). If this principle is correct, plurality of theoretical systems is a natural outcome. Many differing theories might accomplish the same goals of science even as diverse cultures help satisfy common human needs.

5. The recognition of the viability and usefulness of alternative interpretations does not necessarily entail sinking in relativistic quicksand (Paranjpe, 1993). Although in principle there is no limit on the number of alternative theories that could be generated, in practice there has been only a finite number of serious alternatives. Also, there are reliable ways of assessing competing claims of alternative theories in terms of criteria such as accuracy and predictability, although as suggested by Kuhn (1970), "There is no neutral algorithm for theory-choice" (p 200). As long as this is the situation, we might accept a plurality of emics and try to understand one another, even as we must, as members of pluralistic societies, strive to foster mutual understanding and live in harmony.

References

Allport, G.W. (1955) *Becoming: Basic considerations for a psychology of personality.*
    New Haven, CT: Yale University Press.
Bacon, F. (1905). Novum Organum. In J.M. Robertson (Ed.), *The Philosophical works of Francis Bacon.* London: George Routledge & Sons. (Original work published 1620).

Barnes, B., & Bloor, D. (1982). Relativism, rationalism and the sociology of knowledge. In M. Hollis & S. Lukes (Eds.), *Rationality and relativism*. Oxford: Basil Blackwell.

Bernstein, R.J. (1988). *Beyond objectivism and realism: Science, hermeneutics, and praxis*. Philadelphia, PA: University of Pennsylvania Press. (First published 1983).

Berry, J.W. (1969). On cross-cultural comparability. *International Journal of Psychology, 4*, 119-128.

Berry, J.W. (1989). Imposed etics-emics-derived etics: The operationalization of a compelling idea. *International Journal of Psychology, 24*, 721-735.

Berry, J.W., Poortinga, Y.H., Segall, M., & Dasen, P.R. (1992). *Cross-cultural psychology: Research and applications*. Cambridge, U.K.: Cambridge University Press.

Carnap, R. (1949). Logical foundations of the Unity of Science. In H. Feigl & W. Sellars (Eds.), *Readings in philosophical analysis*. New York: Appleton-Century-Crofts. (Original work published 1938)

Geertz, C. (1984). Anti anti-relativism. *American Anthropologist, 86*, 263-278.

Gergen, K.J. 1985. The social constructionist movement in modern psychology. *American Psychologist, 40*, 266-275.

Habermas, J. (1971). *Knowledge and human interests*. (J.J. Shapiro, Trans.). Boston: Beacon Press. (Original work published 1968)

Harre, R. (1987). The social construction of selves. In K. Yardley & T. Honess (Eds.), *Self and identity: Psychosocial perspectives*. New York: John Wiley & Sons.

Jahoda, G. (1983). The cross-cultural emperor's conceptual clothes: The emic-etic issue revisited. In J.B. Deregowski, S. Dziurawiec, & R.C. Annis (Eds.), *Expiscations in cross-cultural psychology* (pp. 19-38). Lisse: Swets and Zeitlinger.

Jones, W.T. (1969). *A history of Western philosophy* (2nd ed., 4 vols.). New York: Harcourt, Brace & World.

Kelly, G.A. (1955). *The psychology of personal constructs* (2 Vols.). New York: W.W. Norton.

Krausz, M. (Ed.). (1989). *Relativism: Interpretation and confrontation*. Notre Dame, IN: University of Notre Dame Press.

Kuhn, T. (1970). *The structure of scientific revolutions* (2nd ed.). Chicago: University of Chicago Press.

Kuhn, T. (1977). *The essential tension: Selected studies in scientific tradition and change*. Chicago: University of Chicago Press.

Leahey, T.H. (1987). *A history of psychology: Main currents in psychological thought* (2nd ed.). Englewood Cliffs, NJ: Prentice-Hall.

Mach, E. (1960). *Space and geometry in the light of physiological, psychological and physical inquiry*. (T.J. McCormack, Trans.). La Salle, IL: Open Court. (Original work published 1901-1903)

McRae, R. (1973). Unity of science from Plato to Kant. In P. Wiener (Ed.), *Dictionary of the history of ideas*. New York: Charles Scribner.

Misra, G., & Gergen, K.J. (1993a). On the place of culture in psychological science. *International Journal of Psychology, 28*, 225-243.

Misra, G., & Gergen, K.J. (1993b). Beyond scientific colonialism: A reply to Poortinga and Triandis. *International Journal of Psychology, 28*, 251-254.

Paranjpe, A.C. (1993). The fear of relativism in post-positivist psychology. In H.J. Stam, B. Kaplan, L. Mos, & W. Thorngate (Eds.), *Recent trends in theoretical psychology* (Vol. 3, pp. 77-83). Springer Verlag.

Piaget, J. (1954). *The construction of reality in the child.* (M. Cook, Trans.) New York: Ballantine.

Poortinga, Y.H. (1993). *Is there no water in the experimental bath water? A comment on Misra and Gergen.*

Quine, W. v. O. (1964). *From a logical point of view* (2nd ed.). Cambridge, MA: Harvard University Press. (Original work published 1953).

Rabinow, P., & Sullivan, W.M. (Eds.). (1987). *Interpretive social sciences: A second look.* Berkeley, CA: University of California Press. (Original work published 1979).

Rorty, R. (1979). *Philosophy and the mirror of nature.* Princeton, N.J.: Princeton University Press.

Rorty, R. (1991). *Objectivity, relativism and truth: Philosophical papers* (Vol. 1). Cambridge, U.K.: Cambridge University Press.

Scheler, M. (1970). On the positivistic philosophy of the history of knowledge and its Law of Three Stages. In J.E. Curtis & J.W. Petras (Eds.), *The sociology of knowledge.* New York: Praeger. (Date of original publication unknown)

Segall, M.H. (1993). Cultural psychology: Reaction to some claims and assertions of dubious validity. *Cross-Cultural Psychology Bulletin, 27,* 2-4.

Shweder, (1990). Cultural psychology - What is it? In Stigler, J.W., Shweder, R.A., & Herdt (Eds.), *Cultural psychology: Essays on comparative human development* (pp. 1-43). Cambridge, U.K.: Cambridge University Press.

Skinner, B.F. (1953). *Science and human behavior.* New York: Free Press.

Staats, A. W. (1983). *Psychology's crisis of disunity: Philosophy and method for a unified science.* New York: Praeger.

Taylor, C. (1980). Understanding in human sciences. *Review of Metaphysics, 34,* 3-23.

Triandis, H.C. (1978). Some universals of social behavior. *Personality and Social Psychology Bulletin, 4,* 1-16.

Triandis, H.C. (1993). Comments on Misra and Gergen's: Place of culture in psychological science. *International Journal of Psycholgy, 28,* 249-250.

Van de Vijver, F.J.R., & Poortinga, Y.H. (1982). *Journal of Cross-Cultural Psychology, 13,* 387-408.

Cross-Cultural Psychology in Germany

Hede Helfrich
University of Regensburg, Regensberg, Germany

Cross-cultural psychology, i.e. the study of human behavior and mental processes under diverse cultural conditions, subsumes two major approaches, the *nomothetic* or *etic* and the *idiographic* or *emic*. The *nomothetically*-oriented researcher views phenomena from a trans- or meta-cultural perspective. "Culture" is stressed as a set of independent variables whose influence on individual behavior is investigated. The attempt is made to represent differences among cultures using a common set of scales. The *idiographically*-oriented researcher, on the other hand, stresses the self-actualization of the individual and the culture-bound definition of psychological phenomena. He/she attempts to view phenomena through the eyes of the subjects under study.

Contingent with the distinction between the *nomothetic* and the *idiographic* approaches is the question of *universality* versus *relativity* in the nature of psychological phenomena across cultures. In Germany, the bulk of recent research on cross-cultural psychology follows the nomothetic approach, with the universalistic viewpoint dominating, but there is a new trend to discuss the controversy explicitly and to work toward a synthesis of the two dialectic positions. Although this current discussion owes much to Anglo-American research, its historical origins can be traced back to German intellectual roots.

The present article on cross-cultural psychology in Germany is divided into three parts. The first part attempts to sketch the historical development of cross-cultural psychology in Germany and its current institutionalization. The second part provides a survey of research on selected topics, including general and social psychology as well as developmental aspects. The third part attempts to draw a conclusion and to outline the prospects of German cross-cultural psychology.

Historical Background and Institutional Framework

As H. Ebbinghaus (1908) said, referring to psychology, cross-cultural psychology in Germany has a short history, but a long past. While the history of cross-cultural psychology starts after the Second World War and is heavily influenced by research carried out in Anglo-American countries, the past can be traced back to the German language philosophers J.G. Herder (1744-1803) and W. von Humboldt. (1767-1835). According to them, cultural communities shape the thoughts of their members by means of a common language, so that people speaking different languages experience the world in different ways (Jahoda, 1990, p. 328). This notion of an intrinsic relationship between language and thought has developed two distinct theoretical branches: one relativistic and the other universalistic. The former was elaborated by the German anthropologist and student of American Indian languages Boas (1848-1942). The heirs of this tradition were the American linguists Sapir (1844-1939) and his student Whorf (1897-1941). Their ideas are familiar to us as the Sapir-Whorf hypothesis, which states that the thinking of an individual is "calibrated" by the particular language to which he or she is exposed. In recent years, this radically relativistic position has been revived by the American linguist Lakoff (1987) who asserts that language and cognition are intrinsically linked, so that "indigenous" definitions of all cognitive phenomena follow.

The universalistic position was advanced by Lazarus and Steinthal (1860), who regarded the cultural community as constitutive for the emergence of the individual mind. Unlike the relativists, they postulated laws of historical development which are universal to all peoples. This means that societies and their languages evolve along a universally valid sequence of developmental stages. Differences between cultures exist because the respective languages are at different evolutionary stages. This idea of universal rules affecting the development of the language and thought of a particular culture was revived by the American researchers Berlin and Kay (1969).

Wundt (1832-1920), the "father of Experimental Psychology" (Hall, 1914), while sharply criticizing Lazarus and Steinthal as too vague and too speculative, shared their conviction that "higher mental processes" emerge as a result of historical development. He distinguished between elementary psychological processes which could be studied experimentally and higher forms of thought which could be analyzed only with reference to their development in language, myth and custom (Wundt, 1904; 1913). During the 1950's and 60's, when Skinnerian behaviorism imported from America dominated German psychology, Wundt's general thesis about the socio-historical contingencies of individual thinking and behavior was relegated to anthropologists, ethnologists, and sociologists (Graumann, 1992). But when the cognitive shift in the seventies led to the era of "social representations" (Moscovici, 1981; 1982; 1990) in the 80's, Wundt's ideas were picked up by German social psychology (Sommer, 1988).

From a methodological viewpoint, the work of the German anthropologist Thurnwald (1924) deserves attention. Thurnwald drew a distinction between a *general* and a *differential* folk psychology (Thomas, 1993a). This distinction mirrors the two-fold goal of modern cross-cultural psychology; on the one hand, the focus lies on psychic features and processes common to all mankind, and, on the other hand, on modifications of those features and processes contingent on specific cultural conditions.

Regrettably, contemporary cross-cultural psychology in Germany has lost the connection to its own earlier research tradition. The reasons for this loss may be mainly attributed to the "brain-drain" which occurred during the national-socialistic era of Hitler, when many German psychologists emigrated to the United States. After World War II, when psychology was reestablished in Germany, it was dominated by American behaviorism. Cross-cultural psychology survived mainly by importing research from the Anglo-American countries and making half-hearted attempts to establish its own credibility in the field through independent research. This lack of "self-reliance" is not unlike that found among natural scientists in developing countries (Moghaddam & Taylor, 1986). As a consequence, cross-cultural psychology is not integrated into the mainstream of psychology and also not included in university curricula. Most German cross-cultural studies are conducted and presented in the context of various other fields of psychology but are rarely discussed in the context of cross-cultural psychology. A notable exception is the reader on cross-cultural psychology edited by A. Thomas (1993). in which the issue of an "appropriate psychology" (Moghaddam & Taylor, 1986) is addressed, at least, implicitly.

Review of Selected Areas

The following review covers research carried out by German psychologists in the cross-cultural area. Topics include general psychological aspects such as cognition, emotion and motivation, developmental aspects such as the development of moral judgment and the development of attachment, social and organizational psychological aspects such as individualism versus collectivism, diagnostic aspects such as personality and psychopathology, and intercultural aspects such as intercultural encounters and acculturation.

The bulk of research is consistent with the nomothetic approach. The goal of this research is two-fold. On the one hand, it seeks structures and processes common to all mankind, and, on the other hand, it looks for modifications of those structures and processes. While the first goal is directed toward finding psychological laws of universal validity, the second is directed towards isolating cultural conditions as causal determinants of individual forms of thinking and behavior. Prototypical for research in line with the first goal is the application of Piagetian tasks in different cultures in order to test the universal validity of Piaget's genetic epistemology. In Germany, this approach is put forward with respect to Kohlberg's stages of moral judgment (Eckensberger, 1993). Prototypical for research in line with the second goal is the comparison of motives, abilities and personality traits between different cultures by means of psychometric tests, behavioral data, or tasks within quasi-experimental settings. An example of German research in this domain is a cross-cultural comparison of inclination to aggression (Kornadt, 1993).

Perception and Cognition

Perception and cognition are not among the research topics favored by German cross-cultural researchers. The reviews on perception (Thomas & Helfrich, 1993; Liebing & Ohler 1993) are based on empirical research performed mainly in Anglo-American countries. Two reasons may account for this lack in interest; one is methodological and the other theoretical in nature. The methodological reason is that "culture" is an organismic instead of an experimental factor (Helfrich, 1993), with the implication that cultural differences are always confounded by other differences such as differences in formal education (Pettigrew & van de Vijver, 1990) or industrialization (Inkeles & Smith, 1974) or even biological differences (Bornstein, 1975). Some of these confusions may be eliminated by a conceptualization at a basic or *intra*-cultural level rather than at a *cross*-cultural level. An example is the relationship between color naming and color perception. Two studies (Zimmer, 1982; Helfrich, 1994a) take advantage of the fact that, within the same culture, linguistic category boundaries between different colors are not fixed at certain points on the wavelength continuum but show considerable variation among individuals. Both studies support a universal perceptual process which means that perception is independent of the availability of color terms and the location of given linguistic category boundaries. However, by using signal detection theory, it can be shown (Helfrich, 1994a) that color discrimination is a two-stage process consisting of sensory discrimination on the one hand and cognitive organization of perceived differences on the other. While the first stage is independent of language, the second is not. Sensory differences crossing the individual linguistic category boundary are weighted cognitively higher than differences of the same magnitude within one category.

The theoretical reason for the lack of empirical studies in perception and cognition concerns the discussion on universality and relativity in the nature of psychological phenomena across cultures (Grossmann, 1993a). Although the generally held position in Germany is that there is interaction between universal basic processes and culture-specific adaptive developments, exceptions may be found especially in the field of cognition. Following the American linguist Lakoff (1987), Liebing & Ohler (1993) maintain a radically relativistic position by claiming an "indigenous" definition of all cognitive phenomena. Such an extreme relativism logically leads away from comparisons across cultures.

Aggression Motive

Based on a motivation theory of aggression (Kornadt, 1982), Kornadt (1990/91; 1993) investigated how and why the aggression motive is differently developed in different cultures. Using questionnaires, projective tests and scenario techniques, he compared levels in both overt aggression and aggression inhibition across five cultures. The results indicated that Eastern cultures (Japan, Bali, Batak) exhibit a lower aggression level than Western cultures (Germany, Switzerland) and that low overt aggression is not compensated by high aggression inhibition. The unequal distribution across cultures challenges instinct based theories of aggression, e.g., Freud (1940) and Lorenz (1963), in which expression of aggression is seen as natural, necessary, and even therapeutic. Kornadt attributed the differences in aggression level to learned differences in interpreting frustrating events. Western people tend to regard frustrations as intentionally caused and unjustified, while Eastern people tend to react with feelings of shame and guilt. Comparing child-rearing practices, Kornadt concluded that the culture-specific reaction tendencies are acquired by culture-specific forms of mother-child interaction. The mother serves as model for the child in interpreting frustrating events. If she feels predominantly frustrated by the child's misbehavior and interprets it as being caused by bad intentions, the child will learn to use similar interpretations. A further important factor is the way the mother enforces rules and norms. The mother in Eastern cultures shows a flexible response set. She makes her demand clear, but if the child is not able or not willing to obey, she reduces her demand step by step, and, in this way, she maintains a basic feeling of harmony between herself and the child. In Western cultures, on the other hand, mothers anger more easily, causing reactive behavior and an escalation of conflict (Kornadt, 1990/91, p. 163).

On the other hand, the results obtained by Kornadt are not as culture-specific as they appear at first glance. An early American study by Sears, Maccoby and Levin (1957) examined the relationship between parents' behavior and the aggression level expressed by their children. Parental behavior was rated as either *high* or *low* on two dimensions: *permissivity* and *punitivity*. Children's behavior was classified as either *high* or *low aggressive*. Children's aggression level was lowest, when parents set rules and make their demands clear (low permissive), but do not punish severely (low punitive). This parental behavior pattern neatly matches the Japanese child-rearing style described by Kornadt.

In line with Sears et al.'s (1957) early research is recent research by Rohner and his colleagues (Rohner, Hahn & Koehn, 1992; Ajdukovic, 1990) on parental practices and child behavior research, which shows such great differences even within each culture, that it is difficult to draw a clear distinction between Western and Eastern styles of parental behavior.

Expression of Emotions

Following the tradition of Darwin (1872) and Ekman (1984), ethologists Grammer & Eibl-Eibesfeldt (1993) attempted to establish that, despite differences in surface structure, the expression of emotions has a universal deep structure. Grammer & Eibl-Eibesfeldt analyzed the temporal patterns of facial expressions across different cultures and concluded that there are universal basic patterns controlled by culture-specific display rules. These rules determine in which situations patterns may undergo a ritualization, indicated by exaggeration and repetition of pattern elements. The ritualization may mask the decodability for persons from other cultures who misinterpret the context situation, and, for this reason, are biased in their inferences. However, when the emotional expression is produced intentionally, it can be interpreted universally. For instance, standardized expressions of emotions in the Japanese Kabuki theater are understood by Germans in the same way as by Japanese (Morishita & Siegfried, 1991).

While ethologists emphasize the innate basis and the universal structure in the perception and production of emotions, developmental psychologists focus on culture-specific patterns. Specifically they stress the learned interpretation of social situations leading to different emotional reactions. Trommsdorf (1993a) compared Japanese and German children in the emotional reactions to persons needing help. The Japanese children showed more socially disengaged emotions such as anxiety and distress in comparison with the German children, who showed more socially engaged emotions such as empathy. Trommsdorf attributed the observed differences to different cognitive "working models" acquired as a consequence of differential socialization practices.

Developmental Psychology

Developmental psychology is an model field where the two-fold goal of cross-cultural psychology is apparent. On the one hand, psychological laws found in Western societies are challenged or validated. On the other hand, systematic cultural influences on basic features and processes are investigated.

Prototypical for the focus on general laws is the application of Piagetian tasks in different cultures in order to test the universal validity of Piaget's genetic epistemology. In Germany, this approach is advanced in relation to Kohlberg's stages of moral judgment (Eckensberger, 1993). Eckensberger's meta-analysis of Kohlbergian studies across cultures leads to two main conclusions. First, the bulk of results support the cross-cultural validity of a universal hierarchy of stages, since most of the longitudinal studies demonstrate that neither regressions back to lower stages nor progressions to higher stages with omissions of lower ones occur. Second, the existence or non-existence of the highest stages depends on global cultural features such as industrialization, conservative versus liberal education or urban versus rural environment, but is not contingent on the dichotomy of Western versus non-Western (Gielen, Cruickhank, Johnston, Swanzey and Avellani, 1986).

Typical of the focus on cultural modification of basic psychological features and processes are cross-cultural developmental studies currently being carried out in Japan (Grossmann, 1993a; Trommsdorf, 1993a, b) and on the Trobriand Islands (Grossmann, 1993b). Characteristic of this research is its interactionist perspective. The term *interaction* has a multiple meaning in this context. First, it refers to an interaction between a universal biological endowment and variable parental behavior. This kind of interaction has been studied primarily in the context of attachment theory (Grossmann, 1993a). Second, it refers to child socialization as an interaction between mother and child implying a mutual shaping of infant's and mother's behavior. This kind of interaction has been studied primarily with respect to the development of aggression and prosocial behavior (Kornadt, 1990/91; Trommsdorf, 1993a). Third, it refers to the interaction between the child's individual personality and the specific culture to which he or she is exposed. This approach proceeds from the assumption that "culture" is not an inevitable treatment but, instead, the individual has choices. However, there are constraints with respect to the range of individual choices (Eckensberger, 1992; Trommsdorf, 1993 b). These constraints are a function of the interaction between the personality type of the individual and culturally prefigured orientations. For example, an *allocentric* individual has greater a chance of self-actualization in a culture with a *collectivistic* orientation, while an *idiocentric* individual has a better chance in an *individualistic* culture (Triandis, 1994).

Social Psychology

Although one would assume that social behavior is the kind of behavior most strongly affected by cultural influences, most text books and readers in social psychology refer to a population which is almost exclusively white and has been educated in the Western context. German textbooks are no exception. Although cross-cultural research on social behavior has been carried out in Germany, the results have not affected mainstream social psychology. However, in a more indirect way, a cross-cultural perspective has been introduced to German social psychology. This perspective has been adopted from the French social psychologist Serge Moscovici, who challenged the American findings on *majority influence* and created a psychology of *minority influence* (Moscovici, 1980; 1985). Moscovici's theories and empirical results have been incorporated into German social psychology; still lacking are the incorporation of culture as an integral part of social psychological theories and the inclusion of cultural variables in empirical studies (Moghaddam, Taylor & Wright, 1993).

Some research has been conducted by Kornadt, Trommosdorf and Kobayashi (1994) and by Helfrich (1994b) on social interaction as reflected in the use of verbal patterns. Based on the speech action model developed by Herrmann and his coworkers (Herrmann & Grabowski, 1994), Helfrich

(1994b) investigated verbal patterns in social acts of making requests in Japan and compared them with the results obtained by Herrmann in Germany. A request can be expressed either directly or indirectly. In order to achieve the intended goal, the speaker has to make a compromise between directness and indirectness. A request uttered too directly may cause opposition, but, on the other hand, a request uttered too indirectly may not be understood at all. Whether directness or indirectness is more effective, depends on the specific situation. In Germany, requests are made more directly when a person of higher rank asks a favor of a person of lower rank than is the case in situations where both persons are of the same rank. Based on Hofstede's characterization of Japan as a collectivistic society (Hofstede, 1980; 1993; Yamaguchi, 1994) as compared to Germany as an individualistic society, one can make two alternative predictions for Japanese subjects. With respect to the hierarchical orientation of collectivistic societies, directness should be more pronounced in situations with asymmetric ranks than in those with symmetric ranks. On the other hand, with respect to the harmony orientation of leadership style which in Japan is assumed to emphasize maintaining good relations rather than emphasize subordinate performance (Bond, 1994; Smith, Peterson, Misumi & Bond 1992), one should expect no differences in directness between asymmetric and symmetric situations. The latter prediction would be in line with guides for social interaction with Japanese people (Hall & Hall, 1985) stating that hierarchic relations are reflected in verbal address forms but not in the directness of content. The two alternative predictions were tested in Helfrich's study. The test material consisted of scenarios in different social situations which required a request (in Japanese). In two separate trials, the test subjects were asked either to produce an appropriate request or to choose one of several alternatives. The results indicated that hierarchical rank has a decisive effect on the frequency of direct and indirect requests. Given the appropriate legitimization, expectations of others are indeed expressed with clarity and directness, in other words, a higher-ranking Japanese person speaks with his subordinates just as directly as would be the case in Germany.

Organizational Psychology

Since Hofstede's (1980) study on work-related values, cultural explanations of organizational behavior have become popular. Wilpert, a German cross-cultural researcher in the field of organizational psychology, points out two shortcomings in Hofstede's analysis (Wilpert, 1993). First, according to Wilpert, Hofstede's definition of the term "culture" is restricted to a cognitive orientation system, without consideration of cultural artifacts such as organizational structures and legal regulations. Second, Wilpert finds fault with the restriction of the analysis to values and attitudes of the interviewees without consideration of real decision behavior. From his own research, Wilpert draws the conclusion that personal factors, such as competence, and situational factors, such as social environment and organizational atmosphere, are much more important than national factors with respect to their influence on decision behavior.

Personality and Psychopathology

Motivated by the reunification of East and West Germany in 1990 and the psychological problems emerging as a consequence, German cross-cultural research on personality traits and syndromes and on behavior patterns focuses on the comparison between East and West Germany (Harenberg, 1991; Maaz, 1990). Such analyses are not without problems because of the differential evaluations of the traits to be investigated (Thomas, 1994). Sometimes personality traits with similar denotative meaning are connotated differently depending on the cultural identity of both research subjects and researcher. Therefore, it seems difficult to separate researcher biases from real differences. Occasionally, this bias itself has become to a subject of research.

Silbereisen, Schwartz and Kracke (1993) compared problem behavior in adolescents across cultures. The term "problem behavior" refers to abnormal behavior such as suicide or alcohol and drug abuse as well as to behavior not appropriate for juvenile age such as premature use of alcohol, premature sexual behavior or violation of parental and school rules. Although, according to Silbereisen, the current state of art does not yet allow the establishing of a universal framework, some hypotheses of universal validity may now be suggested. One such example is the role of value orientations in a culture. Problem behaviors tend to reflect highly valued cultural norms. Thus, the motives for suicides center around school problems in Japan and on problems with relationships to parents or friends in Germany thus reflecting the relative importance of the respective areas in the two cultures. Maladaptations result from an over-identification with cultural norms and standards rather than from an under-identification. This is illustrated by a comparison of clinical sample populations in Thailand and in the USA In Thailand, where education stresses harmony, modesty and courtesy, problems which indicate personal over-control, such as anxiety or sleeplessness, are prevalent. On the other hand, in the USA, where education stresses self-assertiveness, problems with lack of control, such as aggressive behavior and delinquency, are prevalent. Therefore, over-identification seems to cause more problems than under-identification - at least with respect to pathological problems.

Intercultural Encounters and Acculturation

As a result of the internationalization of trade, which affects Germany as an export-oriented country to a particularly high degree, and a concomitant interest in foreign exchange, a great deal of research on intercultural encounters has been undertaken (Thomas, 1993b). Much of this research focuses on foreign manager exchanges with East Asian countries. Using the "critical incidents" technique originally developed by Flanagan (1954), Thomas (1991) identified typical conflicts occurring in real encounters. Content analyses of the reports obtained from managers of the respective countries led to the identification of central "culture standards" which form the basis for the development of a "culture assimilator" training program. The evaluation of the effectiveness of the training is the last step within this research program.

The increase of the number of migrants and refugees in the past and the increase of hostility towards foreigners in contemporary Germany has led to a great amount of research in acculturation and living in multicultural social environments. Most of the research in this area has been carried out by the German Institute for Educational Research. Two main research projects are worth mentioning: first a project exploring the effect of teachers' experience in acculturation through a comparison between "novices" and "experts" in multicultural teaching situations; and, second, a project on expectations and experiences of "Aussiedler" children (Graudenz & Römhild, 1990), German children who have moved with their parents from Eastern Europe to Germany.

Conclusion and Prospects

The dynamics of cross-cultural psychology derive from the basic dilemma between the nomothetic or etic approach on the one hand and the idiographic or emic on the other. A worldwide tendency toward the idiographic approach can be observed. In fact, the term *indigenization* has been coined to characterize this tendency. As Kagitcibasi & Berry (1989) pointed out, two major trends in indigenization can be distinguished. One major trend is a theory-driven approach exhibited in interpretative anthropology and social constructivism. The second major trend is the emphasis on a problem-centered rather than a theory-driven approach. The latter trend, commonly associated with Third World psychology, can also be observed in Germany. This is supported by the interest in international encounters as well as in acculturation having its roots in pragmatic concerns and needs.

The former trend, with its "indigenous" or "emic" definition of all psychological phenomena and consequent radical cultural relativism, is only rarely defended in Germany (Liebing & Ohler, 1993; Schiefenhövel, 1993).

More commonly, such a radically indigenous view is sharply criticized (Bender-Szymanski & Hesse, 1987; 1988a, b; Kornadt, 1993; Kornadt et al., 1994). Instead, there is a tendency in current German cross-cultural psychology to work toward a synthesis of the nomothetic and idiographic positions, a trend which has four aspects.

Figure 1. Principle of Triarchic Resonance

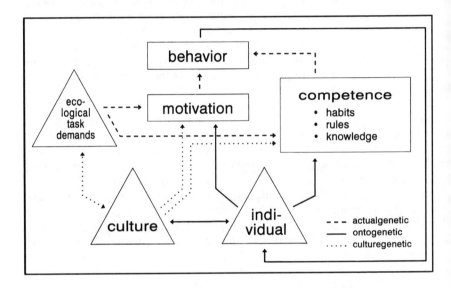

(1) A synthesis of *cross-cultural* psychology and *cultural* psychology: This approach is characterized by the assumption that, at the level of behavior, we are faced with culture-specific manifestations whose relation to underlying universal processes has yet to be clarified (Graumann, 1992; Helfrich, 1993; Trommsdorf, 1993b).

(2) A dialectic relationship between individual autonomy and cultural determination: This approach proceeds from the notion that cultural influences interact dynamically with individual self-actualization tendencies (Eckensberger, 1992; Trommsdorf, 1993 b).

(3) A dialectic relationship between biological and cultural evolution (Grossmann, 1993a): The *emic-etic* controversy is rejected as fruitless from the perspective of evolutionary theories of culture (Durham, 1990). In this framework, both biological and cultural transmission have their place. The capacity to learn is genetically transmitted, but *what* is learned, is transmitted culturally.

(4) A dialectic relationship between temporal changes at different levels (Eckensberger, 1992; Helfrich, 1995): Conceptualization must integrate

different levels of *temporal change* into a common model. This approach is illustrated by the *principle of triarchic resonance* (Helfrich, 1995; see Figure 1), which describes the dynamic relationship between three entities and their respective level of temporal change. The three entities are: *individual, culture,* and *ecological task demands*; the three levels of temporal change are: *ontogenetic* development, *culturegenetic* development or cultural change, and *actualgenetic* development of ecological task demands.

In summary, future research is directed to productively overcome the fundamental distinction between the emic and the emic approach.

References

Ajdukovic, M. (1990). Differences in parent's rearing style between female and male predelinquent youth. *Psychologische Beiträge, 32,* 7-15.
Bender-Szymanski, D., & Hesse, H.-G. (1987). *Migrantenforschung. Studien und Dokumentationen zur vergleichenden Bildungsforschung:* Vol. 28. Köln: Böhlau.
Bender-Szymanski, D., & Hesse, H.-G. (1988a). Migranten als Forschungsobjekt. *Empirische Pädagogik, 2,* 269-287.
Bender-Szymanski, D., & Hesse, H.-G. (1988b). Abschließende Stellungnahme zur Kontroversdiskussion "Migrantenforschung". *Empirische Pädagogik, 2,* 285-287.
Berlin, B., & Kay, P. (1969). *Basic color terms.* Berkeley: University of California Press.
Bond, M.H. (1994). Into the heart of collectivism: A personal and scientific journey. In U. Kim, H.C. Triandis, C. Kagitcibasi, S.-C. Choi, & G. Yoon (Eds.), *Individualism and collectivism* (pp. 66-76). Thousand Oaks, CA: Sage.
Bornstein, M.H. (1975). The influence of visual perception on culture. *American Anthropologist, 77,* 774-798.
Darwin, C. (1872). *The expression of emotions in man and animals. London:* Murray.
Durham, W.H. (1990). Advances in evolutionary culture theory. *Annual Review of Anthropology, 19,* 187-210.
Ebbinghaus, H. (1908). *Abriß der Psychologie.* Leipzig: Veit.
Eckensberger, L. (1992). *Agency, action and culture: Three basic concepts for psychology in general, and for cross-cultural psychology in specific.* Proceedings of the IACCP-Conference, Katmandu, 1992.
Eckensberger, L. (1993). Moralische Urteile als handlungsleitende normative Regelsysteme im Spiegel der kulturvergleichenden Forschung. In A. Thomas (Ed.), *Einführung in die kulturvergleichende Psychologie* (pp. 259-295). Göttingen: Hogrefe.
Ekman, P. (1984). Expression and the nature of emotion. In K.R. Scherer & P. Ekman (Eds.), *Approaches to emotion* (pp. 319-343). Hillsdale: Lawrence Erlbaum.
Flanagan, J.C. (1954). The critical incident technique. *Psychological Bulletin, 51,* 327-358.
Freud, S. (1940). *Gesammelte Werke.* London: Imago, Frankfurt: Fischer.

Gielen, U.P., Cruickshank, H., Johnston, A., Swanzey, B., & Avellani, J. (1986). The development of moral reasoning in Belize, Trinidad-Tobago and the USA. *Behavior Science Research, 20*(1-4), 178-207.

Grammer, K., & Eibl-Eibesfeldt, I. (1993). Emotionspsychologische Aspekte im Kulturenvergleich. In A. Thomas (Ed.), *Einführung in die kulturvergleichende Psychologie* (pp. 298-322). Göttingen: Hogrefe.

Graudenz, I., & Römhild, R. (1990). Kulturkontakt unter Deutschen. Zur interaktiven Identitätsarbeit von Spätaussiedlern. *Bildung und Erziehung, 43*, 313-324.

Graumann, C.F. (1992). Einführung in eine Geschichte der Sozialpsychologie. In W. Stroebe, M. Hewstone, J.-P. Codol, & G.M. Stephenson. (Eds.), *Sozialpsychologie* (2nd ed., pp. 3-20). Berlin: Springer.

Grossmann, K.E. (1993a). Universalismus und kultureller Relativismus psychologischer Erkenntnisse. In A. Thomas (Ed.), *Einführung in die kulturvergleichende Psychologie* (pp. 53-80). Göttingen: Hogrefe.

Grossmann, K.E. (1993b). Aspekte universeller und kulturspezifischer Entwicklung von Kindern: Überlegungen und Feldbeobachtungen auf einer Trobriand-Insel. In H. Mandl, M. Dreher, & H.-J. Kornadt (Eds.), *Entwicklung und Denken im kulturellen Kontext* (pp. 27-53). Göttingen: Hogrefe.

Hall, E.T., & Hall, M.R. (1985). *Verborgene Signale*. Hamburg: Gruner & Jahr.

Hall, S. (1914). *Die Begründer der modernen Psychologie*. Leipzig: Felix Meiner.

Harenberg, W. (1991). Vereint und verschieden. In Spiegel/ Spezial (I, 1991*). Das Profil der Deutschen* (pp. 10-23). Hamburg: Augstein.

Helfrich, H. (1993). Methodologie kulturvergleichender Forschung. In A. Thomas (Ed.), *Einführung in die kulturvergleichende Psychologie* (pp. 81-102). Göttingen: Hogrefe.

Helfrich, H. (1994a). *On the universality of color perception*. Paper presented at the TEAPP 1994, Munich, Germany, April 1994.

Helfrich, H. (1994b). *Soziale Handlungsmuster im Vergleich zwischen Japan und Deutschland*. Paper presented at the conference of the "Deutsch-Japanische Gesellschaft für Sozialwissenschaften", Osaka, Japan, September, 1994.

Helfrich, H. (1995). *Beyond the dilemma of cross-cultural psychology: resolving the tension between nomothetic and idiographic approaches*. Paper to be presented at the Asian Pacific Regional Conference of Psychology, Guangzhou, China, August 1995.

Herrmann, T., & Grabowski, J. (1994). *Sprechen*. Heidelberg: Spektrum.

Hofstede, G. (1980). *Culture's consequences*. Beverly Hills, Cal.: Sage.

Hofstede, G. (1993) *Interkulturelle Zusammenarbeit*. Wiesbaden: Gabler.

Inkeles, A., & Smith, D.H. (1974). *Becoming modern*. Cambridge, MA.: Harvard University Press.

Jahoda, G. (1990). Cross-cultural psychology: Its European roots. In P.J.D. Drenth, J.A. Sergeant, & R.J. Takens (Eds.), *European perspectives in psychology* (Vol. 3, pp. 323-338). Chichester: John Wiley.

Kagitcibasi, C., & Berry, J.W. (1989). Cross-cultural psychology: Current research and trends. *Annual Review of Psychology, 493*-531.

Kornadt, H.-J. (1982). *Aggressionsmotiv und Aggressionshemmung* (Vols. 1-2). Bern: Huber.

Kornadt, H.-J. (1990/91). Aggression motive and its developmental conditions in Eastern and Western cultures. In N. Bleichroth & P.J.D. Drenth (Eds.), *Contemporary issues in cross-cultural psychology* (pp. 155-167). Lisse: Swets & Zeitlinger.

Kornadt, H.-J. (1993). Kulturvergleichende Motivationsforschung. In A. Thomas (Ed.), *Einführung in die kulturvergleichende Psychologie* (pp. 181-216). Göttingen: Hogrefe.

Kornadt, H.-J., Trommsdorf, G., & Kobayashi, R.B. (1994). "Mein Hund hat mich bestorben" - sprachlicher Ausdruck von Gefühlen im deutsch-japanischen Vergleich. In H.-J. Kornadt, J. Grabowski, & R. Mangold-Allwin (Eds.), *Sprache und Kognition* (pp. 233-250). Heidelberg: Spektrum.

Lakoff, G. (1987). *Women, fire, and dangerous things. What categories reveal about the mind*. Chicago: University of Chicago Press.

Lazarus, M., & Steinthal, H. (Eds.) (1860). *Zeitschrift für Völkerpsychologie und Sprachwissenschaft.*

Liebing, U., & Ohler, P. (1993). Aspekte und Probleme des kognitionspsychologischen Kulturvergleiches. In A. Thomas (Ed.), *Einführung in die kulturvergleichende Psychologie* (pp. 217-258). Göttingen: Hogrefe.

Lorenz, K. (1963). *On aggression*. New York: Harcourt Brace Jovanovich.

Maaz, H.-J. (1990). *Der Gefühlsstau*. Berlin: Argon.

Moghaddam, F.M., & Taylor, D.M. (1986). What constitutes an "appropriate psychology" for the developing world? *International Journal of Psychology, 21*, 253-267.

Moghaddam, F.M., Taylor, D.M., & Wright, S.C. (1993). *Social psychology in cross-cultural perspective*. New York: Freeman.

Morishita, H., & Siegfried, W. (1991). *Erkennen von Emotionen in der Mimik von Kabuki-Schauspielern*. Unpublished manuscript.

Moscovici, S. (1980). Toward a theory of conversion behavior. In L. Berkowitz (Ed.), *Advances in experimental social psychology* (Vol. 13). New York: Academic Press.

Moscovici, S. (1981). On social representations. In J.P. Forgas (Ed.), *Social cognition: Perspectives in everyday understanding* (pp. 181-209). London: Academic Press.

Moscovici, S. (1982). The coming era of reprentations. In J.P. Codol & J.P. Leyens (Eds.), *Cognitive analysis of social behavior* (pp. 115-150). The Hague: Nijhoff.

Moscovici, S. (1985). Social influence and conformity. In G. Lindzey & E. Aronson (Eds.), *Handbook of social psychology* (3rd ed., vol. 2). New York: Random House.

Moscovici, S. (1990). *Versuch über die menschliche Geschichte der Natur*. Frankfurt.

Pettigrew, T.F., & van de Vijver, F.J.R. (1990). Thinking both bigger and smaller: finding the basic level for cross-cultural psychology. In P.J.D. Drenth, J.A. Sergeant, & R.J. Takens (Eds.), *European perspectives in psychology* (Vol. 3, pp. 339-353). Chichester: John Wiley.

Rohner, R.P., Hahn, B.C., & Koehn, U. (1992). Occupational mobility, length of residence, and perceived maternal warmth among Korean immigrant families. *Journal of Cross-Cultural Psychology, 23,* 366-376.

Schiefenhövel, W. (1993). Pragmatismus und Utopie als reaktionen auf kulturellen Wandel - Beispiele aus Melanesien. In A. Thomas (Ed.), *Einführung in die kulturvergleichende Psychologie* (pp. 323-337). Göttingen: Hogrefe.

Sears, R.R., Maccoby, E.E., & Levin, H. (1957). *Patterns of child rearing.* Evanston, Ill.: Row, Peterson.

Silbereisen, R.K., Schwarz, B., & Kracke, B. (1993). Problemverhalten Jugendlicher im Kulturvergleich. In A. Thomas (Ed.), *Einführung in die kulturvergleichende Psychologie* (pp. 339-357). Göttingen: Hogrefe.

SmithP.-B., Peterson, M., Misumi, J., & Bond, M. (1992). A cross-cultural test of the Japanese PM leadership theory. *Applied Psychology, 41,* 5-19.

Sommer, M.C. (1988). Von der "kognitiven Wende" zur "Ära sozialer Repräsentationen"? Die Sozialpsychologie Serge Moscovici. In G. Jüttemann (Ed.), *Wegbereiter der Historischen Psychologie* (pp. 370-377). München, Weinheim: Beltz, Pschychologie-Verl.-Union.

Thomas, A. (1991). Psychologische Wirksamkeit von Kulturstandards im interkulturellen Handeln. In A. Thomas (Ed.), *Kulturstandards in der internationalen Begegnung* (pp. 55-69). SSIP-Bulletin 61. Saarbrücken: Breitenbach.

Thomas, A. (Ed.) (1993). *Kulturvergleichende Psychologie.* Göttingen: Hogrefe.

Thomas, A. (1993a). Entwicklungslinien und Erkenntniswert kulturvergleichender Psychologie. In A. Thomas (Ed.), *Einführung in die kulturvergleichende Psychologie* (pp. 27-51). Göttingen: Hogrefe.

Thomas, A. (1993b). Psychologie interkulturellen Lernens und Handelns. In A. Thomas (Ed.), *Einführung in die kulturvergleichende Psychologie* (pp. 377-424). Göttingen: Hogrefe.

Thomas, A. (1994). Kulturelle Divergenzen in der deutsch-deutschen Wirtschaftskooperation. In T. Kornbichler & C.-J. Hartwig (Eds.), *Kommunikationskultur und Arbeitswelt* (pp. 42-52). Berlin: Akademie.

Thomas, A., & Helfrich, H. (1993). Wahrnehmungspsychologische Aspekte im Kulturvergleich. In A. Thomas (Ed.), *Einführung in die kulturvergleichende Psychologie* (pp. 145-180). Göttingen: Hogrefe.

Thurnwald, R. (1924). Zum gegenwärtigen Stand der Völkerpsychologie. *Kölner Vierteljahreshefte für Soziologie,* IV, 32-43.

Triandis, H.C. (1994). Theoretical and methodological approaches to the study of collectivism and individualism. In U. Kim, H.C. Triandis, C. Kagitcibasi, S.-C. Choi, & G. Yoon (Eds.), *Individualism and collectivism* (pp. 41-51). Newbury Park, CA: Sage.

Trommsdorf, G. (1993a). Kulturvergleich von Emotionen beim prosozialen Handeln. In H. Mandl, M. Dreher, & H.-J. Kornadt (Eds.), *Entwicklung und Denken im kulturellen Kontext. Göttingen* (pp. 3-25). Hogrefe.

Trommsdorf, G. (1993b). Entwicklung im Kulturvergleich. In A. Thomas (Ed.), *Einführung in die kulturvergleichende Psychologie* (pp. 103-143). Göttingen: Hogrefe.

Wilpert, B. (1993). Führung und Partizipation im interkulturellen Vergleich. In A. Thomas (Ed.), *Einführung in die kulturvergleichende Psychologie* (pp. 359-375). Göttingen: Hogrefe.

Wundt, W. (1904). *Völkerpsychologie.* Leipzig: Engelmann.

Wundt, W. (1913). *Elemente der Völkerpsychologie - Grundlinien einer psychologischen Entwicklungsgeschichte der Menschheit.* Leipzig: Kröner.

Yamaguchi, S. (1994). Collectivism among the Japanese: A perspective from the self. In U. Kim, H.C. Triandis, C. Kagitcibasi, S.-C. Choi, & G. Yoon (Eds.), *Individualism and collectivism* (pp. 175-188). Newbury Park, CA: Sage.

Zimmer, A.C. (1982). What really is turquoise? A note on the evolution of color terms. *Psychological Research, 44,* 213-230.

Social Psychology and Culture: Reminders from Africa and Asia

Stuart C. Carr
University of Newcastle, Newcastle, NSW, Australia

Arguably, Leon Festinger's major psychological propositions are that people are driven: a) to resolve inconsistency (1957), b) to outperform comparable rivals (1954), and c) to verify reality socially (1950). The conceptual "glue" which binds these propositions together and initially justifies my considering them as one cohesive set of ideas consists of (a) their common emphasis on *cognitive motivation* (taken as people striving to make sense out of their experiences), and (b) the fact that they each seem to have had a wide and profound impact on the discipline of psychology in general, and on mainstream (Western) social psychology in particular.

Let us take a cursory glance at this impact. Moscovici's (1976) theory of social change, for instance, designates consistency as the vital lever by which minorities can compel the majority to internalize their messages (Moscovici, 1980), while one of the axioms of attitude scaling is internal consistency (McIver & Carmines, 1981). Indeed, cognitive dissonance is perhaps the best known and most frequently researched theory in social psychology (Wheeler, Deci, Reis, & Zuckerman, 1978). The Needs for Achievement and Power have as premises the will to compete (McClelland, 1987), while a central tenet in Tajfel's (1978) theory of social change is that members of the ingroup will strive to place themselves above their rivals. A great deal of contemporary social psychology rests on the principle that people are motivated to maintain a positive self-evaluation (Turner, 1991), namely by what Festinger (1954) termed a "unidirectional drive upwards". With regard to dependence on others in defining reality (Festinger, 1950), *reference* groups are today widely recognised as exerting a pervasive influence on our social lives (Sears, Peplau, & Taylor, 1991).

Thus, Festinger has provided contemporary social psychology with some of its cornerstones. It is therefore somewhat surprising, when one examines the *cross-cultural* literature in social psychology, to find that his name is indexed either infrequently (e.g., Smith & Bond, 1993) or not at all (e.g., Brislin, 1993). This surprise increases when one considers that the "conception" of modern cross-cultural psychology (Gergen, 1973) was marked by scepticism about - and a call to test - the cross-cultural validity of Festinger's work. Festinger himself (1954) was certainly aware of this need. Yet Gergen (1994) is still calling for empirical evidence relating to his thesis of cultural relativism. In support of some counter calls that social psychology may have much to offer non-Western societies (Carr, in press; Foster & Louw-Potgieter, 1991), the first objective in this paper is to address the outstanding issue with *empirical evidence* from radically *non*-Western worlds in Tropical Africa and Pacific Asia. Will Festinger's key constructs hold up?

Granted that these constructs are central to Western social psychology, they may provide critical tests between the continuing arguments in favor of "cultural" (Davidson, 1992; Misra & Gergen, 1993a, b), "inter-cultural" (Krewer & Jahoda, 1993), and "cross-cultural" approaches to the study of groups in diverse cultures (Poortinga, 1993; Triandis, 1993). Logically, either a successful, or a partial, or a failed crossing of these basic constructs from "Western" to "non-Western" worlds would also bring basic support for one of the three perspectives. The final objective in this paper is therefore to offer some empirical contribution towards a debate which has, arguably, occupied internationally-focused social psychology for long enough (Carr, MacLachlan, Zimba, & Bowa, in press-a; Carr & MacLachlan, in press-c; MacLachlan & Carr, in press-b).

Cognitive Tolerance

In a recent study of psychiatric admissions in Malawi, the patients frequently attributed their situation at one stroke to both witchcraft and modern medical causes (MacLachlan, Nyirenda, & Nyando, in press). A similar observation has been made that Western-trained nurses sometimes use protective spells to defend themselves against the envious wishes of colleagues vying with them for promotion (Carr & MacLachlan, 1993). Realizing that such pluralism might undermine an existing Western, "rationalist" model for managing health care in Malawi, these *apparent* contradictions have been investigated more closely. Quantitative surveys of traditional and modern medical beliefs about health have been conducted with regard to malaria and schistosomiasis (Ager, Carr, MacLachlan, & Kaneka-Chilongo, in press); "mental illness" (Pangani, Carr, MacLachlan, & Ager, in press); epilepsy (Shaba, MacLachlan, Carr, & Ager, in press); and AIDS (MacLachlan & Carr, in press-a).

In each case, the extent to which people endorsed one type of cause, prevention or treatment was not negatively correlated with their level of endorsement for what might be seen as the *alternative* system of belief. Malawians are apparently very *tolerant* of different health care systems. Carr and MacLachlan (in press b) and MacLachlan and Carr (1994) have described how there may be more "cognitive tolerance" in Malawi than in the West. We discuss the possibility that Festinger's (1957) "cognitive dissonance" theory may not apply in the Malawian context. In order to "refute" dissonance theory in Malawi, we argue, it will be necessary to rule out the possibilities: (1) that people simply failed to notice any inconsistency, and (2) that the issues were simply not salient enough to invoke dissonance (see, Chaiken & Baldwin, 1981). Ultimately, the use of physiological measures of arousal may enable an assessment of the emotional consequences (*or otherwise*) of perceived inconsistency (Croyle & Cooper, 1983).

Cognitive tolerance is not confined to Malawi. It has been recorded in other countries in and around Tropical Africa. These include Ghana (Jahoda, 1970), Sierra Leone (Wober, 1993), Nigeria (Wober, 1971), Kenya (Porter, Allen, & Thompson, 1991), Zimbabwe (Elliot, Pitts, & McMaster, 1992), Swaziland, South Africa, and East and West Africa generally (Wober, 1993). Although it has in the past been viewed as symptomatic of a culture in transition (e.g., Dawson, 1969a,b; Peltzer, 1987), cognitive tolerance may alternatively be viewed as a very adaptive response to the occupational, health, and environmental vagaries of the Third World (see Porter et al., 1991). Schumaker (1990) in fact argues that the ability to hold disparate cognitions in mind is an adaptive reaction to reality in *any* culture, suggesting further that "developed" nations may have much to learn from their "developing" counterparts.

Pluralism is also being increasingly documented in Pacific Asia. In Japan for example, the ability to tolerate inconsistency is both socialized and considered as a mark of maturity (Gui-Young Hong, 1992, cited in Moghaddam, Taylor, & Wright, 1993). Carr, Archer, and Malmberg (1994) have asked Japanese managers to estimate what percentage of people would seek cause, prevention and treatment of illness in (a) modern medical and/or (b) non-medical (spiritual/mystical) factors. If tolerance is the exception rather than the rule, then belief in "modern medicine" should exclude belief in "non-medical" phenomena and vice-versa. There would be a statistically significant negative correlation between variables (a) and (b). In fact, although this did occur with regard to treatment, there was no relationship between (a) and (b) for either cause or prevention. Speaking for themselves, almost one in ten of the sample confided that they held (a) and (b) type beliefs at the same time.

This somewhat ambiguous finding may become clearer in the light of a broader, regional perspective. Among Pacific Asian cultures (Hofstede, 1980), the Japanese are the least "uncertainty tolerant" - which has as a key feature a strong preference for an "absolute truth" (Hofstede & Bond, 1988). Since pluralistic "tolerance" seems to be the opposite of Hofstede's (1980) "uncertainty avoidance," then it would be reasonable to expect less ambiguous tendencies towards cognitive tolerance among Japan's Asian neighbours. For example, although Singaporeans live under a highly certain political regime, they have been found to be among the least "uncertainty-avoiding" peoples (Hofstede & Bond, 1988), and Bishop (1994, in press) has discovered a clear tolerance of both modern medical and traditional Chinese health care practices. Anecdotally, the same is heard of Papua New Guinea (McLoughlin, 1994) and Fiji (Schultz, 1994).

In Tropical Africa, there is at least one highly practical development implication of the tolerance finding. As in many other developing African countries, Malawi's Ministry of Health is attempting to "develop" a western-style health system, with traditional healers perhaps occupying at best a relatively marginal position. In Pacific Asia, Australian Aboriginal traditional healers are increasingly finding themselves in a similar predicament (Dudgeon, 1993). In Malawi however, locally-based psychologists have been effectively *marshalling the psychological evidence* to argue for the re-integration of traditional healers (Carr & MacLachlan, in press-b; MacLachlan, 1993; MacLachlan & Carr, 1993; MacLachlan & Carr, 1994). Such an integration has already taken place in Zimbabwe, where the former Professor of Sociology - now Vice Chancellor of the National University - is also President of the Tribal Healer's Association (Carr & Munro, 1994).

In addition, since it is widely accepted among clinical psychologists and medical doctors in the western mold that an understanding of a patient's perspective may of itself enhance recovery (Sarafino, 1990), the psychologists' case for incorporating the traditional healer has a therapeutic dimension (Carr & MacLachlan, in press-b; MacLachlan & Carr, 1993; MacLachlan & Carr, 1994). There may furthermore be prevention benefits accruing to such integration. Regarding AIDS for example, young Malawians - even with the most intensive exposure to western education - are far from ruling out the numerous traditional healers as "credible" sources of valuable preventative advice (MacLachlan & Carr, in press-a). Ager et al. (in press) argue that cognitive tolerance might undermine attitude theories which assume "rational" decision-making, such as the Theory of Planned Behavior (Ajzen, 1985). This may partly explain why AIDS prevention programs have failed badly in Malawi, where there are reports of four hundred new HIV infections daily (Liomba, 1994; World Health Organization, 1993). The Malawian campaign has relied heavily on the "rational" ethos of media channels, to the exclusion of the "traditional" sources of advice (Carr & MacLachlan, in press-a).

One of the key elements in leading western attitude theories, and indeed the focus of a great deal of western social psychology, is attitude scaling. In turn, one of the common assumptions in such scaling is that people inherently *want* or *need* to be consistent; a valid scale is one that displays high "internal consistency", or inter-item correlation, which is then taken as a sign that these items are all (validly) measuring the same thing (McIver & Carmines, 1981). Yet, if the notion of "cognitive *tolerance*" is valid, then high internal consistency would signal exactly the *reverse*. The attitude scale would be rendered *in*valid! Therefore, in non-western contexts such as Tropical Africa and Pacific Asia, alternative methods of assessing attitudes (e.g., Liggett, 1983) may have to be found (Carr, Munro, & Bishop, under review).

At this stage, there are still a number of question marks over the real extent to which processes of dissonance reduction and cognitive tolerance are valid in both non-Western and Western cultural settings. As Aldous Huxley once remarked, "The only completely consistent consistent people are the dead"! (Ajzen, 1988, p. 25) What is nevertheless clear is that the issue is worth investigating, on both theoretical and practical grounds. In the past perhaps, there has been something of a tendency to dismiss or overlook the non-

Western *cultural* perspective. To that extent, our first conclusion might be that *cultural* psychology is much needed in Tropical Africa and in Pacific Asia.

Motivational Gravity

Those "tolerant" professional nurses in Malawi (Carr & MacLachlan, 1993) were evidently hungry for "success," their personal ambition indicating a highly *competitive* atmosphere in the western-style workplace. In fact, anecdotal accounts of subordinate and coworker sabotage are common currency in Malawi (Carr & MacLachlan, 1993), and workers will allegedly turn down promotions or go for a spell of protection against *chizimba* (witchcraft) before taking up the appointment (Bowa & MacLachlan, 1994). Although such a competitive aura might seem to augur well for the cross-cultural validity of Festinger's (1954) "unidirectional drive upwards", choosing a *traditional* means of protection, against a *traditional* form of attack, suggests some kind of *inter*-cultural phenomenon, namely Western-workstyle, self-promotion running headlong into traditional cultural values and social forces (Carr, 1994).

In fact, managers in non-western worlds may know very well the wider, inter-cultural influences that may envelop or engulf unidirectional and upward motives (see, Blunt, 1983; Bond, 1988b; Carr & MacLachlan, 1993; Triandis, 1989). In many African and Asian societies, there are "collectivist" norms (Hofstede, 1980), in respect to which the emphasis is placed on belongingness, occupying one's proper place, serving group goals, reciprocity, humility, and social achievement (Markus & Kitayana, 1991). Thus for example, preserving "face" may be so important that managers are reluctant to employ individual criticism as a means of re-motivating inadequate individual performance (Abdullah, 1992; Namandwa, 1992).

Ipso facto, the existence of these traditional norms creates the potential for inter-cultural conflicts within non-western organizations (Munro, 1986b). For instance, there may be a conflict of interests between, a) the demands and opportunities of a system favoring individualistic behavior and b) collectivist values (Krewer & Jahoda, 1993). Peers and superiors collectively acting to curtail too much reward being bestowed on any one individual would be one conceivable manifestation of such a clash between organizational and wider social culture. It is possible to draw the same sort of theoretical inter-cultural analysis using the motives of high "power distance" and "uncertainty avoidance" (Hofstede, 1980) that may characterize organizations in countries like Kenya, Nigeria, and Malawi (see, Srinivas, 1994a). An empirical assessment was warranted.

Although some Australian and Japanese studies have recently investigated - and found - envy towards high achievers (Feather, 1994; Feather & McKee, 1993), they have focused on interrelationships between student attributions, values, and attitudes towards socially prominent and therefore remote figures like politicians. Tangible and deleterious behavior towards closer (and more comparable) organizational subordinates and coworkers has received very little attention. In the first organizational study (Carr, MacLachlan, Zimba, & Bowa, in press-b), some kind of guiding theoretical framework was needed. When thinking of formal organizations as distinct from wider informal society, we often think first of hierarchy, and this key feature of power structure renders the possibility of a taxonomy of organizational cultures (Carr & MacLachlan, under review). Along one axis, superiors may either envy or encourage their subordinates, while on a second axis, coworkers may either envy or encourage their peers. These axes create a four-fold classification system. Although Feather (1994) reminds us that we do not - yet - have any sophisticated theory of the psychology of envy, such taxonomies are recognized to provide the first step in new theory-building (Gould, 1994; Pryor, 1993).

A combination of Malawian managers and student managers were presented with a number of brief written scenarios. These depicted situations involving a worker who was performing very well, e.g., coming up with bright ideas, or regularly earning a bonus. They were based on real events described by Malawian psychology lecturers who had themselves been managers. Using a graded scale, the subjects were asked to estimate which reaction - encouragement or sabotage - would prevail among both management and fellow workers. Against the fact that ninety-nine percent of the sample agreed with the statement, "people in general want to do better than others," there were sharp predictions of "push down" by superiors and "pull down" by coworker peers.

It was especially notable that the managers - with more direct experience of the Malawian workplace - were the ones who made these predictions more strongly. In fact, in response to the question, "Should you encourage others to do better than yourself?" the managers were significantly *less* likely to respond "Yes". Fifty-eight percent of the sample responded "yes" - thirty-five percent managers and sixty-six percent students. Such findings lend weight to an inference that Malawi's organizations are indeed characterized by high negative "motivational gravity" (Carr, MacLachlan, Zimba, & Bowa, in press-b).

When the explanations for responses to the issue of encouraging others were content analyzed, the predominant themes were fierce and individualistic competition, with indications of a mixture of "power distance" and a preoccupation with job security ("threat to own position"). "Collectivist" reserving of encouragement for family and friends was relatively low in frequency, suggesting alienation. In its place, individualist competition was so intense that it would stifle any potential benefits - the Malawian sample fell into the worst possible category of the taxonomy of organizational cultures, namely "push down - pull down."

In an Asian study (Carr, Malmberg, & Archer, 1994), Japanese managers were presented with just two scenarios, one involving the reactions to achievement by (a) coworkers and (b) superiors. The target figures were presented as prominent in terms of: (a) frequently receiving a bonus, being awarded a trip overseas, or being promoted ahead of time, and (b) putting forward bright ideas, or taking a correspondence course, or being very keen. The subjects were then asked to estimate how many out of ten typical Japanese coworkers and superiors would somehow offer encouragement ("e.g., congratulations") or discouragement (e.g., "ostracism, mockery").

The overall tendency was consistently towards "encouragement." Thus, the Japanese sample would be classified as "pull up - push up." This contrasts with findings outside the workplace in wider Japanese society (see, Feather & McKee, 1993), where a popular adage states, "the nail that sticks up gets pounded down." Thus, in the Carr et al. workplace sample, there may have been effective management of negative motivational gravity.

Certainly, the content of explanations for personal reactions to high individual achievers in the workplace failed to reveal any envious or fear-driven intentions towards fellow employees. Instead, there was an apparently harmonious and healthy combination of competition tempered by classic "positive interdependence" (Johnson & Johnson, 1991). This finding converges perfectly with other research in Japanese organizations (Clarke, 1979). As one Japanese manager succinctly remarked: "Competition is good. It leads to progress."

What is the likely scope for negative motivational gravity in Tropical Africa and Pacific Asia? The empirical evidence for it converges with observations from a variety of other sources both within and beyond Malawi (Wober, 1993). The national language in Malawi - Chichewa - has no word meaning "congratulations" (Zambezi Mission Inc., 1986), despite being spoken by over ten million people in Malawi and in neighboring Zambia (Hullquist, 1988). Gule Wamkulu society includes punishing those individuals who overstep their position (Carr & MacLachlan, under review). A Malawian undergraduate recently wrote, "We, Africans, we don't accept the superiority of the other without feeling jealousy" (Gondwe, 1993).

In related empirical studies, Jones (1988) found that Malawian managers - like those from other developing societies - were not keen on sharing ideas or power with subordinates. Seddon (1985) found similar tendencies in Nigeria and Kenya. Like Carr and MacLachlan (under review), Jones (1988) also found that Malawian managers were preoccupied with job security. Again, a similar tendency has been found in Kenya and Liberia (Blunt, 1983). In Malawi, Kenya and Nigeria, managers may undervalue subordinate capacities to avoid threats to their own position; desire personal power and prestige; make frequent transfers of personnel to prevent them building any power base of their own; discourage upward communication; and demotivate the junior levels, thereby bringing overload on themselves (Kiggundu, 1991). Munro (1986a) found "fear of success" among Zimbabwean students.

In Pacific Asia, there are also anecdotal accounts of negative motivational gravity. In Papua New Guinea, the achieving individualist is named

"Shoe-sock man" and is frequently victimized (McLoughlin, 1994), while his Hong Kong homologue may provoke "Red Eye disease" (Bond, 1994). Envy in the workplace also causes managerial problems in India (Nadkarni, 1994), in South Korea (Lee, 1994), and in Fiji (Shultz, 1994). "Fear of success" has also been detected socially, for example among Chinese and Indians in Malaysia (Wan-Rafaei, 1984) and among Singaporean women (Kaur & Ward, 1992). For their part, Anglo Australians widely refer to the "tall poppy" being cut down (Conway, 1971, Feather, 1994). We have already seen that beyond the workplace, Japanese preferred to see high achievers fall rather than be rewarded - even more so than their Australian counterparts (see Feather, 1994; Feather & McKee, 1993). Lazarevic (1992) found the term "Big Noting" being applied by Aboriginal Australian students to describe academic conspicuousness.

Taken as a whole, the evidence provides us with several relatively conservative inferences about negative motivational gravity (Carr, MacLachlan, & Schultz, in press). These are: (1) it is widespread, being found across both Tropical Africa and Pacific Asia; (2) it has some respectable traditional foundations; and (3) it often arises as a collectivistic response to the expression of an individualistic "unidirectional drive upwards" at work. Western-style work structuring might therefore inter-act with more traditional social forces to produce the high levels of negative motivational gravity observed, for instance, in the Malawian study (see also, Clarke, 1981). The *full* extent to which Festinger's unidirectional drive upwards may require restatement to accommodate local cultures is perhaps best illustrated by briefly detailing the practical and theoretical implications of the empirical findings.

The first point is that there may be financial costs to motivational gravity in workplace settings. The bulk of the empirical work comes from "equalitarian" Anglo Australia. In support of a widespread stereotype (Dixson, 1993; Grey, 1993; Taft & Day, 1988), Feather (1994) has found a "favor fall" motive applying to hypothetical persons, real political figures, social attitudes, and value priorities. As new Australian research extends to formal organizational groups (Carr & Chidgey, 1994), the economic significance of negative motivational gravity is becoming increasingly clear. For instance, a restaurant complex recently went bankrupt after it became unmanageable due to the gravity provoked by promoting former work "mates" above one another (McLoughlin & Carr, 1994). The danger in Australian organizations may emanate from coworkers more than superiors (i.e., "pull down" but not "push down"). Superiors instead might simply ignore high achievements, i.e. withhold positive reinforcement (Carr & Chidgey, 1994). This is partly perhaps why the introduction of teambuilding and feedback systems are having critical impacts on profitability in Australian industry (respectively, Carnegie & Butlin, 1993; Vecchio, Hearn, & Southey, 1992). Australian advertisements for organizational recruitment often seek to recruit the "quiet achiever" (e.g., Carlton, 1993).

To the extent that African and Asian societies may resemble each other, for example placing value on "collectivism" and "power distance" (Hofstede, 1980), then empirical cases where negative motivational gravity is managed successfully, such as in Japan, may provide practical solutions which generalize to other non-Western countries. The Japanese organization of work is based on the traditional *ie* or family metaphor in place of the machine metaphor of the west (Kashima & Callan, in press). For instance, Japanese superiors are overtly expected to fulfil a paternalistic function towards their subordinates. In such ways, company familism (*keiei-kazoku-shugi*) or a belief in the one-enterprise family (*kigyo-ikka*) serve to integrate workers and worker groups within the Japanese factory. Since such "positive interdependent" teamwork seems to deflect the sentiment that "the nail that sticks out gets pounded down," then traditional family and group metaphors may provide general purpose keys to specific intitiatives for managing negative motivational gravity.

If negative motivational gravity is empirically identified within an organization, one basic tactic would be to creatively write the job description in group-oriented, family terms. This would also provide victims of motivational gravity with a mechanism for expressing grievances, and possibly a greater sense of job security, as would demonopolizing supervisory control. The classification system might also be applied to develop recruitment and selection strategies, by matching candidates to the organizational culture, e.g., seeking out "quiet achievers" for a "push down - pull down" culture, or "high flyers" for "pull up - push up." Feather (1994) has found empirical links between relatively stable personal characteristics and encouragement towards high achievers. These traits include self-esteem and work ethic (positive correlation), and tradition-orientation and belief in equality (negative correlation). Thus, some personality profiles may "fit" the organizational culture of any particular company better than others.

Suggestively, Hofstede (1980) included the following item in his list of "uncertainty avoidance" items: "How often would you say your immediate manager is concerned about helping you get ahead?" This inclusion implies that motivational gravity may correlate (positively) with "uncertainty avoidance." Although the Malawian findings of job insecurity support this hypothesis, the Japanese are both high on uncertainty avoidance and good managers (apparently) of motivational gravity. Hence, empirical research in different cultures must decide whether "uncertainty avoidance" would be a useful selection construct (see, Bond, Leung, & Schwartz, 1992).

With regard to developing existing human resources, management may of course wish to train motivational gravity out of its workforce, and there are obvious possibilities for survey feedback and groupwork (Carnegie & Butlin, 1993). Team-building might be accomplished by techniques such as instructional training which stresses the beneficial impact of *positive* reinforcement (Vecchio et al, 1992); by "outward bound" activities and simulation "management games" (Srinivas, 1994b); by music (Nadkarni, 1994); or by "jigsaw" methods (Aronson, Blaney, Stephan, Sikes, & Snapp, 1978) to cultivate feelings of positive interdependence (Johnson & Johnson, 1991). The latter strategy might be particularly appropriate, by affording the cohabitation of both collectivism (interlocking team roles) and individualism (each worker has their own particular project).

Wober (1993) has suggested that group discussion may be employed in African organizations to reverse motivational gravity. Empirical evidence from Australia gives this idea a promising ring. Feather (1994) finds ambivalence in Australians' attitudes towards high achievers. "Favor fall" and "favor reward" are separate factors. Conway (1971) has traced such ambivalence to identification with traditional authority coupled to an historical rejection by and therefore of that very authority. Making a more general point, Reeve (1992) reminds us that individual "success" often attracts both admiration and resentment.

Whichever is the case, the tendency towards admiration may, upon measurement, prove to be stronger than its opposite. In this event, the tendency for group discussion to polarize group norms (Lamm & Myers, 1978) may produce a clearer dominance of *support* for high achievers over *resentment* against them (a "Net Gain" hypothesis). Carr (1987) also found that group discussion tends to enhance logical consistency in group opinions, thereby suggesting that group discussion may favor the development of admiration of achievements - at the expense of resentment. Assuming that some kind of intercultural tension often exists in non-Western organizations, perhaps such group techniques will be useful to counteract any motivational gravity?

Survey feedback and techniques such as focus groups (for example) could serve to explore (organizational) culture-consistent ways of capitalizing upon it (see Templer, Beaty, & Hofmeyr, 1992). If for instance an organization is characterised by "push down - pull down," then it may be productive to nurture a "conscience professionnelle" as an alternative to encouraging people to work hard as individuals (Munro, 1986b). Professionalism incorporates traditional values like cooperation, teamwork, and "face" before the group. Again, to the extent that non-Western organizations share with the Japanese a tradition of "collectivist" and "power distance" values, then they might profit from quality circles and associated work practices such as TQC (McConnell, 1986).

With regard to occupational health, the existence of motivational gravity suggests that the quality of working life may be partly determined by the degree of "fit" between western-type work practices and non-western, traditional cultural values. In direct and dramatic support for this proposition, a recent sample survey of over one hundred admissions to Malawi's major psychiatric facility has revealed that no less than *forty percent* of these patients attributed their condition *to motivational gravity* - in the majority of cases stemming from over-achievement at work (MacLachlan et al., in press). As one victim of motivational gravity characteristically explained, "I was bewitched by my workmates...because I work hard..." Thus, in addition to a financial bottom line, motivational gravity may carry a very high *human* cost.

In sum, in non-Western workplaces, Western systems which are designed to encourage rather than suppress the expression of a self-promoting "unidirectional drive upwards" may be creating inter-cultural tensions and conflicts, both at group and intrapsychic levels. In further support of these possibilities, a number of observers in Africa and Asia have discussed the extent to which traditional collectivism involves a heightened distinction between ingroup and outgroup. Compared to individualists, collectivists are often both more loyal to the ingroup and more competitive against the outgroup in the workplace (Abdullah, 1992; Hayajneh, Dwairi, & Udeh, 1994; Hui, 1990; Leung & Wu, 1990; Munro, 1986b). Thus, our second conclusion might be that more understanding of *inter* cultural psychology (Krewer & Jahoda, 1993) is needed in Tropical Africa and Pacific Asia.

Cultural Reactance?

There remains Festinger's influential idea that Westerners are obliged to verify much of reality by *social* rather than physical means, an obligation which then motivates them to affiliate with "reference" groups. Hofstede's (1980) study implies that such affiliative behavior will be even more important in non-Western cultures, i.e., those that stress "collectivist" values. This might lull us into assuming, implicitly perhaps, that behavior will follow the same basic *affiliative* pattern across cultures. By taking the influence of reference groups somewhat for granted, we might not see any real need for cross-cultural comparisons in levels of affiliative behavior.

As we have now seen however, workers in "collectivist" countries frequently do not *behave* collectivistically at all! They may even be *anti*-affiliative in the work context. In education too, Clarke (1981) found that Malawian students were intentionally averse to working "collectively." He hypothesized that when Western institutions provide individuals with opportunities for temporary escape from collectivist institutions like the extended family, then "collectivist" societies can in fact - sometimes - produce some of the most individualistic, i.e., "reactant" behavior (Brehm & Brehm, 1981). Similarly perhaps, a highly "individualistic" society like North America may sometimes generate the motivation to find fulfillment through *group* therapies (Keats, 1994). With regard to workplace behavior then, anti-conformity or reactance might partly explain why we find such individualistic, anti-affiliative behavior among the Malawian workers.

A sample of Japanese managers perceived one in two people as being motivated to define reality by means of a reference group, whereas two out of three described themselves as loners, explaining that it is important to be oneself and to recognize that reality is diverse (Carr & Malmberg, in preparation). One recalls too the uncertainty *tolerance* of Singaporeans living within the uncertainty *avoiding* Singaporean *system*. Anglo Australians are supposed to prize individualism (Hofstede, 1980). Yet when we look more closely at their behavioral intentions (Chidgey & Carr, in preparation), the majority regularly consult and depend on "reference groups." These preliminary results concur with other studies which have found that Australians are relatively high on Need for Affiliation (Messinesi, cited in Petzall, Selvarajah, & Willis, 1991). Similarly, much Australian behavior revolves around "mateship" and *team* sports.

The lesson of these findings, perhaps, is that we cannot simply assume that behavior will occur as it "ought" to occur according to differences in "cultural values" (Hofstede, 1980). Socially-oriented behavior can evidently occur where it is *least* expected. In that sense, affiliative behavior may be an extremely *robust* phenomenon. At the same time however, *non*-affiliative behavior may also occur where it is least expected, and in that sense we can *never* afford to take affiliative behavior for granted. To the extent that individual or affiliative behavior does *not* automatically follow from cultural values, our third conclusion might be that maintaining a *cross*-cultural approach is likely to continue to serve important functions in advancing our understanding of actual behavior.

Conclusion

Once an emphasis on *measurement* is adopted, we are drawn to the conclusion that all three approaches to social psychology - cultural, inter-cultural, and cross-cultural - possess considerable merit. This conclusion is all the more compelling because it applies within one person's ideas, which also happen to be at the core of our discipline. Such plurality would seem to undermine further "uncertainty-avoiding" debate about "which" particular approach to adopt. Measuring social processes and behavior also brings considerable practical advantages as well as theoretical renewal. As we have seen, the relatively macroscopic measurement approach of the 1980s (e.g. Hofstede, 1980; Bond, 1988a) has given us some vital compass points. Next, since a good deal of social psychology can intervene between the cultural value and an individual's behavior, social psychologists might begin to chart that particular territory. Let us rise to Gergen's (1973, 1994) challenges, and construct an empirical case that social psychology is *not* merely "history."

References

Abdullah, A. (1992). *Understanding the Malaysian workforce*. Kuala Lumpur: Malaysian Institute of Management.

Ager, A. K., Carr, S. C., MacLachlan, M., & Kaneka-Chilongo, B. (in press). Beliefs about cause, risk reduction, and treatment for schistosomiasis and malaria in rural Malawi. *Malawi Medical Journal*.

Ajzen, I. (1985). From intentions to actions: A theory of planned behavior. In J. Kuhl & J. Beckmann (Eds.). *Action control: From cognition to behavior*. Heidelberg: Springer.

Aronson, E., Blaney, N., Stephan, C., Sikes, J., & Snapp, M. (1978). *The jigsaw classroom*. Beverly Hills: Sage.

Bishop, G. D. (1994). *Health beliefs and the use of "modern" vs. "traditional" medicine*. Third Afro-Asian Psychological Congress, August 22-26, Kuala Lumpur, Malaysia.

Bishop, G. D. (in press). Disease prototypes: Their nature and function. In G. Petrillo (Ed.), *The social psychology of health*. Naples: Liguori.

Blunt, P. (1983). *Organizational theory and behaviour: An African perspective*. New York: Longman.

Bond, M. H. (1988a). Finding universal dimensions of individual variation in multicultural studies of values: The Rokeach and Chinese Value Surveys. *Journal of Personality and Social Psychology, 55*, 1009-1015.

Bond, M. H. (Ed.). (1988b). *The cross-cultural challenge to social psychology*. Beverly Hills: Sage.

Bond, M. H. (1994). *Personal correspondence*, April 1.

Bond, M.H., Leung, K., & Schwartz, S. (1992). Explaining choices in procedural and distributive justice across cultures. *International Journal of Psychology, 27*, 211-225.

Bowa, M., & MacLachlan, M. (1994). *No congratulations in Chichewa: Deterring achievement motivation in Malawi*. Research and Development III: University of Malawi.

Brehm, S.S., & Brehm, J.W. (1981). *Psychological reactance*. New York: Academic Press.

Brislin, R.W. (1993). *Understanding culture's influence on behavior*. Fort Worth: Harcourt Brace Jovanovich.

Carlton, P. (1993). Training coordinator: Someone who wants to make a difference. *The Weekend Australian*, October 2-3, 54.

Carnegie, R., & Butlin, M. (1993). *Managing the innovating enterprise*. Melbourne: Business Council of Australia.

Carr, S.C. (1987). *Images of deviance and social influence*. Unpublished PhD thesis, Stirling University, Australia.

Carr, S.C. (1994). *Generating velocity for overcoming motivational gravity in LDC business organizations*. Penang, Malaysia: Third Annual World Business Congress, June 16-18.

Carr, S.C. (in press). Social psychology and the management of aid. In S. C. Carr & J. F. Schumaker (Eds.), *Psychology and the developing world*. New York: Praeger.

Carr, S.C., Archer, L., & Malmberg, R. (1994). *Japanese beliefs about medicine: A healthy cognitive tolerance?* Newcastle: University of Newcastle.

Carr, S.C., & Chidgey, J. (1994). *Executive consultancy report*. Newcastle: Pristus Consulting Group.

Carr, S.C., & MacLachlan, M. (1993). Asserting psychology in Malawi. *The Psychologist, 6*, 413-419.

Carr, S. C., & MacLachlan, M. (in press-a). Family health care in Malawi: The sustainable community alternative for AIDS management. *Ife Psychologia*.

Carr, S. C., & MacLachlan, M. (in press-b). Managing tropical health: Psychology for development? *British Medical Anthropology Review*.

Carr, S. C., & MacLachlan, M. (in press-c). Towards a Malawian psychology. *The Journal of Psychology in Africa*.

Carr, S. C., & MacLachlan, M. (under review). The Motivational Gravity Grid: Dynamic force fields at work. *Psychological Science*.

Carr, S. C., MacLachlan, M., Zimba, C., & Bowa, M. (in press-a). Community aid abroad: A Malawian perspective. *The Journal of Social Psychology*.

Carr, S. C., MacLachlan, M., Zimba, C., & Bowa, M. (in press-b). Managing motivational gravity in Malawi. *The Journal of Social Psychology*.

Carr, S. C., MacLachlan, M., & Schultz, R. (in press). Pacific Asia Psychology: Ideas for Development? *South Pacific Journal of Psychology*.

Carr, S.C., & Malmberg, R. (in preparation). *The changing face of collectivism in contemporary Japan*. Newcastle: University of Newcastle.

Carr, S.C., Malmberg, R., & Archer, L. (1994). *Positive interdependence counteracting motivational gravity: World's best practice in Japan?* Newcastle: University of Newcastle.

Carr, S.C., & Munro, D. (1994). A New style of psychology for development studies. *Development Bulletin, 32*, 60-63.

Carr, S.C., Munro, D., & Bishop, G. D. (Under review). *Attitude assessment in non-Western countries: Critical modifications to Likert scaling*.

Chaiken, S., & Baldwin, M.W. (1981). Affective-cognitive consistency and the effect of salient behavioral information on the self-perception of attitudes. *Journal of Personality and Social Psychology, 41*, 1-12.

Chidgey, J., & Carr, S. C. (in preparation). *The affiliative tendencies of Australians in the workplace*. Newcastle: University of Newcastle.

Clarke, R. (1979). *The Japanese company*. New Haven: Yale University Press.

Clarke, R. (1981). *Independent learning in an African country, with special reference to the Certificate of Adult Studies in the University of Malawi*. Unpublished PhD thesis, University of Manchester.

Conway, R. (1971). *The great Australian stupor*. Melbourne: Sun Books Pty.

Croyle, R.T., & Cooper, J. (1983). Dissonance arousal: Physiological evidence. *Journal of Personality and Social Psychology, 45*, 782-791.

Dawson, J.L.M. (1969a). Attitudinal consistency and conflict in West Africa. *International Journal of Psychology, 4*, 39-54.

Dawson, J.L.M. (1969b). Theoretical and research bases of bio-psychology. *University of Hong Kong Gaz, 3*, 1-10.

Davidson, G. (1992). Toward an applied Aboriginal psychology. *South Pacific Journal of Psychology, 5*, 1-20.

Dixson, M. (1993). A nation in thrall to the third deadly sin. *The Weekend Australian*, May 26-27, 23.

Dudgeon, P. (1993). *Aborigines and western psychology*. Gold Coast: Australian Psychological Society Annual Conference, September 29-October 2.

Elliot, E., Pitts, M., & McMaster, J. (1992). Nurses' views of parasuicide in a developing country. *The International Journal of Social Psychiatry, 38*, 273-279.

Feather, N. T. (1994). Attitudes toward high achievers and reactions to their fall: Theory and research concerning tall poppies. *Advances in Experimental Social Psychology, 26*, 1-73.

Feather, N. T., & McKee, I. R. (1993). Global self-esteem and attitudes toward the high achiever for Australian and Japanese students. *Social Psychological Quarterly, 56*, 65-76.

Festinger, L. (1950). Informal social communication. *Psychological Review, 57*, 271-282.

Festinger, L. (1954). A theory of social comparison processes. *Human Relations, 1*, 117-140.

Festinger, L. (1957). *A theory of cognitive dissonance.* Stanford: Stanford University Press.

Foster, D., & Louw-Potgieter, J. (1991). *Social psychology in South Africa.* Johannesburg: Lexicon.

Gergen, K. J. (1973). Social psychology as history. *Journal of Personality and Social Psychology, 26*, 309-320.

Gergen, K. J. (1994). *Toward transformation in social knowledge.* Trowbridge: Sage.

Gondwe, S. C. (1993). *An outline of Festinger's key theoretical ideas: Do they help us explain social stability and change?* Zomba: University of Malawi.

Gould, S. J. (1994). Pride of place: Science without taxonomy is blind. *The Sciences, March/April*, 38-39.

Grey, J. (1993). Tall-Poppy soldier never cut down. *The Weekend Australian*, November 6-7, 8.

Hayajneh, A. F., Dwairi, M. A., & Udeh, I. E. (1994). Nepotism as a dilemma for managing human resources overseas: Its impact on employees, management, and organizations. *Journal of Transnational Management Development, 1*, 51-74.

Hofstede, G. (1980). *Culture's consequences: International differences in work related values.* Beverly Hills: Sage.

Hofstede, G., & Bond, M. H. (1988). The Confucius connection: From cultural roots to economic growth. *Organizational Dynamics, 16*, 4-21.

Hui, C. H. (1990). Work attitudes, leadership styles, and managerial behaviors in different cultures. In R. W. Brislin (Ed.), *Applied cross-cultural psychology* (pp. 186-208). Newbury Park, California: Sage.

Hullquist, C. G. (1988). *Simply Chichewa.* Makwasa: Malamulo Publishing House.

Jahoda, G. (1970). Supernatural beliefs and changing cognitive structures among Ghanaian university students. *Journal of Cross-Cultural Psychology, 1*, 115-130.

Johnson, D.W., & Johnson, F.P. (1991). *Joining together: Group theory and group skills.* Englewood Cliffs: New Jersey.

Jones, M. (1988). Managerial thinking: An African perspective. *Journal of Management Studies, 25*, 481-505.

Kashima, Y., & Callan, V.J. (in press). The Japanese work group. *Handbook of Industrial and Organizational Psychology, 4*, 1-67.

Kaur, R., & Ward, C. (1992). Cross-cultural construct validity study of "fear of success": A Singaporean case study. In S. Iwawaki, Y. Kashima, & K. Leung. (Eds.). *Innovations in cross-cultural psychology* (pp. 214-222). Lisse: Swets & Zeitlinger.

Keats, D. (1994). *Personal correspondence*, May 31.

Kiggundu, M. N. (1991). The challenge of management development in Sub-Saharan Africa. *Journal of Management Development, 10*, 32-47.

Krewer, B., & Jahoda, G. (1993). Psychologie et culture: Vers une solution du "Babel"? *International Journal of Psychology, 28*, 367-375.

Lamm, H., & Myers, D. G. (1978). Group induced polarization of attitudes and behavior. *Advances in Experimental Social Psychology, 11*, 145-187.

Lazarevic, R. (1992). *The self-esteem of rural and urban Aboriginal students in New South Wales*. Master's thesis, University of Newcastle, Australia.

Lee, J. K. (1994). *Personal correspondence*, June 18.

Leung, K., & Wu, P. G. (1990). Dispute processing: A cross-cultural analysis. In R.W. Brislin (Ed.), *Applied cross-cultural psychology* (pp. 209-231). Newbury Park, California: Sage.

Liggett, J. (1983). Some practical problems of assessment in developing countries. In F. Blackler (Ed.), *Social psychology and developing countries* (pp. 71-85). Chichester: Wiley.

Liomba, N.G. (1994). Statistics for HIV/AIDS in Malawi. Lilongwe, Malawi: Ministry of Health.

MacLachlan, M. (1993). Mental health in Malawi: Which way forward? *Journal of Mental Health, 2*, 271-274.

MacLachlan, M., & Carr, S. C. (1993). *Tropical tolerance in Malawi*. Ministry of Health. Zomba: University of Malawi Press.

MacLachlan, M., & Carr, S. C. (1994). From dissonance to tolerance: Toward managing health in tropical cultures? *Psychology and Developing Societies, 6*, 119-129.

MacLachlan, M., & Carr, S. C. (in press-a). Managing the AIDS crisis in Africa: In support of pluralism. *Journal of Management in Medicine*.

MacLachlan, M., & Carr, S. C. (in press-b). Pathways to a psychology for development: Reconstituting, restating, refuting and realizing. *Psychology and Developing Societies*.

MacLachlan, M., Nyirenda, T., & Nyando, M. C. (in press). Attributions for admission to Zomba Mental Hospital: Implications for the development of mental health services in Malawi. *International Journal of Social Psychiatry*.

Markus, H. R., & Kitayana, S. (1991). Culture and the self: Implications for cognition, emotion and motivation. *Psychological Review, 98*, 224-253.

McClelland, D. C. (1987). Characteristics of successful entrepreneurs. *Journal of Creative Behaviour, 21*, 219-233.

McConnell, J. (1986). *The seven tools of TQC*. Sydney: Enterprise Australia Publications.

McIver, J. P., & Carmines, E. G. (1981). *Unidimensional scaling*. Beverly Hills: Sage.

McLoughlin, D. (1994). *Personal correspondence*, September 3.

McLoughlin, D., & Carr, S.C. (1994). *The Buick Bar & Grill*. Melbourne: University of Melbourne Case Study Services.

Misra, G., & Gergen, K. J. (1993a). Beyond scientific colonialism: A reply to Poortinga and Triandis. *International Journal of Psychology, 28*, 251-254.

Moghaddam, F. M., Taylor, D., & Wright, S. C. (1993). *Social psychology in cross-cultural perspective*. New York: W.H. Freeman.

Moscovici, S. (1976). *Social influence and social change*. London: Academic Press.

Moscovici, S. (1980). Toward a theory of conversion behavior. *Advances in Experimental Social Psychology, 13*, 209-239.

Munro, D. (1986a). The meaning of Eysenck's personality constructs and scales for Zimbabwean male students. *Personality and Individual Differences, 7*, 283-291.

Munro, D. (1986b). Work motivation and values: Problems in and out of Africa. *Australian Journal of Psychology, 38*, 285-296.

Nadkarni, R. P. (1994). *Use of music for transforming the work culture*. Penang, Malaysia: Third World Business Congress, June 16-18.

Namandwa, D. Z. (1992). Personnel and training manager, Agricultural Development and Marketing Corporation (ADMARC), Malawi. *Personal correspondence*.

Pangani, D, Carr, S. C., MacLachlan, M., & Ager, A. K. (in press). Medical versus traditional attributions for psychiatric symptomatology in the tropics: Which reflects greater tolerance? *Medical Science Research*.

Peltzer, K. (1987). *Some contributions of traditional healing practices towards psychosocial health care in Malawi*. Heidelberg: Asanger.

Petzall, S. B., Selvarajah, C. T., & Willis, Q. F. (1991). *Management: A behavioural approach*. Melbourne: Longman Cheshire.

Poortinga, Y. H. (1993). Is there no child in the experimental bathwater? A comment on Misra and Gergen. *International Journal of Psychology, 28*, 245-248.

Porter, D., Allen, B., & Thompson, G. (1991). *Development in practice: Paved with good intentions*. Guidlford: Routledge.

Pryor, R. G. L. (1993). Returning from the wilderness. *Australian Journal of Career Development, September*, 13-17.

Reeve, J. M. (1992). *Understanding motivation and emotion*. Fort Worth: Harcourt Brace Jovanovich.

Sarafino, E. P. (1990). *Health psychology: Biopsychosocial interactions*. New York: Wiley.

Schultz, R. (1994). Personal correspondence, January 27. The University of the South Pacific, Fiji.

Schumaker, J. F. (1990). *Wings of illusion*. Buffalo, NY: Prometheus.

Sears, D. O., Peplau, L. A., & Taylor, S. E. (1991). *Social psychology*. Englewood Cliffs: Prentice-Hall.

Seddon, J. (1985). The development and indigenisation of Third World business: African values in the workplace. In V. Hammond (Ed.), *Current research in management* (pp. 98-109). London: Pinter.

Shaba, B., MacLachlan, M., Carr, S. C., & Ager, A. K. (in press). Palliative versus curative beliefs regarding tropical epilepsy as a function of traditional and medical attributions. *Central African Journal of Medicine*.

Smith, P. B., & Bond, M. H. (1993). Social psychology across cultures. Cambridge: Harvester Wheatsheaf.

Srinivas, K. M. (1994a). Organization development for national development: A review of evidence. In Kanungo, R. N., & Saunders, D. (Eds.), *Employee management in developing countries* (pp. 1-16). Regina: JAI Press.

Srinivas, K. M. (1994b). *The shrinking globe and the challenge of expanding mindsets*. Penang, Malaysia: Third World Business Congress, June 18.

Taft, R., & Day, R. H. (1988). Psychology in Australia. *Annual Review of Psychology, 39*, 375-400.

Tajfel, H. (1978). *Differentiation between social groups*. London: Academic Press.

Templer, A., Beaty, D., & Hofmeyr, K. (1992). The challenge of management development in S. Africa: So little time and so much to do. *Journal of Management Development, 11*, 32-41.

Triandis, H. C. (1989). Cross-cultural studies of individualism and collectivism. In J. J. Berman, (Ed.), *Nebraska Symposium on Motivation*. Lincoln: University of Nebraska Press.

Triandis, H. C. (1993). Comment on Misra and Gergen's: Place of culture in psychological science. *International Journal of Psychology, 28,* 249-250.

Turner, J. C. (1991). *Social influence.* Milton Keynes: Open University Press.

Vecchio, R. P., Hearn, G., & Southey, G. (1992). *Organisational behaviour: Life at work in Australia.* Marrickville: Harcourt Brace Jovanovich.

Wan-Rafaei, A. R. (1984). Achievement motivation and attribution of success in urban Malaysian ethnic groups. In Y. C. Leong, H. K. Chiam, & L. S. M. Chew (Eds.), *Preparation for adulthood: Proceedings of Third Asian Workshop on Child and Adolescent Development* (pp. 266-267). Kuala Lumpur: University of Malaya.

Wheeler, L., Deci, E. L., Reis, H. T., & Zuckerman, M. (1978). *Interpersonal influence.* Boston: Allyn & Bacon.

Wober, M. (1971). Adapting Dawson's traditional versus western attitudes scale and presenting some new information from Africa. *British Journal of Social and Clinical Psychology, 10,* 101-113.

Wober, M. (1993). The right question. Social Psychological and Marketing Research. *Personal correspondence,* September 9.

World Health Organization. (1993). AIDS - Global data. *Weekly Epidemiological Record, 27,* 193-196.

Wright, P.L., Taylor, D.S. (1985). The implications of a skills approach to leadership. *Journal of Management-Development, 4,* 15-28.

Zambezi Mission Inc. (1986). *The student's English-Chichewa dictionary.* Makwasa: Claim.

Acknowledgement

The Japanese study reported in this paper is funded by grant no. 45/280/340 awarded by the Research Management Committee at the University of Newcastle in Australia.

Part II

Consequences of Acculturation

Cognitive Functioning in Relation to some Eco-Cultural and Acculturational Features of Tribal Groups of Bihar

R.C. Mishra, Banaras Hindu University, Varanasi, India
Durganand Sinha, Allahabad University, India
John W. Berry, Queen's University, Kingston, Canada

Description and explanation of psychological differences among human populations as a function of their cultural diversity is one of the major tasks of cross-cultural psychology. The concept of *culture* has been defined in various ways. However, in recent years it has been generally taken to include the notions that are broader than culture (e.g., ecology), and notions that are relatively narrower (e.g., ethnicity) (Berry, 1985). The term *cognition* is used to refer to every process by which individuals obtain and utilize knowledge of objects and events of their environment. Research indicates that in the course of development, children acquire different cognitive operations as a function of their interaction with the environment. There exist a range and network of contexts for human development and behavior. These are exemplified in an eco-cultural model (Berry, 1976, 1987), which proposes ecology and acculturation as two major influences on human behaviour.

Pictures and pictorial materials have been widely used as tasks for the study of cognitive functioning. Recognition of line drawings has been used as a measure of intellectual development on tests like the Stanford-Binet, and picture assembly constitutes an important subtest of the Wechsler scale. A wide range of pictures (e.g., random shapes, matrix patterns, clear and detailed drawings and photographs of animals and objects, representational and conventional pictures involving sequence and symbolism, and impossible pictures) has been used in studies (see Deregowski, 1980). The general conclusion is that the recognition of clear pictures and photographs is not always immediate in cultures without pictorial tradition (Deregowski, Muldrow, & Muldrow, 1972). Studies that have generated cognitive demands by seeking some sort of interpretation of pictures (i.e., description of events in pictures or abstraction of theme) have often provided evidence of misinterpretation, with a greater frequency for rural and unschooled than for urban and schooled subjects (Mishra, 1976, 1987, 1988; Sinha & Mishra, 1982; Winter, 1963).

The differences noted in the interpretation of pictures are explained either in terms of differences in people's awareness of traditions about the representation of objects or scenes (Gombrich, 1977), or in terms of their failure to analyse information contained in the picture (Gibson, 1966). Hudson's (1960, 1967) classical work with South African samples demonstrates that the probability of three dimensional (3-D) perception of pictures increases with school education and acculturation. Using alternative procedures (largely non-verbal) and tests, the proportion of 3-D responders has been found to be greater than that reported by Hudson (Deregowski, 1968; Deregowski & Bentley, 1986; Jahoda & McGurk, 1974; Sinha & Shukla, 1974). However, the conclusion that cultural factors play a vital role in the perception and interpretation of pictures, remains unchallenged.

In studies of cognitive processes, the role of "contact-acculturation" has been outlined. The "contact" factor virtually lies at the heart of the very concept of acculturation (Redfield, Linton, & Herskovits, 1936). In psychological studies, acculturation refers to psychological changes in an individual whose cultural group is undergoing a change through contact with other groups. A broad category of these changes is called "behavioral shifts" (Berry, 1980). Changes in perception and cognition are foremost among them. With continued contact, these tend to be similar to those displayed by the acculturating group (Berry, 1980).

Test acculturation is another factor found associated with differences in performance of cognitive tasks (Berry, van de Koppel, Senechal, Annis, Bahuchet, Cavalli-Sforza, & Witkin, 1986). It refers to the ease and comfort an individual feels in performing a cognitive task in a test situation in the presence of a tester, all being often unfamiliar to the subject. Individuals who have not been to school and are not used to test taking, generally appear to be hesitant to take the test and respond to the test items. The degree of test acculturation often increases with sedentarization and contact of individuals with the outside world. To examine the performance of different cultural groups, and the role of contact-acculturation in it, the effect of test-acculturation needs to be controlled. This has not been done in studies that have used pictorial interpretation tasks for assessing cognition.

The Problem

The present study aims at examining the cognitive functioning of some indigenous people of Bihar, India in relation to their eco-cultural and acculturational characteristics. The groups (Birhor, Asur and Oraon) were selected to exhibit variation in ecological engagements (ranging, in order, from hunting-gathering through a blend of hunting-gathering and agriculture to full fledged agricultural means of subsistence). In each group sampling variation was obtained with respect to people's level of contact-acculturation. It was hypothesized that: (1) As a group Oraons would perform better than Asurs and Birhors on tasks of pictorial perception and interpretation; (2) Individuals with high contact-acculturation would score higher on various tasks than those with low contact-acculturation; (3) The effect of contact-acculturation would be more evident in the performance of Oraons than those of Asurs and Birhors; and (4) the test-acculturation of individuals would be positively correlated with performance on various tasks.

Method

Design and Sample

The study was conducted in Gumla and Hazaribagh districts of Chotanagpur region of Bihar, India with 210 adults belonging to the Birhor (M age = 37.67 years), Asur (M age = 41.36 years) and Oraon (M age = 38.81 years) tribal cultural groups. The level of "contact-acculturation" was assessed with the help of a measure which comprised objective indicators of acculturation such as knowledge of tribal languages, knowledge of Hindi and other languages (e.g., English), possession of household items, (e.g., ornaments, utensils, clothes, and furnitures), means of livelihood, use of modern technology, religion, dressing style, travel experience and exposure to movie. These indices of acculturation were generally found to be highly intercorrelated (rs ranging between .18 to .92) in the three samples, with the average correlation being .68. The average correlations in the Birhor, Asur and Oraon groups were .65, .57, and .72 respectively.

The participants were tested individually for the degree of their contact or change in terms of these indices, and were ascribed to *high-acculturation* (above 50% score on the contact measure) or *low acculturation* (50% or less scores) categories. The sample included both males and females, and was distributed in a 3 (tribe) x 2 (acculturation) factorial design, with 35 high and 35 low acculturation subjects in each cultural group.

The analysis of contact acculturation scores revealed that as a group Oraons were significantly more acculturated (M = 13.88) than Birhors (M = 10.92) and Asurs (M = 10.07). It may also be mentioned that in the case of Oraons, many of the things listed in the contact acculturation scale were voluntarily acquired. In the case of Birhors and Asurs, they appeared to be more as "given" by the government under the general programme of development, and did not indicate their conscious acceptance. This factor should be kept in mind while examining the effect of contact-acculturation on the psychological functioning of these groups.

The *test-acculturation* of participants was measured using a Picture Recognition Task. The task intended to assess the degree to which the participants could recognize familiar objects in pictures, and felt comfortable to speak them out in the presence of a tester in a situation which was rather unfamiliar to many of them. More naming of objects was indicative of a higher level of test-acculturation. The correlations between contact and test acculturation scores were .22, .15 and .38 ($p < .05$) in the Birhor, Asur and Oraon samples respectively.

The Setting and the Cultural Groups

The Chotanagpur region of Bihar consists of a large number of valleys and hills ranging between 600 to 900 meters above sea level. In these valleys and on hilltops live some 29 tribes of distinct origins. The people belonging to these tribes are regarded as the earliest inhabitants of the country. Their life styles, language, *customs*, belief systems, socio-economic and political organizations accord them the status of distinct cultural groups. The 1991 census reports about 6.62 million tribal people in the State of Bihar, which constitutes approximately 7.66% of the total population of the State. In the Gumla and Hazaribagh districts, they constitute 70.80% and 56.41% respectively of the total population of the district (Census of India, 1991).

Among the Bihar tribes, Birhor is the most traditional. It represents largely a nomadic tribe. Exploitation of forest through hunting and gathering constitutes their chief source of livelihood (Vidyarthi & Sahay, 1976). These people have very little contact with other tribes of the same area or with the outside world.

Birhors are generally divided into two groups: the Uthlus (wanderers) and the Jaghis (settlers). The Uthlus move, camp and hunt in forests. The Jaghis have a fixed place to stay, and make frequent forays into the forest for different periods of time. But this distinction is not rigid. There are evidences of Uthlus turning into Jaghis (by settling down) and for Jaghis turning into Uthlus (by abandoning fixed settlements). Today Birhors live in a transitional state. Although their economy is almost entirely dependent on forest products, due to the need for conserving the forest and legislation regarding wild life protection, the Birhors are being "forced" to settle down in colonies built by the Tribal Welfare Department. Though they are being provided with various incentives to take to agriculture, the sedentary existence has not been fully adopted. In the forest, there is hardly any opportunity to encounter pictures, but in the settlements, posters and other kinds of pictorial materials communicating messages of agriculture, health and other development programs can be found. They are also exposed to pictures printed on clothes and to a few cinema posters in the market places.

The Asurs represent one of the minor tribes of Bihar. The earliest account of Asurs (Forbes, 1872) indicates that they were forest people. As iron-smelters and slash and burn cultivators they pursued a nomadic life-style. With a gradual increase in population and laws regarding the conservation of forests, they settled on the hilltops about 50 years ago as rudimentary agriculturalists. The poor agricultural yield is barely enough to feed them for 3-4 months. For the rest of the year, hunting and gathering form a major part of Asur economy, though it is not so basic and regular as in the case of Birhors. Cattle-rearing, fishing and occasional wage employment also contribute to their livelihood.

About fifteen years ago, the contact of Asurs with the outside world was very limited. Two residential schools and bauxite mining industries set up in the region have now exposed them to the outside world and brought about some changes in their traditional life-style. As a settled community, they have developed rules of marriage, divorce, transfer of property and authority. Some families have also gone under the influence of Christianity. As a group Asurs are exposed to pictorial stimuli more than Birhors, but the exposure has been largely of a passive kind.

The Oraons were among the first few settlers at the foothills of Chotanagpur region some 1800 years ago. The word Oraon in Mundari language means "hardworking" or "unwearied", and is used to refer to a person who digs earth the whole night, and does not notice when the dawn has set in. They made extensive clearance for cultivation and became the rulers of land and a dominant tribe of the region. They represent an agricultural community with occupations such as fishing, cattle-rearing and crafts (e.g., mat making) practiced as subsidiary to agriculture. They have a good knowledge of the qualities of soil and of the crops that may be best grown on them. They follow a system of crop rotation, use indigenous manures and chemical fertilizers to increase the yield of crops, have developed their own system of irrigation suited to the terrain, and produce cereals and vegetables for both consumption and sale.

Oraons live in large villages inhabited by a number of families. They have a definite lineage, social structure and authority system. A high degree of control of the general behavior of its members, and in the matters of marriage, divorce and distribution or transfer of property characterizes the Oraon society. Fast expansion of education has brought them close to the outside world. Many educated Oraons have migrated to towns and cities to work in offices, schools, hospitals and industrial organizations. As a group they have moderate to high level of exposure to pictures and other mass media.

Procedure

The subjects were given three cognitive tasks, namely, Pictorial Interpretation Task (PIT), Sequence Perception Task (SPT) and Picture-Model Matching Task (PMMT). The former two tasks were developed by Sinha (1977); the later one was developed by Mishra, Sinha and Berry (in press).

The PIT consists of a set of three pictures. In each, certain objects were so depicted that they were not visible to some of the characters of the picture either due to interposition of objects, or due to not being in the field of vision of the concerned character. The participant was asked to identify the objects and characters depicted in pictures. Then a number of questions were asked regarding the perceptibility/nonperceptibility of objects which were hidden (due to interposition of objects), or were not in the field of vision of the concerned character. Each correct response was given 1 point. Thus, the interpretation score ranged from zero to twelve. Reason for the nonperceptibility of the concerned object was also probed to ascertain whether the participant's response was based on the grasp of spatial cues, or it simply reflected random guessing. When the response indicated spatial orientation (i.e., one object is hiding another, or the object is not in the field of vision of the concerned character), it was classified as *relevant*. The total number of relevant reasons was the score on the *relevance* measure.

Table 1. Mean Scores of Groups on Different Measures of the Tasks

		Birhor		Asur		Oraon	
		HA	LA	HA	LA	HA	LA
PIT Interpretition							
	Mean	5.00	5.09	8.03	4.94	11.06	7.46
	S.D.	4.17	4.49	3.54	4.29	1.17	2.69
Reasoning							
	Mean	5.03	5.00	8.00	4.80	11.00	7.31
	S.D.	4.12	4.44	3.51	4.33	1.22	2.86
Relevence							
	Mean	4.31	4.57	7.86	4.28	10.29	6.83
	S.D.	4.14	4.19	3.48	4.21	1.28	3.02
SPT Arrangement							
	Mean	6.34	6.40	7.74	7.14	7.63	6.54
	S.D.	1.62	1.58	.78	1.50	1.07	1.13
Description							
	Mean	3.90	3.91	3.91	3.74	3.94	3.37
	S.D.	.23	.28	.26	.84	.33	.83
PMMT Matching							
	Mean	12.97	12.89	13.20	11.94	14.57	12.90
	S.D.	1.93	1.80	1.23	2.87	1.10	1.20
Errors : Oblique							
	Mean	.74	.89	.66	1.29	.23	.89
	S.D.	.72	.92	.63	1.16	.59	.89
Frontal							
	Mean	.63	.54	.60	.83	.11	.49
	S.D.	.80	.84	.76	1.13	.32	.81
Side							
	Mean	.66	.69	.54	.97	.09	.74
	S.D.	1.04	.75	.69	1.00	.28	.97

HA = High Acculturation; LA = Low Acculturation

On the SPT, the participant was given a set of four pictures to arrange in a logical sequence of events depicted in those pictures. The testing was carried out in four phases. In the first phase, the participant was asked to arrange the randomly presented pictures in a sequential order. In the second phase, the first picture of the sequence was placed by the researcher; the remaining three were given to the participant in a random order for arranging them in the proper sequence. In the third phase, the first and the last pictures of the series were arranged by the tester; the two intermediary pictures had to be arranged by the participant. In the fourth phase, the participant was told the story of the event depicted in the picture series, and was asked to arrange the pictures in the sequence so that they corresponded to the story.

Testing on successive phases was done only when the participant failed to arrange pictures on an earlier phase. Scores ranged from 4 to 1 depending on the phase on which the participant was successful in arranging pictures in the correct order. Failure at the fourth phase was scored zero. These were called *Arrangement Scores*.

When the pictures had been arranged in a sequence, the participant was asked to describe the events depicted therein. The "description" was categorized as "good", "poor" or "nil" depending on the degree of correspondance between the *standard description* of pictures and the *participant's description*. These categories were given two, one and zero points respectively. Besides a practice series, two test series were used in the study. The score range on the arrangement measure was 0-8, whereas on the description measure, it was 0-4.

Table 2. ANOVA of Various Measures of PIT, SPT and PMMT

Measures	Acculturation	F-ratios Tribe	Acculturation x Tribe
df	(1, 204)	(2, 204)	(2, 204)
PIT			
Interpretation	19.06**	24.08**	5.23**
Reasoning	20.99**	23.43**	5.20**
Relevance	21.14**	23.40**	6.56**
SPT			
Arrangement	8.62**	11.61**	3.21*
Description	9.52**	3.49*	5.22**
PMMT			
Matching	13.85**	6.66**	3.08**
Errors			
Oblique	16.30**	4.19*	2.00
Frontal	2.27	4.63*	1.41
Side	10.22	3.14	2.50

* $p < .05$ ** $p < .01$

On the PMMT the participant was required to match some models of objects with a set of three pictures of each of those models representing frontal (90 degree), side (180 degree) and oblique (45 degree) views. The models were presented in each of these positions randomly. Each time the participant was asked to choose from a set of three pictures the one in which the model object was depicted exactly in the same position as placed in the front. There were five models, and each was matched with a picture in three viewing positions. At total of fifteen judgements were made by a participant. For each correct judgement 1 point was given. Thus, the score range on the test was 0-15. The errors at different viewing positions were also analysed.

Results

The mean scores of high and low acculturated samples of the Birhor, Asur and Oraon cultural groups obtained on the PIT are given in Table 1. The ANOVA results are presented in Table 2. On the interpretation measure, the Oraons as a group scored significantly higher ($M = 9.26$) than Asurs ($M = 6.48$) who scored significantly higher than Birhors ($M = 5.04$). High acculturated individuals ($M = 8.03$) scored higher than low acculturated individuals ($M = 5.83$). The main effects of tribe, acculturation and their interaction were significant, showing the effect of acculturation for the Oraon and Asur samples, but not for the Birhor sample. A similar pattern of results was obtained with regard to *reason* scores. Generally people who were able to interpret pictures were also able to give reasons for their responses; however, Oraons and Asurs, particularly of high acculturation level, provided more relevant reasons than Birhors.

On the SPT, the mean scores of groups (Tables 1 & 2) revealed that on the arrangement measure, Asurs ($M = 7.44$) as a group scored higher than Oraons ($M = 7.08$) and Birhors ($M = 6.37$) with a significant effect of tribe, though Asurs and Oraons did not differ significantly. High acculturated individuals ($M = 7.24$) scored significantly higher than low acculturated ($M = 6.69$). The interaction effect revealed greater effect of acculturation for Oraons as compared to Asurs and Birhors.

The *description* scores revealed a reverse trend. The Birhors scored significantly higher ($M = 3.91$) than Asurs ($M = 3.82$) who scored significantly higher than Oraons ($M = 3.65$), with a significant tribe effect. High acculturated individuals ($M = 3.92$) scored significantly higher than the low acculturated ($M = 3.67$). However, greater variation due to acculturation was noted in the Oraon than in the Birhor or Asur sample.

Analyses on PMMT task (Tables 1 & 2) revealed relatively higher scores for Oraon ($M = 13.73$) as compared to Asur ($M = 12.57$) or Birhor group ($M = 12.93$), with a significant tribe effect. The Oraons differed significantly from Asurs and Birhors, the latter two showing no evidence of significant difference. The scores of high acculturated individuals ($M = 13.58$) were significantly higher than those of the low acculturated ($M = 12.58$). An enhanced effect of acculturation for the Oraon and Asur samples than for the Birhor was evident.

Table 3. Correlation Matrix of Different Measures for the Three Cultural Groups on Interpretation, Arrangement, and Matching

	Cognitive Measures		
	Interpretation	Arrangement	Matching
BIRHOR			
Interpretation	*	.28	.05
Arrangement		*	.42
Matching			*
ASUR			
Interpretation	*	.37	.39
Arrangement		*	.27
Matching			*
ORAONS			
Interpretation	*	.40	.21
Arrangement		*	.58
Matching			*

Significant r .23 and above

Analysis of errors revealed that tribe effect was significant at all the viewing positions. Generally the Asurs made more incorrect judgements than Birhors and Oraons. Excepting for the frontal position, the low acculturated subjects failed more frequently to match pictures with models than high acculturated subjects. This pattern was similar across all the cultural groups.

Table 4. Relationship of Eco-Cultural Background with Cognitive Measures, Controlling for Acculturation Effects

	Cogitive Measure		
Variables	Interpretation	Arrangement	Matching
Culture group	.21	.35	.00
Culture group (controlling for contact-acculturation)	.15	.31	-.06
Culture group (controlling for test-acculturation)	.19	.37	-.02
Culture group (controlling for contact and test-acculturation)	.14	.31	-.07

Significant r .14 and above

Intercorrelations among the three core measures of the tasks (Interpretation, Arrangement on SPT and Matching) were worked out to examine whether the performance exhibited some degree of commonality (Table 3). While all the values of correlations across groups were positive and generally significant, the magnitude of correlations was higher in the Oraon than in the Asur and Birhor groups. Thus, the concerned cognitive processes tended to display some variation in organization across groups.

It has been indicated that with sedentarization of groups not only is there an increase in the level of *contact-acculturation*, but also in the level of *test-acculturation*, which may influence test scores of groups. Hence, correlational analyses were designed to assess initially the relationship of the eco-cultural position of groups with the core measures of cognitive performance without considering the level of *contact* or *test-acculturation*. Later, the correlation between these variables was computed, first by controlling for the effect of *contact-acculturation*, then for *test-acculturation*, and lastly for both *contact* and *test-acculturation* (Table 4).

Table 5. Relationship of Contact-Acculturation with Cognitive Measures, Controlling for Test- Acculturation

Variables	Congitive Measure		
	Interpretation	Arrangement	Matching
Contact-acculturation	.40	.29	.34
Test-acculturation	.39	.17	.29
Contact-acculturation (controlling for test-acculturation)	.12	.09	.11

Significant r .14 and above.

Interpretation and *arrangement* scores showed positive and significant correlations with the eco-cultural position of the groups, whereas *matching* score did not provide any reliable evidence of relationship. The values of correlations showed some variation as the effects of the covariates were partialled out, but the nature of the relationship did not change.

The pattern of correlation between *contact-acculturation* and various cognitive measures was also found to be positive and significant (Table 5). When the effect of *test-acculturation* was partialled out, the values of correlations, though still positive, turned out to be nonsignificant. It appears that besides the long-term eco-cultural features of groups, which influence cognitive functioning of individuals in important ways, relatively more recent features of acculturation also exercise a potent influence on cognitive processes.

Discussion

The findings largely supported the hypotheses. Both eco-cultural features and contact-acculturation of groups had significant influences on the performance of cognitive tasks. The contact and test-acculturation effects tended to override eco-cultural influences, but they could not displace the latter. The effect of eco-cultural features of groups on performance tended to persist even when the effects of contact and test-acculturation were partialled out. On the other hand, test-acculturation effects tended to override the contact-acculturation influences.

The findings generally suggested that the effect of acculturation was significant only for the Oraon and Asur groups. Why did Birhors fail to demonstrate any psychological benefit from contact-acculturation? Three plausible reasons can be advanced in this respect. First, the indicators of contact acculturation used in the scale were probably superficial. They did not indicate any conscious acceptance of things, and it was difficult to know whether Birhors had got psychologically acculturated. Second, the change among Birhors has not reached a "threshold" point to elicit a real and meaningful response at the cognitive level. Third, pictures do not have any functional salience in a nomadic hunting-gathering life, and Birhors altogether lacked organized experiences with pictorial stimuli such as those encountered in schools or in an urban-industrial set up. This lack of experience with pictures did not provide them with an opportunity to master some of the skills. The findings suggested that they were not incapable of perceiving and interpreting all kinds of pictures. In fact, their description of sequential events on the SPT, was more organized than those of the other groups. A strong oral tradition of story-telling among Birhors may be held responsible for their highly developed description skill.

In psychological literature, contact and test-acculturation have been considered as two broad sets of variables influencing cognitive test performance (Berry et al., 1986; Rogoff, 1981). With sedentarization and the development of certain supportive institutions (e.g., schools), the individuals are likely to acquire greater familiarity with pictorial stimuli, and feel more ease and comfortable in test situations due to greater exposure to such situations. These nonspecific factors, called test-acculturation, can improve an individual's level of performance without bringing about any real change in the underlying cognitive structures. The findings of the study do suggest this possibility. The test-acculturation shared some of the variance in the test scores associated with contact-acculturation. When its effect was partialled out, the correlations of contact-acculturation with test scores were considerably reduced.

These findings suggest that there is no risk in interpreting the performance differences of groups in terms of the eco-cultural demands and the adaptations made by individuals. On the other hand, interpretation in terms of acculturation involves some difficulties. There is need to analyse the physical features (objective aspects) of acculturation at a deeper level. At the same time, it is equally important to analyse the psychological features of acculturation and their relationships with the eco-cultural context of groups and individuals.

References

Berry, J.W. (1976). *Human ecology and cognitive style*. New York: Sage Halsted.
Berry, J.W. (1980). Social and cultural change. In H.C. Triandis & R. Brislin (Eds.), *Handbook of cross-cultural psychology* (Vol. 5). Boston: Allyn & Bacon.

Berry, J.W. (1985). Cultural psychology and ethnic psychology: A comparative analysis. In I.R. Lagunes & Y.H. Poortinga (Eds.), *From a different perspective: Studies of behaviour across cultures*. Lisse: Swets & Zeitlinger.

Berry, J.W. (1987). The comparative study of cognitive abilities. In S.H. Irvine & S. Newstead (Eds.), *Intelligence and cognition: Contemporary frames of reference*. Dordrecht: Nijhoff.

Berry, J.W., van de Koppel, J.M.H., Sehechal, C., Annis, R.C., Bahuchet, S., Cavalli-Sforza, L.L., & Witkin, H.A. (1986). *On the edge of the forest: A comparative study of the development of cognitive style in Central Africa*. Lisse: Swets & Zeitlinger.

Census of India (Bihar) (1991). *Patna: Directorate of Census Operations*.

Deregowski, J.B. (1968). Pictorial recognition in subjects from a relatively pictureless environment. *African Social Research*, 5, 356-364.

Deregowski, J.B. (1980). *Illusions, patterns and pictures: A cross-cultural perspective*. London: Academic Press.

Deregowski, J.B., & Betley, A.M. (1986). Perception of pictorial space by Bushmen. *International Journal of Psychology*, 21, 743-752.

Deregowski, J.B., Muldrow, E.S., & Muldrow, W.F. (1972). Pictorial recognition in a remote Ethiopian population. *Perception*, 1, 417-25.

Forbes, L.R. (1872). *Report on the Ryotwaree settlement of the government farms in Palamau*. Calcutta: Government.

Gibson, J.J. (1966). *The senses considered as perceptual systems*. Boston: Houghton Mifflin.

Gombrich, E.H. (1977). *Art and illusion: A study in the psychology of pictorial representation* (5th ed.). Oxford: Phaidon Press.

Hudson, W. (1960). Pictorial depth perception in sub-cultural groups in Africa. *Journal of Social Psychology*, 52, 183-208.

Hudson, W. (1967). The study of the problem of pictorial perception among unacculturated groups. *International Journal of Psychology*, 2, 89-107.

Jahoda, G., & Mc Gurk, H. (1974). Pictorial depth perception: A developmental study. *British Journal of Psychology*, 65, 141-49.

Mishra, R.C. (1976). *Perception and comprehension of certain types of pictorial materials*. Unpublished Doctoral dissertation, Allahabad University.

Mishra, R.C. (1987). A re-examination of socio-cultural differences in the perception of pictorial symbols. *Indian Journal of Current Psychological Research*, 2, 65-73.

Mishra, R.C. (1988). Perception and comprehension of dual scene pictures. *Journal of Psychological Researches*, 32, 121-127.

Mishra, R.C., Sinha, D., & Berry, J.W. (in press). *Ecology, acculturation, and psychological adaptation: A study of Adivasi in Bihar*. New Delhi: Sage.

Redfield, R., Linton, R., & Herskovits, M.J. (1936). Memorandum on the study of acculturation. *American Anthropologist*, 38, 149-152.

Rogoff, B. (1981). Schooling and the development of cognitive skills. In H.C. Triandis & A. Heron (Eds.), *Handbook of cross-cultural psychology* (Vol. 4). Boston: Allyn and Bacon.

Sinha, D. (1977). Some social disadvantages and development of certain perceptual skills. *Indian Journal of Psychology*, 52, 115-132.

Sinha, D., & Mishra, R.C. (1982). Some socio-environmental disadvantages and skill for perception of pictorial symbols. *Journal of Personality and Group Behaviour, 2,* 111-121.

Sinha, D., & Shukla, P. (1974). Deprivation and development of skill for pictorial depth perception. *Journal of Cross-Cultural Psychology, 5,* 434-450.

Vidyarthi, L.P., & Sahay, K.N. (1976). *The dynamics of tribal leadership in Bihar.* Allahabad: Kitab Mahal.

Winter, W. (1963). The perception of safety posters by Bantu industrial workers. *Psychologia Africana, 10,* 127-135.

Ethnic Stereotypes among Polish and German Silesians

Alicja Ostrowska & Danuta Bochenska
University of Opole, Opole, Poland

This study deals with the mutual perception of two groups inhabiting Opole, Silesia, part of Southern Poland, Silesians of German ethnic identity and Silesians of Polish identity.

German and Polish domination in this region changed numerous times in history. Centuries of "germanization" of Silesia were accompanied by the Polish Silesians' struggle against it and culminated in the Silesian Uprisings in the 1920s. Opole was part of Germany until the end of World War II. Post war Poland's policy towards Silesia included resettlement of a large group of German Silesians to Germany, repopulation of the region with Poles from the eastern territories, placing obstacles in the political and public careers of Silesians, not permitting the German language in schools of the region, etc. In the seventies, after an agreement between the Polish and German governments, thousands of Silesians were allowed to immigrate to West Germany. During the recent period of democratic transition in Poland the ethnic problems of Silesia received considerable publicity. The mass media painted a rather dramatic picture of the growing power of the German minority in the region and its possible consequences, mainly, renewed germanization of the region. It is characteristic that the mass media have recently been treating Silesians as a homogenous population with dominant pro-German orientation.

According to Juros (1992), Silesians do not form a homogeneous population in terms of ethnic-cultural identity. Juros analyzed the complexity of the Silesian ethnic identity situation, but also pointed out that various ethnic identities were declared by members of the local Silesian community he studied, even if there were no clear objective criteria for choosing one.

We decided to determine, whether, in the new socio-political situation, there really was, as was commonly believed, tension or conflict between Silesians of German and Silesians of Polish identity. One way to proceed would be to collect data about intergroup perception, specifically ethnic stereotypes held by both groups and interpret them as one indicator of mutual attitudes.

Apart from the tradition of reconstructing auto- and heterostereotypes of both groups (Triandis, Lisansky, Setiadi, Chang, Marin, & Betancourt, 1982; Marjoribanks, & Jordan, 1986), we included an additional kind of stereotype, studied by Bochenska (1994), a *metastereotype*. "An *autostereotype* refers to the attributes that individuals assign to their own group, whereas a *heterostereotype* includes the attributes that a group assigns to members of another group." (Marjoribanks & Jordan, 1986). A *metastereotype* consists of characteristics that members of a group believe another group is likely to ascribe to them.

Since the concept of metastereotypes is not commonly used in studies of intergroup perception some comments are necessary here. The idea that what others think about an individual shapes his/her self-concept has a long tradition. According to George Herbert Mead (1934) it is essential for communication - the fundamental social process - to take the role of the other in viewing both the world and one's self. For Cooley (1902) the self, as a reflection of what others think of us, is not a simple "looking glass reflection" but requires our interpretation of others' perception of us and their probable response to what they observed about us on the basis of their values and attitudes. Although there is comprehensive literature concerning the role of such a "reflected self" in construction of an individual's self-perception, studies of interaction between group self-concept (e.g., ethnic autostereotype) and reflected stereotypes (e.g., ethnic metastereotypes) are not common. Bochenska (1994a, 1994b) analyzed the quality, favorability, uniformity, intensity and certainty of auto-, hetero- and meta-images of American and Polish students and found significant relationships between the auto-, and meta-images. Correlations between some of the attributes in the meta-images and hetero-images and attitudes towards the other group were also noticed. Casas, Ponteretto, & Sweeney, (1987) reconstructed the results of the "stereotyping the stereotyper", i.e., the content of the stereotypes Mexican American married couples believed Americans formed about their ethnic group. Unfortunately, although they emphasized the potential mediating or directional role of the "stereotyped stereotypes" in determining minority groups' auto-, and heterostereotypes, they did not examine any relationships between the former stereotype and the latter two kinds of stereotypes.

Summarizing, the main purpose of the present study was to reconstruct the contents of three kinds of ethnic stereotypes (auto-, hetero-, and metastereotypes) held by Silesians of different declared ethnic identities, and to characterize traditionally studied stereotypes' features, i.e. their direction, intensity, and uniformity (Triandis et al., 1982, Marjoribanks, & Jordan, 1987). In addition, a complex within and between groups analysis of relations between the three kinds of stereotypes should contribute to our better understanding of mutual perception and attitudes of both groups of Silesians.

Method

Subjects

The sample included 128 adult inhabitants of two towns and three villages from the region of Opole, Silesia. They belonged to three groups of different ethnic identity. German Silesian ethnic identity was declared by 46 persons (20 women, 26 men), Polish Silesian identity by 42 persons (18 women, 24 men), and 40 persons emphasized their unique Silesian identity (neither German, nor Polish). The latter group was used in this study only as a source of additional data concerning stereotypes of Polish and German Silesians. Information from the local church organizations and from the German Minority Cultural Association helped the authors to select those who could be considered as subjects in our study and who were then visited in their homes and asked to complete the questionnaire. The only criterion used for classifying subjects to one of the groups was their individual declaration of ethnic identity. Only the data obtained from the persons declaring German Silesian, Polish Silesian and Silesian ethnic identity were used for further analysis. The Polish Silesians were younger ($M = 40$) and better educated than German Silesians ($M = 47$).

Instrument

In order to select the attributes for the questionnaire, 180 students of the Pedagogical University of Opole and high school students from Opole were asked to list features which they thought were most typical of Polish and German Silesians. After a content analysis of over 100 attributes by a group of psychologists, 28 were eventually selected for the final version of the instrument. Some additional questions were also asked in the study, the most concerning the declared ethnic identity of the subjects.

The auto-, hetero-, and metastereotypes were assessed by means of 28 attributes, 20 personal-behavioral characteristics, and eight ethnic-cultural characteristics, with a 11-point scale, for example: *very lazy* (0) to *extremely industrious* (10); *not religious at all* (0) to *extremely religious* (10); *does not maintain Polish tradition at all* (0) to *extremely involved in the maintenance of Polish tradition* (10). For four attributes: *presumptuous, noisy, aggressive*, and *drink alcohol*, the maximum score, 10, indicated the negative pole.

Each Polish and German Silesian was asked to estimate how typical each of the attributes was of his/her own group (autostereotype) and of the other group (heterostereotype). They also characterized their own group in the way they expected the other group would do so (metastereotype). The subjects who declared themselves as simply Silesians were only asked to characterize Polish and German Silesians. The data were collected in the autumn of 1991.

Results

The presentation of the results begins with the description of the auto-, hetero-, and metastereotypes of both groups of Silesians. Then, the relations between the three kinds of stereotypes are analyzed. The results were analyzed with a two-way ANOVA with independent variables Ethnic Identity (2) and Sterotypes (3).

Table 1. *Means of Personal-Behavioral Characteristics (Auto-, Hetero-, and Metastereotypes) among Polish (N = 42) and German (N = 46) Silesians*

Characteristics	Polish Silesians' Stereotypes			German Silesians' Stereotypes			Ps of ANOVA Effects		
	Auto	Hetero	Meta	Auto	Hetero	Meta	STER	ETHNIC	INTE S x E
religious	**8.1**	7.1	<u>7.4</u>	<u>7.9</u>	<u>6.3</u>	6.6	.000	.157	.445
courageous	6.9	5.6	6.4	4.7	7.3	5.1	.010	.145	.004
hospitable	**8.0**	6.4	7.2	**7.4**	**7.7**	6.2	.006	.925	.000
house proud	**8.0**	**8.7**	7.2	**9.3**	5.5	**6.2**	.000	.008	.000
clean	**7.4**	**8.1**	6.5	**8.7**	6.6	6.8	.000	.959	.000
presumptuous	4.9	<u>7.9</u>	6.5	4.5	<u>7.0</u>	**6.2**	.000	.056	.683
materially oriented	**7.9**	**9.0**	**7.8**	**8.7**	7.0	**7.6**	.194	.178	.000
intelligent	7.0	5.9	6.1	6.8	6.4	5.5	.001	.874	.090
noisy	4.6	<u>6.8</u>	5.6	4.0	7.2	5.0	.000	.515	.276
with self-dignity	**8.1**	<u>6.3</u>	6.7	**8.3**	<u>7.1</u>	6.3	.000	.558	.239
industrious	**7.8**	<u>7.9</u>	6.2	**9.0**	5.0	7.0	.000	.432	.000
aggressive	4.8	6.7	5.8	4.1	6.9	5.5	.000	.509	.481
rich	6.2	7.5	5.8	7.1	6.1	7.3	.723	.276	.000
appreciate education	<u>6.9</u>	5.6	6.8	6.1	6.4	5.2	.178	.191	.000
honest	**7.6**	6.3	6.4	**7.8**	5.3	6.4	.000	.479	.075
helpful	**7.5**	5.9	6.7	7.2	5.1	5.8	.000	.156	.625
drink alcohol	5.8	5.9	6.9	4.7	7.0	5.7	.000	.307	.000
family oriented	**8.8**	<u>7.9</u>	**8.0**	**8.6**	7.3	<u>7.2</u>	.000	.115	.431
cheerful	**7.6**	<u>6.7</u>	6.8	7.2	6.9	6.3	.009	.572	.352
polite	**7.2**	6.2	6.4	**7.2**	5.2	5.9	.000	.250	.227
Total I	7.6	6.9	6.8	7.6	6.3	6.3			
Total II	5.0	6.8	6.2	4.3	7.0	5.6			

Note. The uniform characteristics are indicated by boldface, the intense characteristics are underlined. Ster = Stereotype; Ethnic = Ethnic Identity; Inter Sxe = Interaction: Stereotype X Ethnic Identity. Total I - total mean score for 16 positive characteristics, Total II - total mean score for 4 negative characteristics.

Table 1 contains the mean typicality scores for 16 positive and four negative personal-behavioral characteristics in the auto-, hetero-, and metastereotypes of both groups of Silesians. Total mean scores of typicality are presented separately for the positive and negative attributes. The uniform characteristics are indicated by boldface, whereas the intense ones are underlined. Levels of significance of ANOVA effects are indicated. Table 2 contains analogous data concerning ethnic-cultural characteristics.

The Autostereotypes

The Autostereotype of Polish Silesians

All but two attributes in the Polish Silesians autostereotype were in the range of approximately 7.0 to 8.0. The least typical feature, but still, with mean score above 5.0, was *rich* (6.2), whereas *family oriented* was perceived as the most typical among all features (8.8).

The autostereotype was highly intense. The Triandis et al., (1982) indicator of *high intensity* of stereotypes (i.e., 10 per cent frequency of the extreme scores, 0 or 10), was obtained for 15 attributes. Also, the *high uniformity* indicator of more than 75 per cent (or less then 25 per cent) of the answers being in the range of 6-10 was found in 17 attributes.

As expected, the negative attributes were perceived as being significantly less typical of Polish Silesians than the positive ones. The difference between Total I and Total II was significant at $p = .0001$. The mean scores near 5.0 indicated moderate typicality, with *drinking alcohol* slightly above 5.0. The response were neither uniform nor intense.

Table 2. *Means of Ethnic-Cultural Characteristics (Auto-, Hetero-, and Metastereotypes) among Polish (N = 42) and German (N = 46) Silesians*

Characteristics	Polish Silesians' Stereotypes			German Silesians' Stereotypes			Ps of ANOVA Effects		
	Auto	Hetero	Meta	Auto	Hetero	Meta	STER	ETHNIC	INTE S x E
Polish language	7.2	6.3	<u>7.8</u>	7.0	<u>8.8</u>	7.2	.459	.173	.000
German language	6.1	<u>6.4</u>	3.9	6.5	3.9	5.3	.249	.808	.154
Silesian dialect	**8.8**	8.0	<u>7.3</u>	**8.7**	4.4	<u>7.4</u>	.000	.040	.001
Polish tradition	**9.2**	3.6	<u>8.0</u>	**4.0***	7.9	4.1	.718	.116	.000
German tradition	**3.5***	<u>7.8</u>	3.1	**7.3**	<u>2.1</u>	5.6	.155	.515	.000
Silesian tradition	7.6	<u>6.9</u>	7.6	**8.8**	5.0	<u>6.9</u>	.000	.300	.002
Attached to Silesia	**8.1**	<u>6.6</u>	<u>8.3</u>	**8.5**	<u>6.6</u>	<u>6.9</u>	.001	.464	.084
Introduce German methods and language	3.8	<u>8.9</u>	<u>3.4</u>	**9.2**	3.3	<u>7.2</u>	.070	.029	.000

Note. *Asterisk indicates high uniformity of low scores. Other indications as in Table 1.

As far as ethnic-cultural attributes are concerned, the Polish Silesians described themselves as rigidly: *maintaining Polish traditions* (9.2); *speaking Silesian dialect very well* (8.8); *having a feeling of Patriotism towards Silesia* (8.1); *maintaining Silesian tradition* (7.6); *speaking Polish* (7.2); *speaking German at an average level* (6.1); *not maintaining German tradition* (3.5); and opposed to the restoration of German customs, order, methods and language (3.8).

The Autostereotype of German Silesians
The autostereotype of German Silesians was also defined by highly typical positive attributes, with mean scores ranging from 7.0 to 8.0. *Courageous* (M = 4.7) and *appreciate education* (6.0) were the least typical, whereas *house proud* (9.0); *industrious* (9.0); *clean* (8.7;and *materially oriented* (7.9) were perceived as most typical. The German Silesians, like the Polish Silesians, perceived negative attributes as significantly less typical of themselves than the positive ones (the difference between Total I and Total II was significant at $p=.0001$). As far as the personal-behavioral characteristics were concerned, their autostereotype was highly *uniform* for 17 attributes and *intense* for 14 attributes.

Regarding the ethnic-cultural characteristics, the German Silesians described themselves as: *approving of the introduction of German customs, order, and language* (9.2); *cultivating Silesian tradition* (8.8); *speaking the Silesian dialect very well* (8.67); *feeling of patriotism towards Silesia* (8.5); *maintaining German traditions* (7.3); *speaking Polish* (7.0); *speaking German at the average level* (6.5); and *not cultivating Polish traditions* (4.0).

The Heterostereotypes

The Polish Silesians' Heterostereotype of German Silesians
The Polish Silesians defined German Silesians as highly (means between 8.0-9.0) *materially oriented, house proud, clean, industrious, family oriented*, and slightly above the moderate level (scores 5,5 - 6.0) for the other attributes, with *appreciate education* (5.6), *courageous* (5.6), *intelligent* (5.9), and *helpful* (5.9) obtaining the lowest scores. *High uniformity* was obtained for eight attributes and *intensity* for 15.

According to the Polish Silesians, German Silesians typically: *wanted the introduction of German customs, order and language* (8.4); *spoke the Silesian dialect well* (8.0); *maintained German traditions* (7.8); but were only just above the average at *cultivating Silesian traditions* (6.9); *feeling of Patriotism towards Silesia, speaking German* (6.4); *speaking Polish* (6.3). They were described as *not cultivating Polish traditions* (3.6).

The German Silesians' Heterosterotype of Polish Silesians
The German Silesians described Polish Silesians as being at the moderate level (scores about 5.0) *polite, helpful, honest, industrious,* and *house proud,* and as much above the average (scores about 7.0) *courageous, materially oriented, hospitable, with self-dignity,* and *family oriented.* In contrast

with the Polish Silesians, the answers of German Silesian were not uniform and obtained high intensity for only seven attributes.

According to German Silesians, Polish Silesians *spoke Polish very well* (8.8); and *maintained Polish tradition* (7.9). They *did not approve of the introduction of German customs and language* (3.3); *did not speak German very well* (3.9); and *did not maintain German tradition* (2.1). They *spoke the Silesian dialect* (4.4); and *maintained Silesian tradition* (5.0) at a moderate level.

The Metastereotypes

The Metastereotype of the Polish Silesians
The Polish Silesians expected that German Silesians would perceive them most favorably for: *family oriented* (8.0), *materially oriented* (7.8), and *religious* (7.4), and least favorably for *rich* (5.8). As far as the ethnic-cultural features were concerned, the Polish Silesians predicted that they would be perceived by German Silesians as: *having a feeling of Patriotism towards Silesia* (8.3); *maintaining Polish tradition* (8.0); *speaking Polish* (7.8); and *the Silesian dialect well* (7.3); and *maintaining Silesian tradition* (7.6). According to their metastereotype Germans would also picture them as: *not maintaining German tradition* (3.05); *not approving of the introduction of German customs and language* (3.4); and as *not speaking German well* (3.9). The metastereotype was uniform for only four characteristics and intense for nine of them.

The Metastereotype of German Silesians
The German Silesians expected that Polish Silesians would describe them as highly: *materially oriented* (7.6); and *rich* (7.3); and only at the average level *appreciative of education* (5.2); *courageous* (5.1); and *intelligent* (5.5). German Silesians would be also characterized as: *speaking Polish well* (7.2); *opting for the introduction of German customs and language* (7.2); only moderately *maintaining Polish traditions* (4.2); and, surprisingly, as *speaking German* at the average level (5.3). The metastereotype was not uniform for any of the attributes and intense for eight of them.

Within and Between Groups Comparisons of the Auto-, Hetero-, and Metastereotypes

The German Silesians' *autostereotype* differed significantly from the the Polish Silesians' for five positive attributes and one negative attribute. *House proud* ($p < .0001$), *industrious* ($p < .0001$), *clean* ($p < .002$), were estimated at a significantly higher level, and *courageous* ($p < .0001$), *appreciate education* ($p < .$) at a lower level than the same characteristics in Polish Silesians' autostereotype. The German Silesians perceived that *drinking alcohol* was less typical of them than it was of the Polish Silesians according to their autostereotype. They also differed from the Polish Silesians in terms of *cultivation of Polish and German tradition*, and *attitudes towards the introduction of German customs and language*.

It was typical of both groups that the heterostereotypes they formed about the other group were significantly less positive than their own autostereotypes for most of the attributes, indicating strong in-group favoritism. The Polish Silesians perceived the German Silesians in a less favorable way than themselves for 11 positive and three negative personal-behavioral characteristics (p's ranged from .0001 to .01). However, they also pictured the other group in a significantly more favorable way than their own for four characteristics: *house proud* ($p < .01$), *clean* ($p < .04$), *materially oriented* ($p < .0001$) and *rich* ($p < .001$).

The German Silesians described the Polish Silesians in a less positive way for nine positive and four negative attributes (p's ranged from .0001 to .02). They perceived Polish Silesians in a significantly more favourable way than themselves only for *courageous* ($p < .0001$).

Total I for autostereotype minus Total I for *heterostereotype*, and Total II for *autostereotype* minus Total II for *heterostereotype*) between both groups (M's, respectively, .7, and 1.8 for the Polish and 1.3 and 2.7 for the German Silesians) indicate that the total in-group favoritism was larger for the German Silesians for both the positive and negative features (ps, respectively, .008 and .000).The t-test between the Polish and German Silesians' total typicality scores Total I (6.9 vs. 6.3) and Total II (6.8 vs. 7.0) revealed that the heterostereotype defined by the German Silesians was significantly less favorable than that defined by the Polish Silesians for the positive attributes ($p < .030$). The effect of favoritism of the German Silesians was also found in the group of Silesians who defined their ethnic identity as simply Silesian and who were asked in our study to reconstruct their heterostereotypes of Polish and German Silesians. The Silesians' heterostereotype of Polish Silesians was significantly (t-test, ps from .000 to .05) less positive than their heterostereotype of German Silesians for 18 out of the 28 attributes. Polish Silesians were described in a significantly more favorable way than German Silesians only for *courageous* ($p < .001$), *knowledge of Polish* ($p < .001$), and *maintainance of Polish tradition* ($p < .0001$).

It is important to notice that ANOVA's for each attribute in a 2 (ethnic group) by 3 (type of stereotype) design, with the last variable treated as repeated measures, revealed significant interaction effects for nine of the attributes (Table 1 and 2). As is shown in Figure 1, which illustrates a typical interaction, e.g., *hospitable*, an interaction was also found between the auto- and heterostereotypes characteristics of both groups. This means that for these attributes the members of the group who defined their autostereotype more positively than the other group, also formed the heterostereotype of the other group less positively than their own autostereotype, whereas the group with the less positive autostereotype formed the heterostereotype of the other group more favorably than their own autostereotype. We believe that this effect may suggest the existence of commonly accepted ethnic stereotypes of Polish and German Silesians. The Polish Silesians described German Silesians as *cleaner* ($p < .038$), *more materially oriented* ($p < .0001$), and *house proud* ($p < .01$) than their own group whereas the German Silesians perceived Polish Silesians less favorably than their own group for these attributes (*ps*, respectively, .0001, .020, .0001) . The German Silesians perceived Polish Silesians as more *courageous* than German Silesians ($p = .0001$), and the Polish Silesians described German Silesians as less *courageous* than themselves ($p = .0001$).

Figure 1. The mean typicality scores for hospitable in the Polish and German Silesians' auto-, hetero-, and metastereotypes

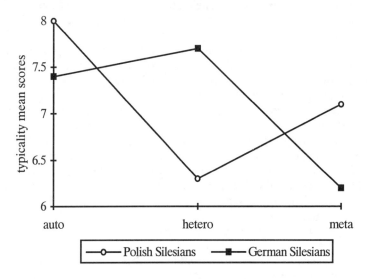

If we add the results concerning the metastereotypes to what we already described regarding both groups' auto-, and heterostereotypes relationship, our interpretation of mutual perception and attitudes of Polish and German Silesians becomes more complex and accurate.

The comparison of the auto- and metastereotypes indicated that for most of the features both groups formed metastereotypes less positive than the autostereotypes they held (Tables 1 and 2). The Polish Silesians formed less favorable metastereotypes than their autostereotype for 14 personal-behavioral characteristics (*ps* from .000 to .017), whereas the German Silesians for 16 attributes (*ps* from .000 to .042). There was however, a tendency to expect from the other group the highest and the lowest estimations of typicality for the features ascribed to one's own group also as the most and the least typical. Despite the significant differences between the auto- and metastereotypes attributes, this effect may suggest existence of commonly accepted (or at least, commonly known) stereotypes of both groups. To some extent it may indicate existence of the "kernel of truth" in the discussed stereotypes.

The German Silesians had less positive metastereotypes than the Polish Silesians, i.e., they expected that Polish Silesians would describe them in a less favorable way than the Polish Silesians expected German Silesians would (the differences between the Total I scores and between the Total II scores in the metastereotypes of the Polish and German Silesians was significant at $p < .033$, and $.02$; respectively).

Finally, comparison of the Total I and II scores in the meta- and heterostereotypes of both groups revealed that for the positive characteristics, the metastereotypes formed by the German Silesians were less positive than the actual heterostereotypes formed by the Polish Silesians (6.3 vs. 6.9; $p < .01$), whereas the Polish Silesians expected more favorable description than the German Silesians actually gave (6.8 vs. 6.3, $p < .01$). For the negative characteristics, both groups tended to underestimate what the other group thought about them (the differences significant at $p < .009$ level for Polish Silesians and at $p < .000$ for the German Silesians). This underestimation was stronger on the German Silesians part ($p < .006$).

Conclusions

The analysis of ethnic stereotypes of the Polish and German Silesians, and especially, the within and between group comparison of three kinds of stereotypes, allows some conclusions to be made about mutual perception and attitudes of two groups of Silesians.

The data indicated a typical in-group favoritism. A high accuracy of intergroup perception, for the most "stereotypic" features, was registered which might suggest the "kernel of truth" symptoms in both groups of Silesians' stereotypes. As it was mentioned earlier, this tendency was indicated by significant interaction effects in the analysis of variance for these attributes which were estimated as highly (or lowly) typical, both, in the autostereotype of one group, and the heterosterotype formed about that group by the other group, and, in addition, in the metastereotype of the first group, whereas the other group defined significantly different (opposite) auto-, hetero-, and metastereotypes. We did not however, register any "mirror image" effects comparable to those described by Salazar & Marin (1977).

Upon such intergroup agreement, according to the Polish Silesians' auto-, and metastereotype and, to the German Silesians' heterosterotype, Polish Silesians are highly *courageous, hospitable, well educated;* they also *maintain Polish tradition,* and *drink alcohol.*

German Silesians are, accordingly, very *clean, industrious, house proud, rich;* they also *maintain German tradition* and *opt for an introduction of German customs and language.*

Both groups formed positive autostereotypes, and relatively positive hetero-, and metastereotypes (even the negative attributes were estimated as only moderately typical of the group). These results seem to contradict common expectations of Poles, supported by the mass media image of the situation in Silesia, that there is a strong conflict between the Polish and German inhabitants of Silesia, caused by German Silesians' group interests. If such a conflict existed, the heterostereotypes, and perhaps also the metastereotypes would have been much less positive.

However, we found indications of some tension between both groups, or more specifically, of ethnocentric attitudes, symptomatic of such tension. Detection of these tendencies was made possible only by the introduction of metastereotype assessment into our study. As far as the personal-behavioral characteristics were concerned, both groups expected that the other group saw them in a less positive way than they described themselves in their autostereotypes. This tendency was stronger for German Silesians, being, from the entire nation's perspective, a minority group. Also, comparison of the metastereoptyes and the actual heterostereotypes formed by the other group revealed that the German Silesians tend to predict their less favorable perception on the Polish Silesians part. However, it is important to notice again, that the ethnocentric trends described here were found for the typicality indices being within positive range, which limits our diagnosis.

References

Bochenska, D. (1994). *The national auto-, hetero-, and metaimages among Polish and American university students.* Paper presented at the XII International Congress of Cross-Cultural Psychology, Pamplona, Spain, 1994.

Bochenska, D. (1994). *The role of metaimages in an international perception.* Unpublished manuscript.

Casas, J.M., Ponteretto, J.G., & Sweeney, M. (1987). Stereotyping the stereotyper. A Mexican American perspective. *Journal of Cross-Cultural Psychology, 18,* 1, 45-57.

Cooley, C.H. (1962[1902]). *Social organization.* New York: Scribners.

Juros, A. (1992). Tozsamosc negatywna Slazakow jako forma obrony wlasnej tozsamosci spolecznej. [Negative identity of Silesians as a form of defense of social identity]. In Chlewinski, Z. & Kurcz, I. (Eds.), *Stereotypy i uprzedzenia [Stereotypes and prejudices]* (pp. 87-103). Warszawa: Kolokwia Psychologiczne, Instytut Psychologii PAN.

Marjoribanks K., & Jordan, D.F. (1986). Stereotyping among Aboriginal and Anglo-Australians. The uniformity, intensity, direction, and quality of auto-, and heterostereotypes. *Journal of Cross-Cultural Psychology, 17,* 1, 17-28.

Mead, G.H. (1934). *Mind, self and society.* Chicago: University of Chicago Press.

Salazar, J.M., & Marin, G. (1977). National stereotypes as a function of conflict and territorial proximity: A test of the mirror image hypothesis. *Journal of Social Psychology, 101*, 13-19.

Triandis, H.C., Lisansky, J., Setiadi B., Chang, B., Marin, G., & Betancourt, H. (1982). Stereotyping among Hispanics and Anglos. The uniformity, intensity, direction, and quality of auto-, and heterostereotypes. *Journal of Cross-Cultural Psychology, 13,* 4, 409-426.

Note

Requests for reprints should be send to Danuta Bochenska, Psychology Department, University of Opole, ul. Oleska 48, 45-951 Opole, Poland. E-mail: danutab@sparc-1.uni.opole.pl

Acculturation and Ethnic Identity: The Remigration of Ethnic Greeks to Greece

James Georgas & Dona Papastylianou
The University of Athens, Athens, Greece

Psychological adaptation refers to changes at individual and group levels that accompany migration (Berry, 1994). Acculturation refers to culture change resulting from continuous, immediate contact between two cultural groups (Redfield, Linton, & Herskovits, 1936). Psychological acculturation refers to changes in an individual whose cultural group is undergoing acculturation (Berry, 1980, 1994; Graves, 1967). One of the psychological consequences of psychological acculturation is changes in ethnic identity.

A significant contribution to the theory of psychological acculturation has been Berry's concept of acculturation strategies. By this term, Berry, employing a model of conflict reduction, argues that there are a variety of adjustment tactics characteristic of the process of psychological acculturation (Berry, 1976, 1980, 1984). The psychological changes in the individual undergoing acculturation are related to the conflict: the degree of maintenance of one's ethnic identity versus the degree of seeking inter-ethnic contact and adopting elements of the ethnic identity of the host society. Berry has described four acculturation strategies. Assimilation refers to the "melting pot" concept of relinquishing one's ethnic identity and identifying completely with the host culture. Integration refers to maintaining synchronously aspects of one's ethnic identity and the identity of the host culture. Separation refers to maintenance of one's ethnic identity and rejection of identification with the host culture. Marginalization refers to the individual's simultaneous rejection of the cultural identity of one's ethnic identity and the host culture identity, and is essentially an example of psychological withdrawal from groups.

Ethnic Identity

Ethnic identity has its roots in the concept of *identity* and the mechanism of *identification*, one of the most basic defense mechanisms of personality formation of Freudian theory (1964). The process of matching a mental representation - part of subjective imagery - with physical reality leads to the construction of the ego. The process of identification of the child with the parent during psychosexual development reduces psychic tension or anxiety, leading to the incorporation of the ego-ideal or superego ideal (the parent) into one's ego or self. Erikson (1956) employed the concept of identity as a fundamental part of his theory (1963, 1968) in that identification bonds the self with aspects of the social environment during successive life stages. As such, the concepts *identity crisis* and *identity confusion* indicate that identity changes occur as a result of conflict between one's role and the social environment, leading to either role-confusion or more differentiated identity. From the point of view of social psychology, Tajfel's Social Identity Theory proposed that a group member is personally motivated to acquire and maintain a positive social identity. In addition, aspects of one's positive self esteem are derived from being a member of groups. (Tajfel, 1981; Tajfel & Turner, 1979).

Weinreich (1985, 1988) proposes a theoretical framework, in which he adds the effects of one's cultural tradition as part of the process of ethnic identity formation. Ethnic identity is defined as "...that part of the totality of one's self-construal of past ancestry and future aspiration in relation to ethnicity" (1988, p.158); in other words, the assimilation of ideas, behaviors, attitudes and the language code that are transferred from generation to generation through socialization.

The Context Variables

We will present in an outline form some relevant aspects of the context variables of the home countries, of the host country, of types of acculturating groups, and the group motivations for migration. A detailed description of Greek indigenous psychology can be found in Georgas (1993).

During the past decade, hundreds of thousands of ethnic Greeks from different countries have been migrating to Greece. One category is remigration, which refers to Greek immigrants to the United States, Canada, Australia, Germany and other European nations, who have decided to remigrate to Greece. A second category refers to the Pontic Greeks of the former Soviet Union. These are ethnic Greeks, descendants of the ancient Hellenic communities of the southern shores of the Black Sea, who were scattered by Stalin to different areas of the Soviet Union. These ethnic Greeks have retained the Greek culture, language, religion, customs, throughout 20 centuries, but have never lived in Greece. These characteristics are of course very similar to the Jews of the Soviet Union and the Diaspora. Following the breakup of the Soviet Union, the Pontics have been migrating to Greece. A third category refers to the ethnic Greeks from Albania, mostly from Northern Epirus, an area on the border of Northern Greece which became part of Albania after the breakup of the Ottoman Empire at the beginning of the century. After the fall of the Communist regime, many ethnic Greeks from Albania migrated to Greece for economic reasons, many of them employed as migrant agriculturer workers or laborers. There are numerous studies in the Greek literature regarding these groups (General Secretariat of Emigrant Greeks, 1990; Haritos-Fatouros & Dikaiou, 1986; Dikaiou,1994; Kasimati, 1992).

One of the interesting aspects of the sample is that each of the above categories has the same ethnic identity, however each group comes from a different culture, and each group has different historical and social reasons for migrated to Greece, although one commonality in their motivation is obviously economic. Georgas and Papastylianou (1994) have studied the stereotypes of migrant students from these three categories toward homoethnic Greeks and vice-versa.

Figure 1. Context Variables of the Home Countries, of the Host Country, Types of Acculturating Groups, and Group Motivations for Migration

Home Country
Ethnic Composition
Polyethnic: USA, Canada, Australia, former Soviet Union
Homogeneous: Albania
Economic Conditions
Stable, high GNP: USA, Canada, Australia
Unstable, low GNP: former Soviet Union, Albania
Host Country: Greece
Ethnic Composition: Homoethnic
Economic Conditions: Relatively stable, medium GNP
Immigration policy toward Ethnic Greeks: Assimilation
Types of Acculturating Groups
Soviet Union Greeks: immigrants and refugees
Albanian Greeks: refugees
Greek-Americans: remigrants
Greek-Canadians: remigrants
Greek-Australians: remigrants
Motivation for Migration
Soviet Union Greeks: voluntary
Albanian Greeks: voluntary
Greek-Americans: voluntary and involuntary
Greek-Canadians: voluntary and involuntary
Greek-Australians: voluntary and involuntary

The cultural differences which are normally found in different groups of returning migrants should affect differentially the way the young students conceive matters related to the formation of ethnic identity (e.g., language, the relationships between the mainstream culture and the culture of each group etc.), and to the types of psychological and psychosomatic reactions to the stress of acculturation to the host culture. More specifically, the model of acculturation refers to: (a) the cultural context of their "home country" - defined as the country from which they migrated to the "host country" Greece - (b) group issues related to the motivation for migration, e.g., persecution or bias toward ethnic group, and (c) individual factors, e.g., individual differences in motivation for migration, acculturation strategies, etc. (Berry, 1994).

The trend in current cross-cultural psychology to describe in detail the cultural context and history of a nation, or indigenous psychology (Kim & Berry, 1993), to attempt to define cross-cultural measures of context variables (Georgas & Berry, 1995) is exemplified by a number of studies in this volume, e.g., Manuel & Palacios; Mishra, Sinha, & Berry; Cook; Ostrowska & Bochenska; Boski.

The Ethnic Composition of the home countries may be related to issues such as bias against ethnic groups, as was the case with the Pontic Greeks in the former Soviet Union under Stalin, or with the ethnic Greeks in Albania at the present time. On the other hand, ethnic Greeks in polyethnic cultures such as the USA, Canada, and Australia have positive experiences regarding national policies toward ethnic groups and regarding interpersonal relations with other ethnic groups. The economic conditions of the Host Country and that of the Home Country - Greece in this study - are related to motivation for migration, and also expectations regarding acculturation in the host country. For the migrants from the former Soviet Union and Albania, which have unstable and poor economic conditions, the economic situation in Greece is perceived as a vast improvement. However, for the remigrants from the USA, Canada and Australia, nations with high GNPs, the economic situation in the host country, Greece, is perceived as lower standard of living. The group motivations for migration, which may be more than economic issues, are voluntary for ethnic Greeks from the former Soviet Union and Albania, where the desire to migrate is related to issues of bias against them, and their desire to return to "the homeland". The motivations for the adolescents in this study from the USA, Australia and Canada are mixed, in that *remigration* was the decision of their parents, who were immigrants, and the adolescents did not necessarily wish to migrate to Greece.

Purpose of the Study

This project studied the acculturation process to Greece of the Ethnic Greeks adolescents, from five nations, the former Soviet Union, Albania, USA, Canada, and Australia, in terms of: (a) modes of acculturation tactics, (b) how the specific acculturation tactic affects ethnic identity, and (c) the relationship between duration in Greece and ethnic identity.

Method

Sample

The sample consisted of 420 adolescent students, children of repatriated Greek immigrants and refugees, with mean age 15.5, and range from 12 to 18. One hundred twenty nine were from the former Soviet Union, 148 from Albania, and 143 from the U.S.A., Canada and Australia. The American, Canadian and Australians were grouped together as Anglophones, because previous studies (Papastylianou, 1992) indicated little variation among the Anglophone ethnic groups in the variables employed in this study. All students were attending Greek public secondary schools. One of the critical variables in acculturation is duration of time in the host culture (Georgas & Papastylianou, 1994). In the above sample, 20.8% were in Greece less than 1 year, 20.5% up to 2 years, 29.6% up to 3 years, 18.3% up to 4 years, and 20% more than 4 years. The proportion of each of the ethnic groups within each time period was fairly uniform and not biased toward any specific group.

The Questionnaires

In addition to demographic information the questionnaires consisted of: (a) the acculturation tactics, and (b) ethnic identity. The questionnaires were translated and back-translated into English, Russian and Albanian, and were administered to the students in the schools. Students choose the language - Greek or home country - in which they preferred to respond. Only a small proportion from each ethnic group chose the questionnaire in Greek.

Acculturation Tactics

Acculturation Tactics were measured in two ways. In the first method, four brief descriptions were prepared, one for each of the four acculturation tactics. The respondent was asked to choose the description which described most him/her. The descriptions are as follows.

Integration "Here in Greece I want to integrate the characteristics of the country in which I lived before with those of Greeks, because they are both very valuable."

Assimilation "Here in Greece I want to forget whatever was related with the country in which I lived before, and to become like all the Greeks."

Separation "Here in Greece I am not interested in relationships with Greeks. My relationships with people from my former country are enough for me."

Marginalization "Here in Greece I am not interested in relationships with Greeks nor with people from my former country. I will get along without their help."

The second method employed the Multi-Cultural Ideology Scales of Schmitz (1987) and the Revised form (Schmitz, 1991), which are based on Berry's model. The items were translated, back-translated, adapted, back-translated, adapted etc., a number of times. However, the problem remained as to whether the Schmitz items "worked" in the Greek cultural context, in relation to acculturation tactic they belonged. Therefore, the items were factor analyzed with the PA 2 method, in an attempt to determine if the four tactics emerged, and to which tactics the items loaded. Five factors emerged with criterion the Scree test. They were rotated according to the varimax criterion, and only items with loadings greater than .45 were interpreted. Factor 1 (18.8% explained variance) was a mixed factor with items from Marginalization, Separation and Assimilation. Factor 2 (8.8%) was clearly Assimilation, e.g., "We immigrants must change our customs and adapt to those of Greece". Factor 3 (8.3%) was Integration, e.g., "If we immigrants want to have good relationships with Greeks of Greece, we must accept their customs but also keep ours." Factors 4 (5.7%) and 5 (5.1%) had only two items each, the former mixed and the latter Integration.

The factor analysis of Schmitz's scales, adapted to Greek, did not result in four clear scales, which would correspond to the four acculturation tactics. This which might be partially due to problems of translation into the Greek language, or other problems related to etic concepts. We decided on the following procedure. Five items from the factor analysis were retained which were assigned to the Assimilation and Integration factors. For the Separation and Marginalization tactics, we chose five items for each scale, based on Schmitz's categorization. We then proceeded to conduct two parallel statistical analyses with the variables described below, e.g., ethnic identity, duration of time in Greece, etc.; employing in the first analysis our measurement of acculturation tactics, and in the second analysis Schmitz's system. In comparing the results of the two analyses, we judged that Schmitz's method led to results that were not interpretable, while our method of determining acculturation tactics led to the results that will be described below, which were consistent with the results of previous studies (Georgas & Papstylianou, 1994; Papstylianou, 1992).

Thus, the results are based on our system of determining the acculturation tactics of the respondents.

Ethnic Identity

Ethnic Identity was measured with the question, "How proud do you feel to be called: (a) a "Greek", (b) a "Greek-American", (c) an American/Australian/Canadian", or "Albanian", or "Russian" (The name of the ethnic group on the questionnaire varied in the third category accordingly).

Results

Acculturation Tactics

The frequencies of the four acculturation tactics of the Anglophone, Pontic and Northern Epirus Greeks, employing our method of categorization, are presented in Table 1. Integration was clearly the most preferred acculturation tactic among the three ethnic groups, which is consistent with the findings of Berry (1994), and Schmitz (1992b). Assimilation was found in 18.5% of the total sample, while 15.2% chose Separation. As would be expected from Berry's model and previous findings, Marginalization had the lowest percent (8.1%). A Chi Square test (10, $N = 414$, $p < .014$) analyzed possible differences in the acculturation tactics of the 3 groups. A tentative interpretation of the results in Table 1 is that Northern Epirus students chose Assimilation more than the other groups, that the Anglophone students prefer Separation more than the other groups. These results are consistent with the findings of a previous study (Georgas & Papastylianou, 1992) with samples from the same countries.

Table 1. *Frequencies of Acculturation Tactics of the Anglophone, Pontic and Northern Epirus Greeks*

	Anglo		Pontic		Albania		n
	n	%	n	%	n	%	
Integration	92	63.4	78	60.5	77	53.8	247
Assimilation	15	10.3	19	14.7	35	24.5	69
Separation	28	19.3	14	10.9	22	15.4	64
Marginalization	9	6.0	17	13.2	8	5.6	34
							414

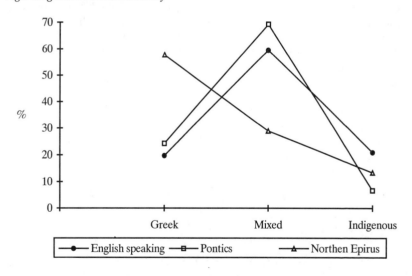

Figure 2. *Relationship between Northern Epirus, Pontic, and Anglophone Greeks regarding Stated Ethnic Identity*

Ethnic Identity

Responses to the question, "How proud do you feel to be called: (a) a "Greek", (b) a "Greek-American", or "Greek-Albanian", etc. (c) an American/ Australian/Canadian", or "Albanian", or "Russian", were cross-tabulated with the respondent's country of origin. The Chi Square Test (4, $N = 410$, $p <$

.0001) indicated systematic differences between the three groups in terms of their declared "Greek", "Mixed", or "country of origin" ethnic identity. The ethnic Greeks from Albania clearly chose "Greek" identity, in comparison to the Anglophone and the Pontic Russians (Figure 2). These latter two groups clearly chose the "mixed" identity, in contrast to the Greeks from Albania. The percent which chose the "country of origin" identity was lower than for "Greek" and "mixed", but within this category, the Anglophones appeared to have a higher percent of home country identification.

Figure 3. Relationship between Stated Ethnic Identity and Acculturation Tactics

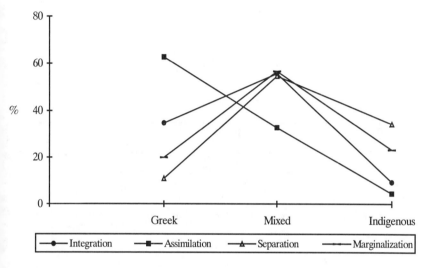

Ethnic Identity and Acculturation Tactics

The next question has to do with the relationship between acculturation tactics and stated ethnic identity, which is a fundamental hypothesis in Berry's theory (Berry & Annis, 1988; Berry, 1994). The data were analyzed with Chi Square (10, $N = 405$, $p < .0001$). There was a clear relationship between stated identity and assimilation (Figure 3), in which, of those who chose assimilation, 62,7% also chose Greek, 32.8% chose Mixed Identity, and only 4.5% chose Home Country identity. However, Mixed Identity was the highest percent of choices for the other three acculturation tactics, Integration (55.7%), Separation (54.7%, and Marginalization (56.7%). On the other hand, if one looks only at Greek Identity and Home Country Identity, it would appear that those who chose Separation have a higher Home Country Identity than Greek Identity, 34.4% vs. 10.9%, which supports the theory. The Integration tactic is a mirror image of the Separation tactic in that exactly the opposite is evident; 9.4% chose Home Country Identity while 34.5% chose Greek Identity. The Marginalization tactic is more balanced.

Figure 4. Relationship between Ethnic Identity and Duration of Time in Greece

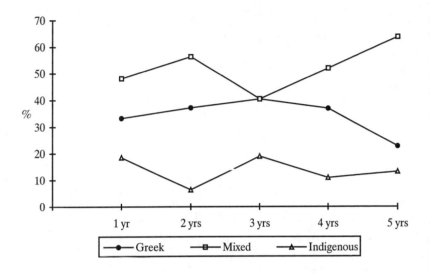

Figure 5. Relationship between Acculturation Tactics and Duration of Time in Greece

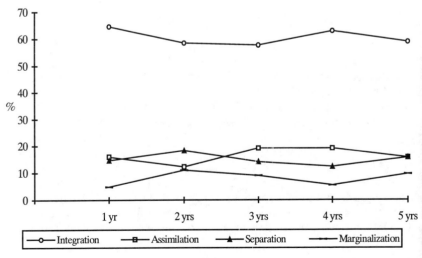

Ethnic Identity and Duration of Time in Greece

Figure 4 illustrates the relationship between duration of time in Greece, separated into five categories - from less than one year to more than four years - and stated ethnic identity Chi Square (8, $N = 404$, p < .048. The Mixed Identity group increased initially, decreased during the third year, and then increased afterward. This U-Curve is consistent with Lysgaard's (1956) description of acculturation phases. On the other hand, the curve begins to decline after 3 years for those with stated Greek Identity. Those who maintained the identity of the Home Country remained stable over the entire period. This supports previous findings that the third year appears to be a critical period in acculturation, which was found by Papastylianou (1992) in her study of English-speaking student remigrants to Greece

Acculturation Tactics and Duration of Time in Greece

Figure 5 presents the relationship between duration of time in Greece and acculturation tactics. The Chi Square (12, N = 398, $p > .05$) was not significant, indicating the acculturation tactics remain at the same level over the five year time period.

Discussion

The findings in this study corroborate two aspects of Berry's acculturation model. The first important finding is the inclination of the remigrants toward Integration. The same results were obtained in our previous study with the same groups of Ethnic Greeks (Georgas & Papastylianou, 1992). That the majority of immigrants chose Integration as the acculturation tactic is a finding that has important implications for acculturation theory. It is a fact that for decades, assimilation was considered to be the primary mode of acculturation, particularly in polyethnic societies. The findings by Berry (1994), and Schmitz (1994), and our findings indicate that psychological reasons related to the self-concept, are more important than social or political pressures to conform to a specific governmental policy, i.e., toward immigrants. This finding was somewhat unexpected with our samples, for the historical, political, and economic reasons outlined above. Of the three Ethnic Greek samples in this study, we would have predicted that the Northern Epirus Greeks of Albania, who have had historic ties with Greece, who live on the border with Greece, who have been persecuted by the previous Communist regime, who are essentially refugees, most of whom illegally emigrated to Greece, the salient acculturation tactic would be Assimilation. Although the percent of Greek-Albanians choosing Integration, 54%, was lower than that of Anglophone Greeks and Pontic Greeks, the fact that the majority of

Greek-Albanians chose Integration - as compared to the 24% who chose assimilation - supports the importance of the psychological aspects of the model, and the important of the integrity of the Self as part of personality.

Indeed, we had predicted the opposite with the Anglophone students; that the predominate acculturation tactic would be Separation, one reason being that with most of these adolescents, their parents decided to remigrate to Greece. In a previous study (Papastylianou, 1992) students from the same group expressed negative attitudes towards the Greek school and the people in Greece, and maintained highly negative stereotypes of Greeks in comparison with people from their country of origin (Georgas & Papastylianou, 1994). Most of these adolescents could be characterized as being subjected to involuntary migration, since the decision to repatriate was their parents'. This again is further evidence of the importance of the acculturation tactic Integration as an element of personality theory, as a means of maintaining the integrity of the Self.

The findings in relation to the stated Ethnic Identity of the three groups contain certain similarities with the above. The extremely restricted conditions of living and the persecution of the Albanian Greeks could help to explain the exceptionally high percent who chose Greek identity.

Regarding the home-country identity, Anglophones had a higher percent than the other two Ethnic Groups, which is consistent with previous findings (Georgas & Papastylianou, 1992). But, Mixed Identity is salient for both the Pontic Greeks and the Greek Americans.

The relationship between acculturative tactics and ethnic identity are also strongly supportive of Berry's acculturation theory, with Assimilation as the most clearly predictive tactic of Ethnic Identity.

As indicated above, the findings regarding acculturation tactics and ethnic identity are relevant to Conflict theory (Berry, 1994), and to aspects of Self theory as recently discussed by Marcus and Kitayama (1991), and by Triandis (1991). The process of acculturation should entail the reduction of anxiety related to the conflicts engendered by the stressful events, which is accompanied by the process of self-evaluation. Integration entails modifying one's construal of ethnic identity by adding to it a different construal of ethnic identity, and seeking means of harmonization. In terms of personality theory, this is a coping strategy - a strategy of harmoniously combining construals of the self from the past with construals of the present self, with the goal of harmonious adaptation in the future. Assimilation is like a defense mechanism in psychodynamic theory, implying repression or suppression of aspects of the Self. To deny or attempt to repress aspects of one's Self-construal has certain negative psychological consequences. Separation is similarly a tactic in which one denies the future and holds onto the past. The defense mechanisms, regression or retrogression, imply maintaining one's construal of the Self, in this case, one's ethnic identity, despite the pressures of adaptation to the host country, a process which also has dysfunctional psychological consequences. Marginalization is clearly a denial of one's construal of social self, with similar negative consequences.

The findings regarding the differential changes in Ethnic Identity over five years also support the above interpretation, in addition to the support they provide to Lysgaard's theory of U-Curve adaptation phases (1956). Decline after the third year for those who stated Greek identity might indicate that disappointments in acculturation, negative incidents with the indigenous population, etc., might lead to the reduction of the Indigenous identity. On the other hand, the increase of Mixed Identity after the third year is an indication of increasing attempts to integrate construals of one's home country identity with the host country identity, which, as indicated above, is a coping strategy, associated with successful psychological adaptation. That the acculturation tactics did not follow the U-Curve adaptation process was not expected.

In closing, the results are supportive of Berry's acculturation theory, and particularly of the relationship of acculturation tactics to construals of Ethnic Identity. Overall, acculturation seems to evoke fundamental adaptations of the ethnic identity and self-construals of individual immigrants. We would argue that the process of acculturation is more related to the individual process of personality development, than to explanations which stress group or social processes, taking into account both the culture of the host country and the culture of the home country.

References

Berry, J.W. (1976). *Human ecology and cognitive style: Comparative studies in cultural and psychological adaptation*. London: Sage.

Berry, J.W. (1980). Acculturation as varieties of adaptation. In A.M. Padilla (Ed.), *Acculturation: Theory, models and some new findings* (pp. 9-26). Boulder: Westview Press.

Berry, J.W. (1984). Cultural relations in plural societies: Alternatives to segregation and their sociopsychological implications. In N. Miller & M. Brewer (Eds.), *Groups in contact*. New York: Academic.

Berry, J.W. (1994). Acculturation and psychological adaptation: An overview. In A.M. Bouvy, F.J.R. van de Vijver, P. Boski, & P. Schmitz (Eds.), *Journeys into cross-cultural psychology* (pp. 129-141). Lisse: Swets & Zeitlinger.

Berry, J.W., & Annis, R.C. (Eds.). (1988). *Ethnic psychology: Research and practice with immigrants, refugees, native peoples, ethnic groups and sojourners*. Lisse: Swets & Zeitlinger.

Dikaiou, M. (1994). Present realities and future prospects among Greek returners. *International Migration, 32*, 29-47.

Erickson, E.H. (1956). The problem of ego identity. *The Journal of the American Psychoanalytic Association*, *4*, 56-121.

Erickson, E.H. (1963). *Childhood and society* (2nd ed.). New York: Norton.

Erickson, E.H. (1968). *Identity: Youth and crisis*. New York: Norton.

Freud, S.(1964). *An outline of psychoanalysis*. In Standard edition (Vol. 23). London: Hogarth. (First German edition, 1940).

General Secretariat of Emigrant Greeks. (1990). *Programma erevnon apodemias-palinnosteses tou ellenikou plethysmou* [Research program in emigration-repatriation of the Greek population]. Athens: General Secretariat of Emigrant Greeks.

Georgas, J. (1993). An ecological-social model for indigenous psychology: The example of Greece. In U. Kim & J.W. Berry. (Eds.), *Indigenous psychologies: Theory, method & experience in cultural context* (pp. 56-78). Beverly Hills: Sage.

Georgas, J., & Berry, J.W. (1995). An ecocultural taxonomy for cross-cultural psychology. *Cross-Cultural Research, 29*, 121-157.

Georgas, J., & Papastylianou, D. (1992, July). Acculturation to Greece of Greek migrants from the Soviet Union and Albania. In J.W. Berry & D. Sinha (Chairs), *Psychological acculturation: The generalizability of theories and findings*. Symposium conducted at the XXV International Congress of Psychology, Brussels, Belgium.

Georgas, J., & Papastylianou, D. (1994). The effect of time on stereotypes: Acculturation of returning migrants to Greece. In A.M. Bouvy, F.J.R. van de Vijver, P. Boski, & P. Schmitz (Eds), *Journeys into Cross-Cultural Psychology*. Lisse: Swets & Zeitlinger.

Graves, T.D. (1967). Psychological acculturation in a tri-ethnic community. *Southwestern Journal of Anthropology, 23*, 337-350.

Haritos-Fatouros, M., & Dikaiou, M. (1986). Counselling migrant persons concerning their children. *International Journal for the Advancement of Counselling, 9*, 301-316.

Kasimati, K. (1992). *Pontioi metanastes apo tin proen Sobietike Enose* [Pontic migrants from the former Soviet Union]. Athens: General Secretariat of Emigrant Greeks.

Kim, U., & Berry, J.W. (Eds.) (1993). *Indigenous psychologies: Theory, method & experience in cultural context*. Beverly Hills: Sage.

Lysgaard, S. (1956). Adjustment in a foreign society: Norwegian Fullbright grantees visiting the United States. *International Social Science Bulletin, 17*, 45-51.

Markus, H.R., & Kitayama, S. (1991). Culture and self: Implications for cognition, emotion and motivation. *Psychological Review, 98*, 224-253.

Papastylianou, A. (1992) *E psychologike prosarmoge ton matheton- paidion palinnostounton* [The psychological adaptation of students-children of immigrants]. Ph.D. Dissertation. University of Athens, Athens, Greece.

Redfield, R., Linton R., & Herskovits M. J. (1936). Memorandum on the study of acculturation. *American Anthropologist, 38*, 149-152.

Schmitz, P.G. (1987). *Acculturation, attitudes and beliefs of immigrants*. Paper presented at the First Regional North American Conference of the International Association for Cross-Cultural Psychology. Kingston, Canada.

Schmitz, P.G. (1991). Basic personality dimensions. Paper presented at the symposium on *Personality research in east and west*. Regional Congress of the International Association for Cross-Cultural Psychology, Debrecen, Hungary.

Schmitz, P.G. (1994). Acculturation and adaptation processes among immigrants in Germany. In A.M. Bouvy, F. J.R. van de Vijver, P. Boski, & P. Schmitz (Eds.), *Journeys into cross-cultural psychology* (pp. 142-157). Lisse: Swets & Zeitlinger.

Schmitz, P.G. (1992b). Acculturation styles and health. In S. Iwawaki, Y. Kashima, & K.S. Leung (Eds.), *Innovations in Cross-Cultural psychology* (pp. 360-370). Lisse: Swets & Zeitlinger.

Tajfel, H., & Turner, J. (1979). An integrative theory of intergroup conflict. In W. Austin & S. Worchel (Eds.), *The social psychology of intergroup relations* (pp. 33-47). Monterey, CA: Brooks/Cole.

Tajfel, H. (1981). *Human groups and social categories*. Cambridge: Cambridge University Press.

Triandis, H.C. (1991). The self and social behavior in differing cultural contexts. *Psychological Review, 96*, 506-520.

Weinreich, P. (1985). Rationality and irrationality in racial and ethnic relations: A metatheoretical framework. *Ethnic and Racial Studies, 8*, 500-515.

Weinreich, P. (1988). The operationalization of ethnic identity. In J.W. Berry & R.C. Annis (Eds.), *Ethnic psychology: Research and practice with immigrants, refugees, native peoples, ethnic groups, and sojourners* (pp. 149-168). Lisse: Swets & Zeitlinger.

Empirical Distinctiveness between Cognitive and Affective Elements of Ethnic Identity and Scales for their Measurement

Dan Hocoy
Queen's University, Kingston, Canada

A distinction that is commonly made among researchers of ethnic identity is one between the *cognitive* and *affective* elements of ethnic identity. In a review of 70 studies of ethnic identity published since 1972, Phinney (1990) found that *ethnic self-identification* (i.e., primarily a cognitive element) is commonly distinguished from *attitudes toward one's ethnic group* (i.e., primarily an affective element), as components of ethnic identity. Cross (1987) also distinguishes *race self-identification* (i.e., cognitive element) from *race esteem* (i.e., affective element), in his schema for Black identity. Berry and Boski (1988), in their construct of national self-identity or ethnic identity, also distinguish between a *criterial component*, which refers to the knowledge and personal relevance of cultural-specific symbols (e.g., language) and a *correlative component*, which refers to the similarity between one's self-description and construction of a nation's prototypical person (i.e., *cognitive* element), from an *affective* component, which refers to feelings toward one's country or ethnic group (i.e., affective element).

The distinction between cognitive and affective elements can be also regarded as a separation of descriptive from evaluative elements of ethnic identity. It is a distinction that becomes necessary when one's identification with an ethnic group is inconsistent with one's regard for that group. For instance, this occurs when an individual consciously regards oneself as being very much a part of a particular ethnic group, but does not regard the group positively.

Although this is a common experience among ethnic individuals in societies perceived to be hostile to one's ethnicity (e.g., Cross, 1987; Fordham & Ogbu, 1986; Hocoy, 1993), the lack of empirical support has led researchers to believe that such a distinction is only theoretical. For instance, Keefe (1992), in a factor analysis of integrated empirical findings from data gathered from the mid-1970s to 1987 from two different cultural groups, the Chicanos and the Appalachians, found that ethnic identity emerged as a general dimension of ethnicity, and that this factor was composed of two elements, the *identification of one's own versus other ethnic groups* (i.e., cognitive element), and the *kind and degree of sentiment attached to each group and its heritage* (i.e., affective element). Although a distinction was made here theoretically, the elements did not demonstrate empirical distinctiveness as they load on the same factor.

The findings of another empirical study also imply a distinction between cognitive and affective elements in ethnic identity, but again do not fully demonstrate this. Hinkle, Taylor, and Lee Fox-Cardamone (1989) examined the factor structure of the Group Identification Scale (Brown, Condor, Mathews, Wade, & Williams, 1986) and obtained one factor reflecting the cognitions of belonging, association or being linked to the group, and another factor reflecting emotional or affective aspects of group membership, among the three distinct, statistically identifiable factors. The existence of an empirical distinction between cognitive and affective elements in relation to ethnic groups, and ethnic identity in general, was not demonstrated in the samples of undergraduate subjects.

In addition to the lack of empirical support, there may be theoretical reasons to doubt the existence of the distinction. Tajfel and Turner's Social Identity Theory (1979) suggests that such a distinction may be merely an abstraction. According to Social Identity Theory, individuals are motivated to seek a positive social identity, in order to derive positive self-esteem. Since individuals are inclined to regard themselves in such a way that is favorable, the cognitive and affective concepts of ethnic identity are expected to be one and the same. Another source of doubt may come from the inability to empirically separate *self-concept* (i.e., cognitive element) from *self-esteem* (i.e., affective element) (Shavelson, Hubner, & Stanton, cited in Marsh, Relich & Smith, 1983), in the area of self-esteem.

Despite these empirical and theoretical problems, there are reasons to believe that a distinction does exist, and can be measured. The inability to capture the distinction in previous empirical research may be the result of an insufficient sensitivity in the scales previously used. The experiential evidence strongly suggests a distinction exists, and it may require the life experience and sensitivity of an ethnic individual to develop adequate scales.

The theoretical doubts raised by SIT and the self-esteem research may not apply to the area of ethnic identity. In regard to SIT, it may be difficult, if not impossible, for individuals of various ethnic backgrounds to *only* regard themselves in a way that is favourable, especially if they cannot easily dissociate themselves from their ethnicity and if their ethnicity is perceived as constantly being disparaged in society. As well, there is no reason to believe what has been found in the area of self-esteem is applicable to ethnic identity.

The purpose of this study was to investigate, via factor analysis, whether the cognitive and affective elements can be distinguished empirically, and in the process, develop scales that reliably do so.

Method

Scale Construction

For the elements to be empirically distinct, the instruments used must be sensitive to the subtle differences between the cognitive and affective elements of ethnic identity, and the concepts must be operationalized in a manner that make the concepts exclusive of one another. With this intent in mind, two scales were developed termed *Ethnic Self-Perception* and *Ethnic Esteem*, measuring respectively, the cognitive and affective elements of ethnic identity. Ethnic Self-Perception was operationally defined as the perception of oneself as an ethnic individual, while Ethnic Esteem was defined as the evaluative attitude toward one's own ethnicity.

Ethnic Self-Perception, unlike Ethnic Esteem, is a new concept and deserves some explanation. Ethnic Self-Perception emphasizes an internalized - although not necessarily desired - personal identity, in contrast to an externally imposed, social identity. Ethnic identity is both a subjective and objective phenomenon (Moodley, 1981), one containing internal and external boundaries (Rosenthal & Hrynevich, 1985). Ethnic Self-Perception concerns the subjective and internal aspects, and involves the degree to which a person, who is visibly seen, and socially identified as being of some ethnic origin, actually sees oneself, as an ethnic individual. Ethnic Self-Perception concerns the sense of oneself as ethnic, over and above one's social categorization of being an ethnic individual by others. It is not necessarily a desired sense of ethnic self-hood, because one can actively reject or have disdain for one's ethnicity. Ethnic Self-Perception pertains to a choiceless appraisal of the degree of one's ethnicity, which may not be personally desired. The rationale for the name "Ethnic Self-Perception" is to emphasize a purely descriptive measure of one's ethnic identification, devoid of any affective aspects. Other terms used in the past, such as "acceptance", "belongingness", were rejected because they can involve affective elements.

The scales were constructed based on conceptualizations from the ethnic identity literature, as well as from the author's own experiences as an ethnic individual. The Likert-like items reflected as many aspects of their respective domains as conceivable, based on previous scales and ethnic identity theory and research. These items were formulated by the author, and carefully worded and selected as to tease out the subtle differences between a cognitive identification with one's ethnicity and one's affective regard for that ethnicity, based on the author's own sensitivity to such a difference. A semantic differential format with an additional 15 items was included to measure Ethnic Esteem. The semantic differential format and the particular items used have been shown to measure evaluative elements (Kumata & Schramm, 1969) in previous research.

The Ethnic Self-Perception item pool was composed of 40 Likert-type items, while the Ethnic Esteem item pool was composed of 40 Likert-type items and 15 semantic differential items.

Because Ethnic Self-Perception is a new concept, three validation items were included in the questionnaire. The first item asked the subject to describe oneself as either Chinese, Chinese-Canadian, or Canadian. The second item asked the subject to indicate the strength of their identification with being

Chinese, Chinese-Canadian, and Canadian. The third item asked the respondent's first language.

Sample

The 140 individuals (77 males and 63 females) of Chinese heritage, were recruited in Kingston and Toronto, Ontario, Canada by the author from telephone lists of various Chinese associations, and at various gatherings and events of these associations. Participants were sampled from a variety of Chinese associations so that a diversity in age, country of origin, dialect of Chinese, years in Canada, and occupation, among subjects would be assured. Subjects ranged in age from 18 to 83 years, with a mean of 32 years. Respondents were born in 12 different countries, the largest group was born in China (40.7%), followed by Hong Kong (24.3%), and Canada (16.4%). Others, in decreasing number were born in Malaysia (4.3%), Trinidad and Tobago (3.6%), Jamaica (3.6%), Taiwan (2.1%), Singapore (1.4%), the U.S.A. (0.7%), Vietnam (0.7%), Portugal (0.7%), and Peru (0.7%). At least four different Chinese dialects were represented. The number of years in Canada ranged from 0 (i.e., arrived within the year) to 72 years, with the average number of years being 10.6 years. An informal assessment confirmed a wide diversity of occupations. The language of the scales used was English, and each subject was personally interviewed and assessed as to his or her fluency and competence in the language before administration of the scales.

Why the Chinese?

The particular visible ethnic group in which the distinction between cognitive and affective elements was examined was the Chinese in Canada. Given the exploratory nature of the research, it was considered necessary to control as many variables as possible and focus on one group, especially since there may be reason to suspect ethnic group differences. In addition, the idea for this study derived from the author's personal experiences, and the validity of its generalizability to other ethnic groups, at this stage at least, is questionable.

Chinese individuals were considered the optimal subjects on which to initially test the distinction between cognitive and affective elements in that, as a group, they should provide greater variation in ethnic identity, given that they possess a greater flexibility in identification than most visible minorities. That is, they may have more freedom to choose not to identify personally with their ethnicity, given their ability to assimilate, and greater acceptance relative other groups (Berry & Kalin, 1993); other groups may be more confined to the physical salience (e.g., skin colour) of their social identities (e.g., East Indian).

Also, research on the Chinese, being an Asian sub-group, may be of some urgency, as self-derogatory attitudes have been found to be more prevalent among Asians than other ethnic groups (Phinney, 1989). For instance, Asian-Americans were more likely than their Black or Hispanic peers (10th graders) to express the desire to belong to a different ethnic group if they had their choice (Phinney 1989). And a focus on the specific Asian sub-groups seems to be "essential" for future research (J. Phinney, personal communication, January 13, 1993).

Measures

Demographic Information
The following demographic variables were measured: age, sex, country of birth, mother's country of birth, father's country of birth, age of settlement in Canada, status in Canada, first language, and last in family to speak Chinese.

Ethnic Self-Rating
Apart from the scale, Ethnic Self-Perception was assessed by the three validation items described before.

Ethnic Self-Perception
In order to measure Ethnic Self-Perception, a 40-item pool that operationalized the conceptual definition of Ethnic Self-Perception was constructed and administered.

Ethnic Esteem
Ethnic Esteem was measured in two formats: a 40-item Likert-type scale and a 15-item semantic differential scale, consisting of items shown to have an evaluative element (Kumata & Schramm, 1969). This development pool was constructed to operationalize the conceptual definition of Ethnic Esteem.

Self-Esteem
The 10-item Rosenberg (1965) Self-Esteem scale was chosen to assess self-esteem in order to allow for comparisons with other studies examining the relationship between ethnic identity and self-esteem, using the same measure (e.g., Phinney, Chavira, & Williamson, 1992; Verkuyten, 1990; White & Burke, 1987). The Rosenberg scale is likely the most widely used self-esteem instrument in ethnic group research.

Method of Analysis

The first stage was a factor analysis of the combined pool of items from the Ethnic Self-Perception and Ethnic Esteem items to determine whether the dimensionality underlying the combined items was consistent with the conceptual distinctiveness assumed by the two scales.

The second step was to refine the scales from lengths of 40 and 55 items to 15, for the sake of brevity of administration. The refinement of the Ethnic

Self-Perception and Ethnic Esteem scales entailed: (1) computing two types of *item-total* correlations; *item-own* scale correlations and *item-other* scale correlations, and (2) obtaining a variant of the Differential Reliability Index (DRI variant; Jackson, 1984), which represents the portion of the item variance associated with its own scale that remains after subtracting the item's variance associated with the other scale. The DRI variant is a measure of the degree to which an item associates with its own scale, over and above the item's association with the other scale. The 15 items with the highest variant DRI's for their own scale were selected to comprise that scale.

Results

Empirical Distinctiveness

To assess the first hypothesis, that Ethnic Self-Perception and Ethnic Esteem are empirically distinct, as operationalized by the item pools developed for the study, a factor analysis of the combined, 95 items of the original Ethnic Self-Perception and Ethnic Esteem item pools was performed. Factor analytic results, regardless of the number of factors extracted, demonstrated the empirical distinctiveness between the theoretical concepts of Ethnic Self-Perception and Ethnic Esteem.

A factor analysis extracting only two factors, theoretically consistent with two distinct constructs underlying the items, resulted in factors that were conceptually distinct. The criterion by which an item was accepted as loading on a factor was .45. Factor I contained 36 items loading greater than the criterion. Of these, 33 items were from the original Ethnic Self-Perception pool, while only 3 were from the original Ethnic Esteem pool, and 2 of the 3 were among the lowest 7 loadings on Factor I. Therefore, Factor I seems clearly to be an Ethnic Self-Perception factor, as it is almost exclusively comprised of items from the Ethnic Self-Perception pool.

Factor II contained 14 items loading greater than the criterion. All 14 items are from the original Ethnic Esteem pool, 11 of which were Likert-like items, while 3 were semantic differential items. Thus, Factor II seems to be an Ethnic Esteem factor, as it is exclusively comprised of items from the Ethnic Esteem pool.

Ethnic Self-Perception Scale

The mean and the standard deviation for the original pool of 40 items administered to the subjects were 4.53 and .94 respectively. Cronbach's *alpha* for the development pool was .94. From this original pool of 40 items, the 15 items that most associated with its own scale, over and above the item's association with the Ethnic Esteem (55-item) scale (i.e., highest Differential

Reliability Index Variants) were selected to comprise the final Ethnic Self-Perception Scale.

The mean and the standard deviation for the Ethnic Self-Perception Scale were 4.20 and 1.00 respectively; Cronbach's *alpha* was .87.

Validity

Correlations with other scales were used as indirect measures of the validity of the newly developed Ethnic Self-Perception scale. One would expect a positive correlation between Ethnic Self-Perception and the degree to which one identified oneself as Chinese, and indeed, a positive correlation (r (138) = .57, $p < .01$) was found. In addition, when given a choice between Chinese, Chinese-Canadian, and Canadian to describe oneself, one would expect a positive correlation between Ethnic Self-Perception and the choice of the category of Chinese to describe oneself. which was found (r (138) = .47, $p < .01$). Also, a positive correlation between Ethnic Self-Perception and those whose first language is Chinese was found (r (138) = .39, $p < .01$).

Ethnic Esteem Scale

The mean and standard deviation for the pool of 40 Likert-type items were respectively, 4.73 and .64. Cronbach's *alpha* for the 40-item Likert-type scale was .90. The mean and the standard deviation for the pool of 15 semantic differential items, were 4.49 and 0.68. Cronbach's *alpha* for the 15-item semantic differential scale was .83.

From this original pool of 40 items, the 15 items that most associated with its own scale, over and above the item's association with Ethnic Self-Perception (40-item) scale were selected to comprise the final Ethnic Esteem Scale. The mean and the standard deviation for the Ethnic Esteem Scale were 4.79 and .68 respectively, and Cronbach's *alpha* for the scale was .81.

Validity

Correlations with other scales were used as indirect measures of the validity of the Ethnic Esteem scale. Other measures of Ethnic Esteem, for instance, the Black Identity Questionnaire (Morse, cited in Paul and Fischer, 1980), and the evaluative orientation regarding ethnic identity scale (Verkuyten, 1990), have obtained significant positive correlations with global self-esteem (e.g., Paul & Fischer, 1980; Verkuyten, 1990). Demonstrating convergent validity, the Ethnic Esteem scale developed here, similarly correlated positively (r (138) = .57, $p < .01$) with global self-esteem.

Discussion

Consistent with the prediction, factor analysis demonstrated the empirical distinctiveness between the instruments measuring Ethnic Self-Perception and Ethnic Esteem. These findings are consistent with a conceptual distinction made in the ethnic identity literature. Given the high *alphas* within item pools, and high separation between pools (i.e., items load on orthogonal factors), it seems that the item pools developed here were successful in operationalizing the conceptual distinction between Ethnic Self-Perception and Ethnic Esteem.

From these items pools, 15-item scales were developed which may be used in future ethnic identity research. The scales can be easily modified for use with other ethnic groups. As a recommendation for future research, Phinney (1990), in her extensive review of the ethnic identity research, concludes that the most serious need in ethnic identity research is to devise reliable and valid measures. And according to Mackie and Brinkerhoff (1984), ethnic identification (measured by Ethnic Self-Perception here), although being a critical concept in ethnic identity research, has received insufficient systematic attention in Canadian research. This neglect has been both theoretical and methodological. Perhaps the operationalization of Ethnic Self-Perception can be seen as a step in remedying this situation, by clarifying the conceptual distinctiveness between ethnic identification and ethnic esteem, and providing a reliable empirical measure of ethnic identification. Phinney (1990) also recommends that general aspects of ethnic identity that apply across groups be distinguished from specific aspects that distinguish groups.

Whether this empirical independence of Ethnic Self-Perception and Ethnic Esteem reflects a general aspect of ethnic identity, or merely a characteristic peculiar to Chinese in Canada is still in question. However, there is no reason to believe that this distinction is exclusive to Chinese individuals as the experiential research has indicated otherwise (e.g., Cross, 1971, 1978), but it still remains a question to be answered in future research. As well, since there was no direct measure of validity for the scales, their construct validity is also yet to be established by future research.

References

Berry, J. W., & Boski, P. (1988). *On becoming a Canadian or remaining a Pole: Stability and change in cognitive and affective components of national self-identity among Polish immigrants in Canada.* Report to Advisory Research Committee, Queen's University, Kingston, Canada.
Berry, J. W., & Kalin, R. (1993). *Multicultural and ethnic attitudes in Canada: An overview of the 1991 national survey.* Paper presented at Canadian Psychological Association Conference, Montreal, Canada.
Brown, R., Condor, S., Mathews, A., Wade, G., & Williams, J. (1986). Explaining intergroup differentiation in an industrial organization. *Journal of Occupational Psychology, 59,* 273-286.
Cross, W. E. (1971). The Negro to Black conversion experience: Towards a psychology of Black liberation. *Black World, 20,* 13-27.
Cross, W. E. (1978). The Cross and Thomas models of psychological Nigrescence. *Journal of Black Psychology, 5,* 13-19.

Cross, W. E. Jr. (1987). A two-factor theory of Black identity: implications for the study of identity development in minority children. In J. S. Phinney & M. J. Rotheram (Eds.), *Children's ethnic socialization* (pp. 117-133). Beverly Hills, CA: Sage.

Fordham, S., & Ogbu, J. U. (1986). Coping with the burden of acting white. *The Urban Review, 18,* 176-206.

Hinkle, S., Taylor, L. A., & Lee Fox-Cardamone, D. (1989). Intragroup identification and intergroup differentiation: A multicomponent approach. *British Journal of Social Psychology, 28,* 305-317.

Hocoy, D. (1993). *Ethnic identity among Chinese in Canada: Its relationship to self-esteem.* Unpublished master's thesis, Dept. of Psychology, Queen's University, Kingston, Canada.

Jackson, D. N. (1984). *Personality Research Form manual* (3rd ed.). Port Huron, MI: Research Psychologists Press.

Keefe, S. E. (1992). Ethnic identity: The domain of perceptions of and attachment to ethnic groups and cultures. *Human Organization, 51,* 35-43.

Kumata, H., & Schramm, W. (1969). A pilot study of cross-cultural meaning. In J. Snider & C. Osgood (Eds.), *Semantic differential technique* (pp. 273-288). Chicago, IL: Aldine Publishing Co.

Lorr, M., & Wunderlich, R. A. (1986). Two objective measures of self-esteem. *Journal of Personality Assessment, 50,* 18-23.

Mackie, M., & Brinkerhoff, M. B. (1984). Measuring ethnic salience. *Canadian Ethnic Studies, 16,* 114-131.

Marsh, H. W., Relich, J. D., Smith, I. D. (1983). Self-concept: The construct validity of interpretations based upon the SDQ. *Journal of Personality and Social Psychology, 45,* 173-187.

Moodley, K. (1981). Canadian ethnicity in comparative perspective: issues in the literature. In J. Dahlie & T. Fernando (Eds.), *Ethnicity, power and politics in Canada* (pp. 6-21). Toronto: Meuthuen.

Paul, M. J., & Fischer, J. L. (1980). Correlates of self-concept among black early adolescents. *Journal of Youth and Adolescence, 9,* 163-173.

Phinney, J. (1989). Stages of ethnic identity development in minority group adolescents. *Journal of Early Adolescence, 9,* 265-277.

Phinney, J. (1990). Ethnic identity in adolescents and adults: A review of the research. *Psychological Bulletin, 108,* 499-514.

Phinney, J. (1993). Personal communication, January 13.

Phinney, J., Chavira, V., & Williamson, L. (1992). Acculturation attitudes and self-esteem among high school and college students. *Youth and Society, 23,* 299-312.

Rosenberg, M. (1965). *Society and the adolescent self-image.* Princeton: Princeton University Press,.

Rosenberg, M. (1979). *Conceiving the self.* New York: Basic Books.

Rosenthal, D. A., & Hrynevich, C. (1985). Ethnicity and ethnic identity: a comparative study of Greek-, Italian-, and Anglo-Australian Adolescents. *International Journal of Psychology, 20,* 723-742.

Shavelson, R. J., Hubner J. J., & Stanton, G. C. (1976). Self-concept: Validation of construct interpretations. *Review of Educational Research, 46,* 407-441.

Tajfel, H., & Turner, J. (1979). An integrative theory of intergroup conflict. In W. Austin & S. Worchel (Eds.), *The social psychology of intergroup relations* (pp. 33-47). Monterey, CA: Brooks/Cole.

Verkuyten, M. (1990). Self-esteem and the evaluation of ethnic identity among Turkish and Dutch adolescents in the Netherlands., *Journal of Social Psychology, 130*, 285-297.

White, C. L., & Burke, P. J. (1987). Ethnic role identity among black and white college students: An interactionist approach. *Sociological Perspectives, 30*, 310-331.

Before and after Cross-Cultural Transition: A Study of New Zealand Volunteers on Field Assignments

Colleen Ward & Antony Kennedy
University of Canterbury, Christchurch, New Zealand

In recent years Ward and colleagues have pursued the investigation of cross-cultural transition and adaptation in a variety of sojourning groups. The conceptual basis of these investigations has been the theoretical and empirical distinction of two related constructs: *psychological* and *socio-cultural adjustment*. The former refers to psychological satisfaction or emotional well-being during a cross-cultural transition. The latter, in contrast, pertains to the ability to "fit in" or to negotiate interactive aspects of life in a new culture.

The conceptual distinction is underpinned by two popular approaches to the study of "culture shock." The first is derived from a clinical perspective, broadly associated with a stress and coping framework, and exemplified in pioneering work by Berry and colleagues (Berry & Kim, 1988). The second is more closely aligned with a social learning perspective, associated with a culture-learning framework, and exemplified by research by Furnham and Bochner (1986). Reflecting the merger of these two theoretical approaches in the study of the process and products of cross-cultural transitions, our research findings have consistently demonstrated that psychological and socio-cultural adjustment are inter-related, but that they are predicted by distinctively different variables. In the main, psychological adjustment is influenced by life changes, social support, and personality factors, such as locus of control or coping with humor. In contrast, socio-cultural adaptation is predicted by behavioral and cognitive variables such as length of residence in the new culture, quantity of interaction with hosts, perceived cultural distance, cultural identity, and language ability (Searle & Ward, 1990; Kennedy, 1994; Ward & Kennedy, 1992, 1993a,b; Ward & Searle, 1991).

Despite the burgeoning program of research, we have acknowledged two major shortcomings of our work on cross-cultural transition and adjustment. The first relates to the paucity of longitudinal studies. The second is concerned with the restricted range of sojourner samples. To circumvent the limitations of previous research, the study reported here entails a longitudinal design - including both pre-departure measures and multiple post-departure assessments - and a new type of sojourning sample - volunteer workers on overseas assignments.

The bulk of current research on cross-cultural transition and adaptation, like our own studies, has been cross-sectional. Even in those less common investigations of adjustment over time, it is extremely rare to find research, such as earlier studies by Kealey (1989) and Zheng & Berry (1991), which includes pre-departure measures. The reasons for this are obvious in terms of the difficulties involved in conducting longitudinal research in multinational settings, including the practical problems associated with overseas mailing and the location of current pre- and post-departure addresses for sojourners (Brabant, Palmer, & Gramling, 1990). However, the advantages are also obvious in terms of securing both information on the variations in adjustment over time and exploring pre-departure predictors of a successful sojourn. In addition to these theoretical and methodological advantages, longitudinal research, especially studies which incorporate pre-departure measures, has practical benefits for international organizations in terms of personnel recruitment, selection and training.

The majority of our previous studies has been undertaken with foreign students, with a smaller number of investigations based on samples of working adults and their spouses. This, again, is characteristic of the broader research area. Yet it has been argued that the patterns of cross-cultural adjustment are likely to vary, in at least some aspects, across different types of sojourners, and that this should be investigated in future research (Ward & Kennedy, 1995). A group of volunteer workers abroad provides a relatively atypical sample in that they are sent overseas to developing countries with the objective of offering a specialized service for a fixed period of time, they are expected to "go native" or live in the local way, as opposed to as pampered expatriates, yet they are frequently left to their own devices with little direct supervision and without any co-national support. As such, they present an interesting case study and allow us to extend the range of variation in our research on cross-cultural transition and adjustment.

The study reported here was designed to contribute to the growing body of literature on cross-cultural transition as well as to service the needs of the sponsoring organization in relation to their selection and training programs. Along these lines, personality factors and expectations have provided a starting point for the prediction of psychological and socio-cultural adaptation. In the first instance, we have considered the investigation of tolerance of ambiguity, self-acceptance and cultural identity. In the second, we have pursued the investigation of pre-departure expectations about social difficulty during field assignments. These are examined in relation to the criterion variables which include job performance and job satisfaction, in addition to our standard measurements of psychological and socio-cultural adjustment.

The proposition that successful adaptation to a new cultural environment would be facilitated by the capacity to withstand ambiguity is not a new idea. Early writers on adjustment during cross-cultural relocation described the successful sojourner as open-minded, tolerant, non-judgemental and flexible (Church, 1982; Hammer, Gudykunst & Wiseman, 1978). Rigid personalities and intolerance of ambiguity have been assumed to precipitate adjustment problems (Brislin, 1981; Locke & Feinsod, 1982; Maretzki, 1969). As sojourners are required to deal with complex and novel environments, tolerance of ambiguity should enhance psychological adjustment. In fact, Cort and King's (1979) study with American travellers in East Africa revealed that intolerance of ambiguity was related to psychological distress.

Self-concept is also likely to affect psychological adjustment during cross-cultural relocations. Of course there is a broader clinical literature which links positive self-concept to psychological well-being (Schlenker, 1985) as well as emerging research on the beneficial effects of perceived self-efficacy (Bandura, 1989). Along these lines, self-acceptance, the positive evaluation of oneself in terms of self-confidence and coping capacity, is likely to be a primary resource in managing cross-cultural transitions. Unfamiliar situations and unusual challenges commonly confronted during cross-cultural relocation are likely to engender adjustment problems, but self-acceptance should decrease the probability of psychological dysfunction. This is supported by evidence from the psychometric literature which has demonstrated that self-acceptance is generally associated with lowered levels of depression (Gough, 1987).

Following from our previous research, we also included an examination of cultural identity. Past studies have shown that a strong identity with culture of origin impedes socio-cultural adaptation (Ward & Kennedy, 1992). This is not surprising, in general, as one who strongly identifies as a New Zealander, may find it more difficult to behave in a culturally-discrepant manner if the situation demands it. However, later research has demonstrated that co-national and host national identification may interact to produce differential effects on psychological and socio-cultural adaptation (Ward & Kennedy, 1994).

Expectations have long been regarded as a crucial factor in determining adjustment during cross-cultural transitions (Gullahorn & Gullahorn, 1963; Klineberg & Hull, 1979). Indeed, many cross-cultural training programs are predicated on the assumption that it is necessary to prepare sojourners for life abroad by teaching them what to expect in a new cultural setting (Landis & Brislin, 1983). However, few studies have empirically examined these assumptions. Hawes and Kealy (1980) reported that expectations of a rewarding sojourn and competent performance on overseas assignments were associated with self-rated satisfaction on international development projects. Similarly, Searle and Ward (1990) reported a significant relationship between expected and actual difficulties in socio-cultural adjustment in Malaysian and Singaporean students in New Zealand. A major problem with these studies, as with Black and Gregersen's (1990) work with international business people, is that the expectation ratings were done retrospectively.

An exception to this trend is work by Weissman and Furnham (1987) who compared expectations and experiences of American sojourners in the United Kingdom. They reported that expectations were similar to experiences, but they failed to demonstrate that discrepancies between expectations and experiences significantly related to psychological adjustment. In one of our later studies Rogers and Ward (1993) examined re-entry expectations of AFS students abroad and found that the expected socio-cultural difficulties were not related to the actual experiences, nor were they directly related to psychological well-being. However, when re-entry adaptation problems were greater than expected, discrepancies between expected and real experiences were linked to higher levels of depression. Altogether the literature seems equivocal as to whether it is accurate or positive expectations which facilitate psychological and social adjustment (Bandura, 1989; Cochrane, 1983; Kealey, 1989). Our study extends this line of research, examining pre-departure expectations about socio-cultural adaptation.

While the longitudinal approach is useful for identifying variables which may predict successful adaptation during cross-cultural transitions, it is also suitable for investigating the pattern of psychological and socio-cultural adjustment over time and the causal direction of the relationship between the two adjustment domains. On the first count we have previously demonstrated socio-cultural adaptation improves steadily over time (approximating a learning curve) but that psychological adjustment is more variable and less predictable (Ward & Kennedy, 1995). This research allows for a more extensive investigation of temporal variation in psychological and socio-cultural adaptation. On the second count we have noted the strong relationship between the two adjustment domains in our past research and have suggested that the causal dimension of the relationship may run in either or both directions. Monitoring the two forms of adjustment over time allows the use of cross-lagged correlations to consider if one type of adjustment is causally predictive of another.

Finally, the analysis of the relationship between psychological and socio-cultural adjustment is also facilitated by the selection of a sample of international volunteers. In this instance the inclusion of this sample in our broader research program allows the exploration of possible differences in the magnitude of the relationship between psychological and socio-cultural adaptation. In this context Ward and Kennedy (1993a,b; 1995) have argued that while the association between psychological and socio-cultural adjustment is robust and reliable, the magnitude of the relationship varies. Among other conditions, the magnitude is increased when sojourners' experiences are characterized by a high incidence of host national contact and opportunity for integration. The strength of the relationship may also be augmented by a host-sojourner match. Along these lines we have demonstrated that, as expected, the magnitude of the relationship between psychological and socio-cultural adaptation was significantly greater in a culturally and ethnically similar group of sojourners (Malaysian students in Singapore) than in a culturally and ethnically dissimilar sample of sojourners (Malaysian and Singaporean students in New Zealand). In a separate study we likewise confirmed that the magnitude of the relationship was greater in an indigenous sample (New Zealand students at home) than in an overseas sample (New Zealand AFS students abroad; Ward & Kennedy, 1993a). In both instances it was postulated that data from a group such as volunteer workers are likely to demonstrate a particularly high correlation between psychological and socio-cultural adaptation in that these sojourners are forced to adapt to the host culture, being sent to perform a job and often existing in conditions without co-national support. This research allows us to compare the relationship between psychological and socio-cultural adjustment in volunteers with a previous sample of sojourning adults from New Zealand.

The primary objective of the research, then, was the study of psychological and socio-cultural adjustment over time. Emphasis is placed on pre-departure predictors of psychological and socio-cultural adjustment, variations in psychological and socio-cultural adjustment over time, and the magnitude of the relationship between psychological and socio-cultural adjustment in a group of volunteers and a comparative sample of New Zealand adults. The hypotheses concerning these variables and their inter-relationships follow. In addition to these concerns, however, organizations have interests in alternative "adjustment" indices, particularly the prediction of work-related variables. In this event, we also examined self-ratings of job performance and job satisfaction; this component of the research was considered exploratory.

Hypotheses:

1. Greater tolerance of ambiguity and self-acceptance will be positively associated with psychological adjustment in the field.
2. A strong identity with culture of origin will be negatively associated with socio-cultural adaptation.
3. Expectations of social difficulty will be related to the experience of social difficulty (socio-cultural adjustment); however, these expectations will be unrelated to psychological adjustment.
4. When socio-cultural difficulties are greater than anticipated, discrepancies between actual and expected difficulties will be associated with poorer psychological adjustment.
5. Poorer psychological adjustment will be apparent during field assignments than before departure; no hypothesis is advanced regarding the differences in psychological adjustment in the two field testings.
6. Socio-cultural adaptation difficulties will decrease during field assignments.
7. The magnitude of the relationship between psychological and socio-cultural adaptation during cross-cultural transition will be significantly greater in the group of VSA volunteers compared to New Zealand employees of a multi-national organization.

Method

Subjects

Fourteen recruits, ten women and four men, for New Zealand's Volunteer Service Abroad (VSA) participated in the study. The VSA program (akin to the American Peace Corps) provides New Zealand volunteer workers for various projects in developing countries. Placements most typically occur for a two year period, and the volunteers are often the only New Zealanders resident at the placement site. Twelve of the participants described themselves as Pakeha (white New Zealanders); one Chinese and one Maori subject also participated in the research. Subjects' mean age was 37.9 years ($SD = 12.5$). Twelve had previously lived abroad.

The volunteers were assigned to work in the following locations: Bhutan, Cook Islands, Namibia, Papua New Guinea, Solomon Islands, Tanzania, Vanuatu and Zimbabwe. Six were employed as teachers, three as managers, two as nurses, two as scientists, and one as an engineer. Although volunteers received pre-departure training of various kinds, with only one exception, they described their pre-departure language ability for the country of assignment as poor.

Procedure

Volunteers completed three questionnaires. The first testing occurred at their pre-departure training in Wellington, New Zealand, within a month before their anticipated field assignments. The subsequent testings involved postal questionnaires which were sent to volunteers approximately two months (returned on average at 9.7 weeks) after arrival in the field and then again after one year of field placement (returned on average at 52.1 weeks).

Participation in the research was anonymous and voluntary although the study was endorsed by VSA. Of the original 33 volunteers who were tested before departure to the field, 14 returned both of the two subsequent questionnaires. This represented 42% of the original sample.

Materials

The original, pre-departure questionnaire included personal and demographic information (e.g., age, previous overseas experience), measurements of personality and identity (tolerance of ambiguity, self-acceptance, and cultural identity), expectations (of social difficulty) and psychological adjustment (depression). The field testings included measurements of socio-cultural adaptation (social difficulty) and psychological adjustment (depression) as well as ratings of job performance and job satisfaction.

Pre-Departure Questionnaire

Tolerance of Ambiguity. Tolerance of ambiguity, or the tendency to perceive ambiguous situations as desirable, was assessed by Budner's (1962) 16-item Intolerance of Ambiguity Scale with a 5-point scale, where higher scores represent greater tolerance of ambiguity.

Self-Acceptance. Self-acceptance, or favorable evaluations of self-confidence and self-efficacy, was assessed by 20 items extracted from the Self-Acceptance subscale of the California Psychological Inventory (Gough, 1987). Subjects select true/false options in response to statements such as, "I set a high standard for myself, and I feel others should do the same," and, "I am certainly lacking in self-confidence." Higher scores indicated greater self-acceptance.

Cultural Identity. This 10-item scale was based on Tajfel's (1981) theory of social identity and pertains to the salience and importance of one's cultural group in relationship to personal identity. The original version of this scale was constructed by Hewstone and Ward (1985) for research on intergroup perceptions; subsequent versions have been used in our research on cross-cultural transition and adaptation and have proven reliable; this includes at least one study with New Zealand adults (Ward & Kennedy, 1992). Subjects rate issues such as the similarity of their own beliefs and values to other members of their culture, similarity and differences between cultural groups, and perceptions of others in terms of group membership, on 7-point bipolar scales.

Social Difficulty Expectations. Expectations about socio-cultural adaptation, more specifically, difficulties in managing daily tasks in a new environment, were tapped by an author-devised scale, based on earlier work by Furnham and Bochner (1982) with the Social Situations Questionnaire. Subjects specify the amount of difficulty they expect to experience in 23 everyday situations, e.g., getting used to the pace of life, using local transport on a 5-point scale. The measurement of actual socio-cultural adaptation problems has proven reliable in samples of New Zealand adults (Ward & Kennedy, 1992, 1994).

Psychological Adjustment. The Zung (1965) Self-Rating Depression scale (ZSDS) was used, as in our past research, as a measurement of psychological adjustment. Subjects rate their affective, physiological and cognitive components of depression on a 4-point scale, with 20 items. This scale has been used extensively in cross-cultural research and has proven reliable and valid in our studies with New Zealand adults (Ward & Kennedy, 1992, 1994).

Field Tests

Both of the subsequent field tests included measurements of psychological adjustment (depression) and socio-cultural adaptation (social difficulty). In addition, subjects evaluated their job performance and job satisfaction. In each of these instances 5-point rating scales (endpoints: poor/ excellent) were used for the evaluation.

Results

The sample size precluded direct reliability estimates for the measurement scales; however, there was some support of scalar validity in the intercorrelations amongst pre-departure measures. For example, strong cultural identity was associated with a low tolerance of ambiguity ($r = .74, p < .01$).

Table 1. Correlations between Pre-Departure Measures and Field Measures of Adjustment

Field Measures	Pre-Departure Measures						
	Age	Tolerance of Ambiguity	Self Acceptance	Cultural Identity	Expected Social Difficulty	Expectation Discrepancies	Depression
Test 1							
Social Difficulty	-.63*	-.41	-.12	-.01	.76**	-.03	.15
Depression	-.55*	-.43	-.11	.17	.45	.01	.58*
Job Performance	.08	-.09	.37	.59*	-.48	-.16	-.34
Job Satisfaction	-.15	-.06	.44	.54*	-.24	.12	-.49
Test 2							
Social Difficulty	-.52	.27	.00	-.02	.76**	.06	.04
Depression	-.16	-.07	.18	.12	.59*	.24	.20
Job Performance	.34	-.05	.31	.28	-.37	-.03	-.28
Job Satisfaction	.06	-.74**	.06	.69**	.04	.26	-.13

* $p < .05$ ** $p < .01$

Subsequent data analyses are organized in four parts. First, correlations between the pre-departure and field test variables are presented. Second, cross-lagged panel correlations conducted on psychological and socio-cultural dimensions of adjustment are described. Third, changes in psychological and socio-cultural adjustment are examined over time. Finally, comparisons are made between the magnitude of the correlation between psychological and socio-cultural adaptation in the VSA sample and a second group of New Zealand sojourners.[2]

The relationship between pre-departure and field test variables are presented in Table 1. Results indicate that pre-departure personality varables are unrelated to psychological and socio-cultural adjustment. However, cultural identity, more specifically, a strong self-perception as a New Zealander, is linked to more positive evaluations of job performance ($r = .59, p < .05$) and job satisfaction ($r = .54, p < .05$) after two months in the field. It is also associated with enhanced job satisfaction after a one-year period ($r = .69, p < .05$).

Figure 1. Cross-Lagged Correlations amongst Adjustment Measures during Field Assignments

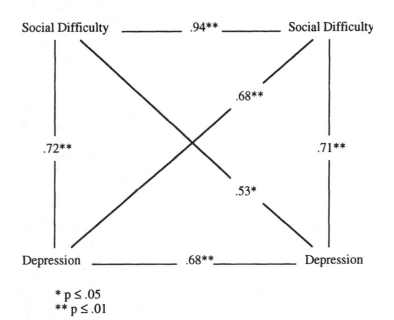

* $p \leq .05$
** $p \leq .01$

Pre-departure expectations bore some relationship to cross-cultural adjustment. More specifically, expectations of social difficulty were predictive of actual socio-cultural adaptation problems in both field testings ($r's = .76, p < .01$); they were also related to higher levels of depression after one year in the field ($r = .59, p < .05$). In addition to expectation scores, a measure of directional discrepancy was calculated on the signed difference between real and expected difficulties (See Rogers & Ward, 1993, for additional methodological information). Directional discrepancies between expected and real socio-cultural difficulties, however, were unrelated to psychological adjustment.[3]

Overall the most powerful predictor of psychological and socio-cultural adjustment was age. Analysis revealed that older vnlunteers experienced less psychological ($r = -.63, p < .05$) and socio-cultural ($r = -.55, p < .05$) adjustment problems early in their field assignments.

Figure 2. Depression before and during Field Assignments

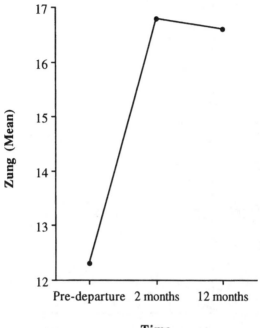

Time

Cross-lagged panel correlations were conducted to examine the direction of causality in the relationship between psychological and socio-cultural adjustment. As expected, psychological and socio-cultural adaptation were highly related at both concurrent field testings ($r > .70$). In addition, there was a significant relationship between the first and second field testings of depression ($r = .68$) and social difficulty ($r = .94$). However, the diagonals of the cross-lagged panels also produced significant relationships; that is, depression at two months in the field was significantly related to social difficulty experienced after one year ($r = .68$); likewise, social difficulty at the two month field test was significantly associated ($r = .53$) with depression after one year (See Figure 1).

Changes in psychological and socio-cultural adjustment were also examined over time. In the first instance a one-way analysis of variance was employed to compare depression before and after cross-cultural transition. Analysis revealed that volunteers experienced significantly more depression after beginning their field assignments than before departure from New Zealand ($M = 12.3$) ; $F (2, 26) = 6.6, p < .005$ (Figure 2). However, further analysis by Newman-Keuls indicated there was no significant difference between the first ($M = 16.8$) and second field testings ($M = 16.6$).

Figure 3. Pre-Departure Expactations of Social Difficulty and Actual Social Difficulty during Field Assignments

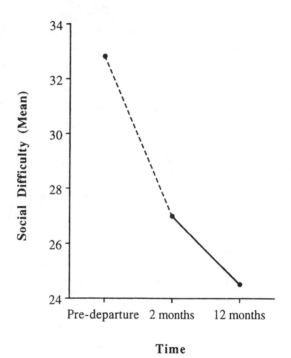

Pre-departure expectations of socio-cultural difficulty and actual social difficulty in the field are presented in Figure 3. As the measurement of actual, rather than expected, socio-cultural adaptation was limited to post-departure assessments, a t-test for paired samples was used for comparative purposes. As expected, results confirmed that socio-cultural adaptation difficulties were greater in the earlier stages of the field assignment; $t(13) = 1.81, p < .05$ ($M_1 = 26.7, M_2 = 24.5$). [4]

Finally, the magnitude of the relationship between psychological and socio-cultural adjustment was compared between two samples of New Zealanders abroad. In both cases there was a significant relationship between the two adjustment indicators (.71 and .72 in the VSA field testings and .36 in the comparison group). A test for significant differences between independent correlations revealed that the magnitude of the correlation was significantly greater in the VSA sample than in the comparison group; $z = 1.67, p < .05$. [5]

Discussion

Blending clinical and social learning perspectives on "culture shock," this study further investigated the prediction of, changes in, and the relationship between, psychological and socio-cultural adjustment during cross-cultural transition. Examining cross-cultural adaptation in a group of international volunteers, the research largely failed to identify pre-departure personality variables which predicted psychological and socio-cultural success in the field. However, the study confirmed that changes in psychological and socio-cultural adjustment varied in a predictable fashion, and that the relationship between the two domains is particularly strong in this type of sojourning group.

The failure to confirm the predictive influence of personality on psychological adjustment may be due to a number of factors. The first possibility is that tolerance of ambiguity and self-acceptance are actually unrelated to future psychological outcomes. Indeed, these domains have not been previously investigated in our cross-sectional research. Before that conclusion may be drawn, however, alternative explanations should be considered.

First, there are issues of scale reliability. The sample size in this study precluded the calculation of scale reliability, and there were no data available on the psychometric properties of these scales with New Zealand samples. In addition, Budner (1962) cited reliability estimates in the range of .50 for the Intolerance of Ambiguity Scale, and Gough (1987) reported a similarly low reliability index ($\alpha = .52$) for Self-acceptance in American samples. If reliability figures are similar for New Zealand samples, this would adversely affect the hypothesized relationship between personality and psychological well-being.

Second, the failure to confirm an association between tolerance of ambiguity, self-acceptance and psychological adjustment in sojourners may relate to the larger problem cited by Church (1982) on the person-situation interaction in cross-cultural research. Indeed, similar arguments about the interactive influence of situational variables were presented by Kealey (1989) whose longitudinal research on sojourner adaptation demonstrated only limited predictive power of personality factors. It is also possible that the relationship between personality and adjustment is diluted in longitudinal and predictive research, as compared to concurrent, cross-sectional studies. Further investigation is required to unravel these interwoven issues.

Although personality variables did not predict adjustment outcomes, demographic characteristics, more specifically age, related to both psychological and socio-cultural adaptation in the field. This is consistent with our earlier work (Armes & Ward, 1989) which indicated that older expatriates in Singapore experienced greater emotional satisfaction and less psychological distress. Similar findings were reported in a recent associated study by Pilcher

(1994) on psychological and socio-cultural adaptation of New Zealand business people abroad.

While cultural identity failed to relate to socio-cultural adaptation as hypothesized, it was predictive of work-related outcomes. More specifically, a strong identity as a New Zealander was associated with more positive self-ratings of job performance and satisfaction. Perhaps this is not surprising as volunteers act as New Zealand ambassadors to developing countries. In this sense their work roles are also connected with their national identities. Alternatively, they may possess a culture-bound perspective on their job performance which inflates self-appraisals. Performance evaluations by others, particularly host nationals, may or may not concur with self-ratings.

Expectations about social difficulties in a new culture were related to social difficulties experienced in the field. This suggests that VSA training packages provide some degree of accurate information about field assignments and living conditions. To a large extent volunteers experienced what they expected, although it appears that they may have anticipated even greater difficulties than actually encountered. Contrary to our hypotheses, discrepancies between expectations and experiences were not significantly related to psychological adjustment. On the whole, the pre-departure variables did not fare well as predictors of psychological and socio-cultural adjustment and of work performance and satisfaction.

The mutual effects of psychological and socio-cultural adjustment on successful cross-cultural adaptation were also explored. Cross-lagged correlational analysis suggested bidirectional causality between the two adjustment outcomes. This is not surprising as the issue of directional causality has also been raised in the mainstream clinical literature where a robust relationship has been demonstrated between depression and impaired social competence (Haley, 1985). In the context of cross-cultural transition one might ask: Do individuals have more social difficulties because they are depressed and/or are they depressed because they have more social difficulties? The answer to both questions appears to be yes.

The findings pertaining to the variations in psychological and socio-cultural adjustment over time are less controversial. As predicted, and in line with a culture-learning approach to cross-cultural relocation, socio-cultural adaptation problems diminish over time. Results further revealed that levels of depression increased after commencement of field assignments. This is consistent with research by Zheng and Berry (1991) on acculturative stress in Chinese sojourners in Canada, and is also expected in light of the clinical literature which links life changes with physical and psychological symptoms of distress (Holmes & Rahe, 1967). Certainly the amount and intensity of life changes engendered in a cross-cultural transition would put sojourners at risk. However, it is interesting to note that there were no differences in depression between the two and 12 month testings. To date our research has failed to demonstrate a discernible pattern of psychological adjustment over time (see also Ward & Kennedy, 1995). Indeed, it may be the case that temporal patterns of psychological adjustment are more strongly affected by extra-cultural factors such as school or work schedules, holidays or home leaves.

Finally, data analysis revealed that the association between psychological and socio-cultural adjustment is significantly greater in a group of VSA volunteers than in a comparable sample of New Zealand adults on overseas assignments. This is consistent with our proposition that the magnitude of the relationship is affected by contextual issues, including the capacity and opportunity for the sojourner to integrate with the host culture. Along these lines, it has been suggested that the relationship between psychological and socio-cultural adaptation is enhanced when sojourners primarily operate in the host culture milieu, as opposed, for example, to isolating themselves in "expatriate bubbles." In comparison with the samples we have studied to date, the level of integration demanded for volunteers is particularly high, and the alternative sources for social support are particularly low. Under these conditions, the primary operating environment, i.e., the host culture, provides not only the major source of feedback on social skills and competence, but also offers the primary, if not exclusive, resources for social support. Consequently, the relationship between psychological and socio-cultural adjustment should be exceptionally robust. Future research should explore additional dimensions which might affect the strength of the relationship between psychological and socio-cultural adjustment during cross-cultural transitions.

Before concluding, the limitations of this research should also be acknowledged. An obvious problem is the small sample size. On the most basic level this presented difficulties for the assessment of the reliability of the psychometric instruments. On a further count, while the 42% return rate compares favorably with return rates in international studies (Jobber & Saunders, 1988), the representativeness of this sample and the capacity to generalize from such a small number of respondents should be called into question. In addition to the sample size, concerns may be expressed about the selection of some of the measurement instruments. The Intolerance of Ambiguity Scale and Self-acceptance subscale of the CPI have not been previously used (to our knowledge) with New Zealand subjects, and their reliability appears to be somewhat low with American samples. On a related count, the validity of self-ratings for job performance evaluations is controversial and subject to limitations as has been discussed extensively in the industrial and organizational literature (e.g., Dunnett, 1976).

In contrast, the strength of this study lies in its longitudinal design, its pre- and post-testings of predictor and criterion variables, and its sample selection. In this light it has made a modest contribution to understanding the process and the product of cross-cultural transition. Despite the practical difficulties, future research should rely more extensively on longitudinal investigations and incorporate the use of more diverse samples in the investigation of cross-cultural transition and adjustment.

References

Armes, K., & Ward, C. (1989). Cross-cultural transitions and sojourner adjustment in Singapore. *Journal of Social Psychology, 12*, 273-275.

Bandura, A. (1989). Human agency in social cognitive theory. *American Psychologist, 44*, 1175-1184.

Berry, J.W., & Kim, U. (1988). Acculturation and mental health. In P. Dasen, J.W. Berry, & N. Sartorius (Eds.), *Health and cross-cultural psychology* (pp. 207-236). London: Sage.

Black, J.S., & Gregersen, H.B. (1990). Expectations, satisfaction and intention to leave of American expatriate managers in Japan. *International Journal of Intercultural Relations, 14*, 485-506.

Brabant, S., Palmer, C.E., & Gramling, R. (1990). Returning home: An empirical investigation of cross-cultural re-entry. *International Journal of Intercultural Relations, 14*, 387-404.

Brislin, R. (1981). *Cross-cultural encounters*. New York: Pergamon.

Budner, S. (1962). Intolerance of ambiguity as a personality variable. *Journal of Personality, 30*, 29-50.

Church, A.T. (1982). Sojourner adjustment. *Psychological Bulletin, 91*, 540-572.

Cochrane, R. (1983). *The social creation of mental illness*. London: Longman.

Cort, D.A., & King, M. (1979). Some correlates of culture shock among American tourists in Africa. *International Journal of Intercultural Relations, 3*, 211-225.

Dunnett, M.D. (Ed.) (1976). *Handbook of industrial and organizational psychology*. Chicago: Rand McNally.

Furnham, A., & Bochner, S. (1982). Social difficulty in a foreign culture: An empirical analysis of culture shock. In S. Bochner (Ed.), *Cultures in contact: Studies in cross-cultural interactions* (pp.161-198). Oxford: Pergamon.

Furnham, A., & Bochner, S. (1986). *Culture shock: Psychological reactions to unfamiliar environments*. London: Methuen.

Gough, H.G. (1987). *California Psychological Inventory: Administrator's Guide*. Palo Alto, CA: Consulting Psychologists Press.

Gullahorn, J.T., & Gullahorn, J.E. (1963). An extension of the U-curve hypothesis. *Journal of Social Issues, 19*, 33-47.

Haley, W.D. (1985). Social skills deficits and self-evaluation among depressed and non-depressed psychiatric inpatients. *Journal of Clinical Psychology, 41*, 162-168.

Hammer, M.R., Gudykunst, W.B., & Wiseman, R.L. (1978). Dimensions of intercultural effectiveness: An exploratory study. *International Journal of Intercultural Relations, 2*, 382-393.

Hawes, F., & Kealey, D. (1980). *Canadians in development: An empirical study of adaptation and effectiveness on overseas assignment*. Ottawa: Canada International Development Agency.

Hewstone, M., & Ward, C. (1985). *Ethnic identity, stereotypes and attributions in Malaysia and Singapore*. Unpublished raw data.

Holmes, T.H., & Rahe, R.H. (1967). The Social Readjustment Rating Scale. *Journal of Psychosomatic Research, 11*, 213-218.

Jobber, D., & Saunders, J. (1988). An experimental investigation into cross-national mail survey response rates. *Journal of International Business Studies, 19*, 483-489.

Kealey, D.J. (1989). A study of cross-cultural effectiveness: Theoretical issues, practical applications. *International Journal of Intercultural Relations, 13*, 387-428.

Kennedy, A.D. (1994). *Personality and psychological adjustment during cross-cultural transitions: A study of the cultural fit proposition*. Unpublished masters' thesis. University of Canterbury, New Zealand.

Klineberg, O., & Hull, W.F. (1979). *At a foreign university: An international study of adaptation and coping*. New York: Praeger.

Landis, D., & Brislin, R. (1983). *Handbook of intercultural training* (Vol. 1). Tarrytown, NY: Pergamon Press.

Locke, S.A., & Feinsod, F.M. (1982). Psychological preparation for young adults travelling abroad. *Adolescence, 17*, 815-819.

Maretzki, T.W. (1969). Transcultural adjustment of Peace Corps volunteers. In J.C. Finney (Ed.), *Culture change, mental health and poverty* (pp. 203-221). Lexington: University of Kentucky Press.

Pilcher, K. (1994). *The work and non-work related adjustment and adaptation of New Zealand business people while on overseas assignment*. Unpublished master's thesis, University of Canterbury, New Zealand.

Rogers, J., & Ward, C. (1993). Expectation-experience discrepancies and psychological adjustment during cross-cultural re-entry. *International Journal of Intercultural Relations, 17*, 185-196.

Schlenker, B.R. (Ed.) (1985). *The self and social life*. New York: McGraw Hill.

Searle, W., & Ward, C. (1990). The prediction of psychological and socio-cultural adjustment during cross-cultural transitions. *International Journal of Intercultural Relations, 12*, 61-71.

Tajfel, H. (1981). *Human groups and social categories*. Cambridge: Cambridge University Press.

Ward, C., & Kennedy, A. (1992). Locus of control, mood disturbance, and social difficulty during cross-cultural transitions. *International Journal of Intercultural Relations, 16*, 175-194.

Ward, C., & Kennedy, A. (1993a). Psychological and socio-cultural adjustment during cross-cultural transitions: A comparison of secondary students overseas and at home. *International Journal of Psychology, 28*, 129-147.

Ward, C., & Kennedy, A. (1993b). Where's the culture in cross-cultural transition? Comparative studies of sojourner adjustment. *Journal of Cross-Cultural Psychology, 24*, 221-249.

Ward, C., & Kennedy, A. (1994). Acculturation strategies, psychological adjustment and socio-cultural competence during cross-cultural transitions. *International Journal of Intercultural Relations, 18*, 329-43.

Ward, C., & Kennedy, A. (1995). Cross-cultures: The relationship between psychological and socio-cultural adjustment. In J. Pandey, P.S. Bhawuk, & D. Sinha (Eds.), *Asian contributions to cross-cultural psychology*. New Delhi: Sage.

Ward, C., & Searle, W. (1991). The impact of value discrepancies and cultural identity on psychological and socio-cultural adjustment of sojourners. *International Journal of Intercultural Relations, 15*, 209-225.

Weissman, D., & Furnham, A. (1987). The expectations and experiences of a sojourning temporary resident abroad: A preliminary study. *Human Relations, 40*, 313-326.

Zheng, X., & Berry, J.W. (1991). Psychological adaptation of Chinese sojourners in Canada. *International Journal of Intercultural Relations, 26*, 451-70.

Zung, W.W.K. (1965). A self-rating depression scale. *Archives of General Psychiatry, 12*, 63-70.

Notes

1. The study was made possible by a grant from the Department of Psychology, University of Canterbury, Christchurch, New Zealand. The authors would like to thank the volunteers and staff, especially Trevor Richards and Peter Brock, of Volunteer Service Abroad, Wellington, New Zealand, for their cooperation and support.

2. The organization allowed its staff to participate in the research on the condition that it would not be identified. However, it is accurate to describe the sojourning group as one which is well provided for, interacts extensively with a small number of co-nationals on the same placement, and frequently encounters host nationals in business settings and through formal social engagements. It is also important to note that like volunteers, this group is expected to represent New Zealand's interests abroad. N = 98; further details are provided in Ward and Kennedy (1994).

3. Absolute discrepancy scores, that is, a measure of realistic expectations, were also calculated, but they failed to relate to any of the criterion variables.

4. one tailed test.

5. one tailed test.

Part III

Cognitive Processes

The Cognitive Structure of Emotions in Indonesia and The Netherlands: A Preliminary Report

Johnny R.J. Fontaine, University of Leuven, Leuven, Belgium
Ype H. Poortinga, Tilburg University, Tilburg, The Netherlands, & University of Leuven, Leuven, Belgium
Bernadette Setiadi & Suprapti S. Markam, Universitas Indonesia, Jakarta, Indonesia

One central focus of concern in recent cross-cultural research is to what extent emotions, in the sense of distinguishable psychological states, are identical across cultures, and to what extent there are essential differences (Mesquita & Frijda, 1992; Russell, 1991). In the 19th century Darwin collected evidence to demonstrate the cross-cultural invariance of emotions and hence their biological origin. With the advent of a more social orientation on behavior emotions came to be seen as the product of the cultural context in which people are living. This point of view is especially influential within anthropology (Heelas, 1986; Lutz, 1988). Within cross-cultural psychology the balance has shifted again to a point of view in which all emotional states are assumed to be based on a set of universal reaction types, but in which systematic differences in the soliciting situations, the expression, the appraisal, and perhaps the cognitive meaning attributed to various states are not excluded, while in addition the frequency of occurrence may differ from culture to culture. This shift was strongly influenced by the empirical research conducted in the tradition of Ekman (Ekman, 1972, 1982; Ekman & Friesen, 1971). It seems likely that discrepancies in findings between schools have much to do with the perspective that researchers take when looking at reality. Nevertheless, some of the obvious contrasts in results and interpretations logically are incompatible and require clarification.

In this chapter we will report preliminary results of a study on the cognitive structure of emotions, as it emerges from the categorization of emotion words. In most cross-cultural studies of emotion use is made of *emotion* terms, but there are serious problems. First of all, there is a lack of criteria to decide whether an *emotion* term has been translated accurately or even can be translated (Russell, 1991; Mesquita & Frijda, 1992). In the absence of exact criteria for translation equivalence it is impossible to decide whether observed cross-cultural differences are merely linguistic or whether there is also non-identity of emotional states. This leads to a more fundamental problem, that the relationship is unclear between emotion *taxonomies* and emotional *experience* (Mesquita & Frijda, 1992). Any analysis of emotions through verbal terms requires a theory which specifies the relationships between emotional *experiences* and emotion *labels*.

In many respects the present study is a replication of the work reported by Shaver, Schwartz, Kirson & O'Connor (1987). These authors based their approach on Rosch's prototype theory (Fehr & Russell, 1984; Rosch, 1973, 1978; Rosch., Merris, Gray, Johnson, & Boyes-Braem, 1976). According to this theory, everyday concepts do not refer to categories which are characterized by necessary and jointly sufficient features. Natural categories are organized around prototypes, a prototype being the clearest example of a category. When an element has many features in common with the prototype, it clearly belongs to the category. There is a gradual change from membership to nonmembership; natural categories have fuzzy boundaries. For less prototypical elements it is often arguable whether or not they belong to the category. Prototype theory is not only concerned with the horizontal dimension of *categorization*, i.e., the segmentation of categories at the same level of inclusiveness, but also with the vertical dimension of categorization, i.e., hierarchical relations among categories. Following Rosch, Shaver et al. (1987) make a distinction between three levels of inclusiveness: the superordinate, the basic and the subordinate level. They write: "Basic level concepts ... convey more, and more specific information about category members than superordinate categories do, and at the same time, they are superior to subordinate-level concepts in identifying major distinctions between categories. Because the basic level offers the best compromise between informativeness and cognitive economy, people seem naturally to prefer basic-level categorization for much of their everyday conversation and thought" (p. 1062). Such a hierarchical ordering could be the sequence of furniture, e.g., chair and rotan chair.

Fehr and Russell (1984) found empirical evidence for the prototypical organization of the emotion domain. In a free listing task more prototypical emotion words were generated with a higher frequency. The fuzzy boundary of the emotion domain was supported in their view by the finding of peripheral words for which it is not clear whether or not they refer to an emotion state. Fehr and Russell also found a correlation between the number of shared attributes that emotion words had in common and ratings of their prototypicality.

Shaver et al., (1987) collected similarity sorting data of American students. A cluster analysis revealed the following hierarchical structure. At a high level they found a distinction between positive and negative emotion words. At intermediate level there were five categories that could be identified as *love, joy, anger, fear* and *sadness*. At a low level they distinguished 25 smaller clusters. For example, their love cluster was divided in the subclusters *affection, lust* and *longing*. The high, medium, and low cluster levels were taken as superordinate, basic and subordinate categories by Shaver et al. At the lowest level a distinction was made between core and noncore subclusters. A core subcluster is more representative of the mediate level cluster it belongs to, in the sense that it contains emotion words which are better examples of the cluster concerned, and thus have a higher degree of prototypicality than the emotion words in noncore subclusters.

The five basic-level clusters are included in other emotion theories as universal emotions (Izard, 1977). According to these theories, 7 ± 2 basic emotions exist which represent universal, probably innate categories, of subjective emotional experience. Other emotions are conceptualized as a specification of a basic emotion, or as a combination of basic emotions - usually one speaks of a blended emotion. Furthermore, different emotion words can designate different intensity levels and fairly specific situational antecedents. Especially for terms referring to a certain level of intensity or to specific situations found in one culture there may not be matching terms in some other culture.

A combination of basic-emotions theories and the prototype approach to emotions offers a framework for formulating relationships between emotional states and cross-cultural similarities and differences in labeling. If there exists a limited set of universal basic emotions, we can expect the basic level and the superordinate level to be cross-culturally identical. On the other hand, following Shaver et al.(1987), one would expect cross-cultural differences at the subordinate level and at the level of separate terms.

In a review, Russell (1991) argued that in the cross-cultural analysis of similarities between emotion words, two or three dimensions are usually found. The first dimension is consistently an *evaluation* - valence or pleasure - dimension. The second dimension is an *activity* - arousal or activation - dimension or a *potency* - power, control or dominance - dimension, depending on the study. In some studies all three dimensions have been found. Sometimes different investigations within the same culture result in different dimensions. Russell (1991) suggests that the selection of words may be important. Potency would emerge when the words tend to emphasize interpersonal contexts, activity would emerge when the words refer more to non-social (noninterpersonal) contexts. In the study of Shaver et al., (1987) all three dimensions were found. A possible explanation is the extensive list of emotion words included in that project. Overall, the findings suggest that the subjective experience of emotions has a connotative meaning structure as identified in Osgood's well-known Atlas of Subjective Meaning for 30 cultures (Osgood, May, & Miron, 1975).

The present study forms part of a comparative project conducted in Indonesia and The Netherlands. One of the problems facing the researchers is whether translated terms have the same meaning, or at least a similar meaning. Another point of concern is the obvious objection against the selection of terms on the basis of existing theories through which implicitly the results of Western research are imposed on other cultural populations. Therefore, we started out with the collection of indigenous emotion words in the two societies. Cross-cultural similarities derived from such indigenous emotion lists are more convincing than similarities derived from a translated list.

More specifically, we tried to replicate Shaver's et al. (1987) American findings in two other societies, The Netherlands and Indonesia (Jakarta), but with locally generated words. Our hypothesis was that in a hierarchical structure we would find (a) the same superordinate clusters (positive versus negative emotion words) and (b) the same basic clusters (love, joy, anger, fear and sadness), but that (c) differences would emerge at the subordinate level. We also expected (d) the same two or three dimensions of meaning (evaluation and activation and/or potency) in both cultures.

Method

Phase 1: Free Listing of Emotion Terms

The aim in this phase of the study was to gather a rather complete set of emotion words for each culture. We used the method described by Fehr and Russell (1984).

Subjects
Sixty Dutch (10 men and 50 women) and 73 Indonesian (15 men and 58 women) psychology students participated. All subjects were volunteers.

Procedure
Each subject was asked to write down as many emotion words s/he could think of during 10 minutes. In both samples all emotion words generated more than three times were selected for further research. In Indonesia the list was extended with 16 emotion words from an already existing list (Markam, unpublished report). The final result was a list of 153 Indonesian emotion words. The Dutch list contained 78 emotion words on the basis of the free listing task. This list was extended with emotion words readily translatable into Dutch that had been found in the Shaver et al., (1987) study and in the Indonesian list, but had not been generated by at least three subjects. The final Dutch list contained 136 emotion words.

Phase 2: Prototypicality Ratings

In this phase empirical ratings were gathered on the perceived prototypicality of each emotion word. In this way we also obtained information about the relevance of the emotion words not generated by the subjects in the previous phase, but included by us in the preliminary lists.

Subjects

Similar samples as in Phase 1 participated, consisting of 71 Indonesian (15 men and 56 women) and 89 Dutch (10 men and 79 women) psychology students.

Procedure

Each subject was asked to rate each emotion term on a scale indicating to which extent it represents an emotion; the scale points were 0 (certainly not), 1 (rather not), 2 (rather) and 3 (certainly). The 120 words with the highest scores in both Indonesia and The Netherlands were selected for further study. In Indonesia words with a scale value of 1.94 or higher were included; in The Netherlands the cutting point was 1.19.

Phase 3: Similarity Sorting

Subjects

The samples were again similar to those in phase 1. In Indonesia $n = 109$ (28 men and 81 women) and in The Netherlands $n = 105$ (46 men and 59 women) students participated. Conditions for the two groups of subjects were the same, except that the Dutch students were paid Dfl 10 for their work.

Procedure

Each of the 120 terms was printed on a small white card. The cards were presented with the same instructions as in the Shaver et al., (1987, p. 1065) study. The subjects were asked to sort the cards in categories of similarity. They could make as many and as small categories as they liked. They were given an hour to complete the task.

Results

As in the research of Shaver et al., (1987) a 120x120 co-occurence matrix was constructed for each subject, with 1 indicating that two terms were placed in the same category and 0 indicating that they were not. These matrices were added across the 105 Dutch subjects and across the 109 Indonesian subjects to form two matrices. The cell entries could range from 0 to 105 for the Dutch and from 0 to 109 for the Indonesian sample.

A preliminary finding was that the two samples differed significantly in mean number of clusters (28.5 in Indonesia and 19.4 in The Netherlands). This had a non-negligible effect on the co-occurence matrices. We can expect that emotion words which are very similar (or dissimilar) are placed often (or never) in the same category. But emotion words which have a somewhat similar meaning are likely to be put more often in the same category in The Netherlands than in Indonesia.

Approximate translations were found easily in Dutch for all Indonesian terms, and in English for both Indonesian and Dutch terms. More precise translations were established with the help of bilinguals, three of whom translated the Indonesian terms into Dutch and two who translated the Dutch terms into Indonesian. An initial criterion for equivalence was that one and the same word had to be mentioned by three of these five persons with none of the five giving a clearly deviating translation. So, *muak* and *walging* [disgust] were included, since muak was translated with walging two times by the Indonesian-Dutch translators, and walging once with muak by the Dutch-Indonesian translators. Another translation for muak was *misselijk* [nauseating] and the Dutch word *afschuw* [horror] was translated as memuakan. These translations were not incompatible with the translation of muak by walging. This resulted in a set of 48 terms which were used for comparative analysis.

Hierarchical structure

To find the hierarchical structure of the emotion domain in each sample both matrices were analyzed after rescaling the maximum frequencies of 109 and 105 to 100. The SAS Proc Cluster Procedure Average Linkage Method (Everitt, (1980) was used. Both in Indonesia and in The Netherlands there was a distinction between positive and negative emotion words at the highest level. Each subcluster is represented by the emotion term with the highest centrality rating; Tables 1 and 2 contain the emotion words of each subcluster. At the basic level a distinction could be made between *love*, *joy*, *anger*, *sadness* and *fear* in both cultures. In the Dutch group a sixth cluster emerged, identified as *guilt*. This somewhat unexpected finding was due to two infrequently used words, i.e. *zondigheid* [sinfulness] and *wroeging* [remorse]. When the analysis was repeated without these, the other emotion words of this cluster, i.e. *schuld* [guilt], *spijt* [regret] and *berouw* [repentance/feel sorry] became a subcluster of the fear cluster. *Zondigheid* [sinfulness] and *wroeging* [remorse] were the least prototypical emotion words of the guilt subcluster (prototypicality rating of the guilt emotion words were: *zondigheid* = 1.19; *wroeging* = 2.14; *schuld* = 2.21; *spijt* = 2.25; *berouw* = 2.35).

Since there was no prior matching of the emotion words it was difficult to compare the subclusters across the two samples. Ad hoc rules for cross-cultural correspondence were applied. With a maximum average distance of .70 there were 30 subclusters in the Indonesian data (Table 1) and 32 subclusters in the Dutch data (Table 2). A distance of .70 means that the emotion terms were placed in the same category by 30 percent of the subjects on average. Considering the subclusters that consisted of three or more words of which at least two belonged to the set of 48 translation equivalent terms, we identified 10 such subclusters (with a total of 30 terms) in the Indonesian and 8 (with a total of 34 terms) in the Dutch data. Some of these subclusters shifted from one basic cluster in Indonesia to another in The Netherlands, or vice versa. A minimum criterion for a shift of a subcluster was that at least two equivalent terms would be involved.

Table 1. Mean Centrality and Emotion Words of Indonesian subclusters

Centr.	Basic-, subcluster[1] and emotion words
	JOY
2.72	**bahagia** *[happiness]*: suka cita *[happiness]*, girang *[joy]*, nikmat *[pleasantness]*, suka *[joy]*, gembira *[cheerfulness]*, riang *[gaiety]*, senang *[happiness]*
2.49	**syahdu** *[quietness]*
2.46	**lega** *[relief]*: puas *[satisfaction]*, plong *[relief]*, bebas *[independence]*
2.42	**tenteram** *[peacefulness]*: tenang *[calmness]*, betah *[to feel at home]*, damai *[peacefulness]*, aman *[peacefulness]*
2.38	**tertarik*** *[attraction]*: terpesona *[fascination]*, kagum *[admiration]*
2.32	**bangga** *[pride]*: tersanjung *[to be flattered]*
2.18	**yakin** *[to be sure]*: optimis *[optimism]*, mantep *[confidence]*, semangat *[enthusiasm]*
	LOVE
2.79	**cinta** *[love]*: kasih *[affection]*, sayang *[love]*
2.51	**rindu*** *[homesickness]*: kehilangan *[to miss]*, kangen *[longing]*
2.36	**tulus** *[honesty]*: rela *[willingness]*, ikhlas *[resignation]*
	ANGER
2.61	**tersinggung** *[to be hurt]*: sakit hati *[to be hurt]*, terhina *[insult]*
2.57	**cemburu** *[jealousy]*: iri *[jealousy]*, dengki *[envy]*, dendam *[resentment]*, sirik *[envy]*
2.54	**benci** *[hate]*: jijik *[disgust]*, muak *[disgust]*, tidak suka *[dislike]*
2.47	**sebal** *[annoyance]*: murka *[fury]*, gondok *[suppressed anger]*, gusar *[anger]*, dongkol *[annoyance]*, geram *[fury]*, jengkel *[irritation]*, marah *[anger]*, kesal *[annoyance]*
	SADNESS
2.56	**sedih** *[sadness]*: pilu *[sorrow]*, pedih *[pain]*, duka *[grief]*, murung *[melancholy]*, sendu *[melancholy]*
2.53	**terharu** *[to be moved]*: kasihan *[pity]*, tersentuh *[to be touched]*, iba *[to be touched]*
2.50	**kecewa** *[disappointment]*: terpukul *[dejection]*
2.45	**bersalah** *[guilt]*: berdosa *[sinfulness]*, menyesal *[regret]*
2.42	**jenuh** *[to be fed up]*: bosan *[boredom]*, jenuh *[to be fed up]*, suntuk *[to be overtaken]*
2.30	**frustrasi** *[frustration]*: putus asa *[desperation]*, pesimis *[pessimism]*, pasrah *[resignation]*
2.24	**kesepian** *[loneliness]*: sepi *[loneliness]*, terkucil *[isolation]*, terasing *[alienation]*, hampa *[emptiness]*, diabaikan *[to be neglected]*, sunyi *[loneliness]*
2.21	**tertekan** *[dejection]*: merana *[misery]*, susah *[worry]*, menderita *[suffering]*, tertindas *[oppression]*
	FEAR
2.63	**cemas** *[worry]*: was-was *[suspicion]*, gelisah *[nervousness]*, kuatir *[worry]*, risau *[worry]*, gundah *[restlessness]*, resah *[anxiety]*
2.56	**rendah-diri** *[inferiority]*: minder [inferiority]
2.39	**bimbang** *[doubt]*: bingung *[confusion]*, ragu-ragu *[uncertainty]*
2.38	**curiga** *[suspicion]*: penasaran *[suppressed anger]*
2.32	**takut** *[fear]*: ngeri *[terror]*, tercekam *[fright]*, tegang *[tenseness]*, serem *[terror]*
2.27	**malu** *[shame]*: gugup *[nervousness]*, sungkan *[reluctance]*, kikuk *[awkwardness]*, segan/*diffidence]*, rikuh *[embarrassment]*, grogi *[groggy]*
1.99	**panik** *[panic]*
1.97	**terkejut** *[startle]*

[1]Each subcluster is named after its emotion word with the highest centrality rating.
*Subclusters with two translation equivalent terms shifting from basic cluster in The Netherlands.

Table 2. Mean Centrality and Emotion Words of Dutch Subclusters

Centr.	Basic-, subcluster[1] and emotion words
	LOVE
2.16	**verliefh.*** *[to be in love]*: genegenh. *[affection]*, tederh. *[tenderness]*, vriendschap *[friendship]*, verlangen *[longing]*, geluk *[happiness]*, passie *[passion]*, liefde *[love]*, aantrekking *[attraction]*, begeerte *[desire]*
2.12	**overgave** *[surrender]*: onderwerping *[submission]*
2.21	**ontroering*** *[to be moved]*: medelijden *[pity]*, medeleven *[sympathy]*, bezorgdh *[worry]*
1.82	**gevleidh.** *[to be flattered]*: trots *[pride]*
1.78	**bewondering** *[admiration]*: fascinatie *[fascination]*
	JOY
2.32	**blijh.** *[joy]*: optimisme *[optimism]*, meligh. *[to be corny]*, opgetogenh. *[elation]*, opgewekth. *[cheerfulness]*, verrukking *[delight]*, uitgelatenh. *[elation]*, plezier *[pleasure]*, verheuging *[joy]*, enthousiasme *[enthusiasm]*, vrolijkh. *[gaiety]*, vreugde *[joy]*, uitbundigh. *[exuberance]*
1.92	**verbazing** *[amazement]*: verrassing *[surprise]*
1.68	**tevredenh.** *[satisfaction]*: kalmte *[calmness]*, vredigh. *[peacefulness]*, berusting *[resignation]*
1.67	**hoop** *[hope]*: verwachting *[anticipation]*
1.52	**opluchting** *[relief]*
1.36	**moed** *[courage]*
2.51	**boosh.** *[anger]*: kwaadh. *[anger]*, agressie *[aggression]*, wrok *[resentment]*, jaloezie *[jealousy]*, wraakzucht *[vengefulness]*, nijd *[envy]*, vijandigh. *[hostility]*, woede *[fury]*, haat *[hate]*
2.43	**ontevredenh.** *[dissatisfaction]*: het beu zijn *[to be fed up]*
2.24	**walging** *[disgust]*: afkeer *[aversion]*, afschuw *[disgust]*, verachting *[contempt]*, afgrijzen *[horror]*
2.11	**ergernis** *[irritation]*: irritatie *[irritation]*, chagrijn *[to feel chagrined]*, ongeduldigh. *[impatience]*, ontstemming *[to be put out]*, humeurigh. *[moodiness]*, ongesteldh. *[indisposition]*, geprikkeldh. *[irritation]*
1.93	**gekwetsth.** *[to be hurt]*: vernedering *[humiliation]*, belediging *[insult]*, gekweldh. *torment]*
1.78	**wantrouwen** *[suspicion]*
1.64	**onverschilligh.** *[indifference]*: verveling *[boredom]*
1.26	**ontzetting** *[horror]*
	FEAR
2.20	**angst** *[fear]*: huivering *[shiver]*, bangh. *[fear]*, bevreesdh. *[apprehension]*, verontrusting *[alarm]*, paniek *[panic]*, schrik *[terror]*
2.17	**gefrustreerdh.** *[frustration]*
2.06	**schroom** *[diffidence]*: schaamte *[shame]*, verlegenh. *[shyness]*
2.06	**onzekerh.** *[uncertainty]*: minderwaardigh. *[inferiority]*
2.00	**stress** *[stress]*: nervositeit *[nervousness]*, zenuwachtigh. *[nervousness]*
1.91	**verwarring** *[confusion]*
	SADNESS
2.30	**verdriet** *[sadness]*: neerslachtigh. *[dejection]*, somberh. *[gloominess]*, teneergeslagenh. *[dejection]*, ongelukkigh. *[unhappiness]*, droevigh. *[sadness]*, depressie *[to be depressed]*, triestigh. *[melancholy]*, treurigh. *[sorrow]*, pessimisme *[pessimism]*, bedroefdh. *[sadness]*
2.18	**teleurstelling** *[disappointment]*
2.13	**wanhoop** *[desperation]*: onmacht *[impotence]*, hopeloosh. *[hopelessness]*
2.00	**lijden** *[to suffer]*
1.86	**heimwee*** *[homesickness]*: verlorenh. *[to feel lost]*, leegte *[emptiness]*, eenzaamh. *[loneliness]*, gemis *[to miss]*
1.64	**zieligh.** *[pitifulness]*
	GUILT
1.86	**berouw** *[repentance]*: wroeging *[remorse]*, schuld *[guilt]*, zondigh. *[sinfulness]*, spijt *[regret]*

[1]Each subcluster is named after its emotion word with the highest centrality rating.
*Subclusters with two translation equivalent terms shifting from basic cluster in Indonesia.

This proved to be the case for two subclusters in the Indonesian data (Table 1). *Kagum* [admiration] and *tertarik* [attraction] were part of the *joy* cluster, but the Dutch equivalents were part of the *love* cluster. The equivalents of *kehilangan* [to miss] and *rindu* [homesickness] shifted from the *love* to the *sadness* cluster in The Netherlands. Of the eight relevant subclusters in the Dutch data there were three that showed a non-correspondence with the Indonesian cluster analysis (Table 2). *Gemis* [to miss] and *heimwee* [homesickness] belonged to the *sadness* cluster in The Netherlands and the Indonesian equivalents shifted to the *love* cluster. In two of the *love* subclusters translation equivalents shifted to the Indonesian subclusters of *joy* (i.e. *aantrekking* [attraction] and *geluk* [happiness]) and sadness (i.e. *medelijden* [pity] and *ontroering* [to be moved]) respectively. It should be noted that all of these shifts involve the *love* cluster. In addition, when the five-cluster solution was followed, the Dutch terms of the subcluster guilt had their equivalents in the Indonesian cluster of sadness.

At the level of the separate emotion words non-correspondence was observed for 16 of the 48 translation equivalent terms, in the sense that these 16 terms did not belong to the same basic cluster in the two data sets (Table 3). It may be noted that the rate of shifts for subclusters (6/18) was about the same as that for separate words, but that this is mainly due to the deviant results for love.

Table 3. Shifting Emotion Words in the Hierarchical Cluster Analyses

Dutch emotion word	Dutch basic cluster	Indonesian basic cluster	Indonesian emotion word	English translation
gevleidheid	LOVE	JOY	tersanjung	[being flattered]
aantrekking	LOVE	JOY	tertarik	[attraction]
geluk	LOVE	JOY	bahagia	[happiness]
trots	LOVE	JOY	bangga	[pride]
bewondering	LOVE	JOY	kagum	[admiration]
medelijden	LOVE	SADNESS	kasihan	[pity]
ontroering	LOVE	SADNESS	terharu	[to be moved]
bezorgdheid	LOVE	FEAR	kuatir	[worry]
spijt	GUILT/FEAR	SADNESS	menyesal	[regret]
zondigheid	GUILT/FEAR	SADNESS	berdosa	[sinfulness]
schuld	GUILT/FEAR	SADNESS	bersalah	[guilt]
wantrouwen	ANGER	FEAR	curiga	[suspicion]
verveling	ANGER	SADNESS	bosan	[boredom]
heimwee	SADNESS	LOVE	rindu	[homesickness]
gemis	SADNESS	LOVE	kehilangan	[to miss]
gefrustreerdh.	FEAR	SADNESS	frustrasi	[frustration]

Dimensional structure

The two 120x120 matrices were subjected to a classical nonmetric multidimensional scaling analysis (using the SAS MDS procedure). Coordinates averaged over the terms in the basic clusters and in the subclusters are presented in Tables 4 and 5.

Table 4. Coordinates of the Indonesian Basic and Subclusters in Two- and Three-Dimensional Space

Sub-/basic cluster	English translation	Two-dimensional coordinates		Three-dimensional coordinates		
		Dim1	Dim2	Dim1	Dim2	Dim3
JOY		-1.58	0.06	-1.71	0.01	0.00
tenteram	[peacefulness]	-1.71	-0.13	-1.88	-0.28	0.22
syahdu	[quietness]	-0.51	-0.78	-0.48	-1.15	0.10
lega	[relief]	-1.50	0.19	-1.71	0.20	-0.34
bahagia	[happiness]	-2.08	0.18	-2.22	0.14	-0.04
bangga	[pride]	-1.09	-0.23	-1.18	-0.45	-0.03
tertarik	[attraction]	-1.00	-0.14	-0.97	-0.23	-0.32
yakin	[to be sure]	-1.44	0.44	-1.61	0.60	0.37
LOVE		-1.20	-0.24	-1.26	-0.52	0.18
cinta	[love]	-1.67	0.22	-1.88	0.01	0.12
rindu	[homesickness]	-0.38	-0.88	-0.27	-1.18	-0.03
tulus	[honesty]	-1.55	-0.05	-1.65	-0.39	0.46
ANGER		0.84	0.87	0.73	1.03	1.07
benci	[hate]	0.97	0.94	0.78	1.05	1.28
sebal	[annoyance]	0.97	0.88	0.83	1.13	0.99
tersinggung	[to be hurt]	1.02	0.17	0.97	0.65	0.46
cemburu	[jealousy]	0.40	1.21	0.38	1.07	1.39
SADNESS		0.39	-0.86	0.59	-0.94	0.18
bersalah	[guilt]	0.88	-1.05	1.11	-0.93	-0.66
kesepian	[loneliness]	0.05	-0.94	0.25	-1.08	0.27
sedih	[sadness]	0.75	-1.02	0.97	-1.13	0.15
tertekan	[dejected]	0.70	-1.01	1.00	-1.04	0.21
kecewa	[disappointment]	-0.46	-0.24	-0.45	-0.14	0.20
frustrasi	[frustration]	0.59	-1.15	0.86	-1.14	-0.09
terharu	[to be moved]	-0.27	-1.18	-0.10	-1.25	-0.15
jenuh	[to be fed up]	0.60	0.13	0.62	-0.23	1.22
FEAR		0.79	0.44	0.77	0.56	-1.08
bimbang	[doubt]	0.95	0.41	0.84	0.67	-1.18
cemas	[worry]	1.07	0.05	1.02	0.38	-1.07
curiga	[suspicion]	0.39	0.94	0.43	1.15	-0.68
malu	[shame]	0.67	1.06	0.58	0.93	-1.40
rendah-diri	[inferiority]	0.81	-0.82	1.05	-0.49	-0.78
takut	[fear]	0.60	0.69	0.65	0.70	-1.13
panik	[panic]	1.05	0.89	0.92	1.12	-1.03
terkejut	[startle]	0.54	-1.24	0.84	-1.37	0.20

Table 5. *Coordinates of the Dutch Basic and Subclusters in Two- and Three-Dimensional Space*

Sub-/basic cluster	English translation	Two-dimensional coordinates		Three-dimensional coordinates		
		Dim1	Dim2	Dim1	Dim2	Dim3
LOVE		-1.14	0.15	-1.26	-0.18	-0.46
verliefdheid	[to be in love]	-1.33	-0.00	-1.43	-0.02	-0.42
bewondering	[admiration]	-1.65	-0.17	-1.84	0.21	-0.38
gevleidheid	[to be flattered]	-1.61	-0.24	-1.83	0.27	-0.04
overgave	[to surrender]	-0.25	0.02	-0.43	-0.05	-0.86
ontroering	[to be moved]	-0.62	0.95	-0.69	-1.09	-0.59
JOY		-1.58	-0.05	-1.71	0.05	0.48
tevredenheid	[satisfaction]	-1.24	0.51	-1.37	-0.59	0.53
blijheid	[joy]	-1.72	-0.25	-1.82	0.24	0.76
opluchting	[relief]	-1.62	0.10	-1.82	-0.10	0.46
verbazing	[surprise]	-1.46	-0.06	-1.57	0.12	-0.31
hoop	[hope]	-1.47	0.11	-1.69	-0.03	-0.22
moed	[courage]	-1.52	-0.10	-1.74	0.18	-0.26
ANGER		0.84	-0.89	0.94	0.99	0.32
walging	[disgust]	0.77	-1.47	0.88	1.71	-0.29
ontzetting	[horror]	0.96	-0.63	1.07	0.70	-0.77
boosheid	[anger]	0.72	-1.42	0.84	1.64	0.06
wantrouwen	[suspicion]	1.09	-0.46	1.25	0.63	-0.42
gekwetstheid	[to be hurt]	0.99	-0.31	1.18	0.37	0.02
ergernis	[irritation]	0.82	-0.79	0.89	0.73	0.98
ontevredenh.	[dissatisfaction]	1.16	-0.42	1.20	0.17	1.05
onverschilligh.	[indifference]	0.90	0.74	0.83	-0.63	1.29
FEAR		0.68	0.53	0.71	-0.44	-0.95
angst	[fear]	0.88	0.42	0.82	-0.32	-1.31
verwarring	[confusion]	0.30	1.00	0.47	-1.22	-0.42
gefrustreerdh.	[frustration]	1.14	-0.28	1.34	0.31	0.31
stress	[stress]	0.71	0.41	0.82	-0.32	-0.65
onzekerheid	[uncertainty]	0.82	0.67	0.97	-0.81	-0.44
schroom	[diffidence]	0.06	0.94	0.06	-0.60	-1.35
SADNESS		0.77	0.86	0.84	-1.06	0.48
verdriet	[sadness]	0.69	0.85	0.72	-1.00	0.68
heimwee	[homesickness]	0.66	1.20	0.78	-1.44	0.35
zieligheid	[pitifulness]	0.98	0.98	1.03	-1.24	0.52
teleurstelling	[disappointment]	1.08	0.65	1.16	-0.89	0.65
lijden	[suffering]	1.08	0.59	1.20	-0.94	0.14
wanhoop	[desperation]	0.97	0.42	1.13	-0.71	-0.00
GUILT		0.58	-0.05	0.53	0.17	-1.39
berouw	[repentence]	0.58	-0.05	0.53	0.17	-1.39

We selected a two and a three dimensional solution. For the Indonesian data the stress measure (using Young's s-stress formula 1; Young & Hammer, 1987) was .35 for the two-dimensional and .26 for the three-dimensional solution. For the Dutch data the stress values were .31 and .23, respectively. In the two cultures the first dimension could be interpreted clearly as an *evaluation* dimension (*love* and *joy* emotions were situated at the one side and *anger, sadness* and *fear* at the other side). In a two-dimensional solution the second dimension was interpreted as an *arousal* dimension (*sadness* at the negative end, *love* and *joy* in the middle, and *fear* and *anger* at the positive end) for the Indonesian data, and as a potency dimension (*anger* at the negative end, *joy* and *love* in the middle, and *fear* and *sadness* at the positive end) for the Dutch data. In a three dimensional solution for the Indonesian data the first dimension could again be interpreted as *evaluation*, the second as *arousal* and the third as *potency*. For the Dutch data only the first and the second dimension had a meaningful interpretation, namely as *evaluation* and *potency*. The interpretation of the third dimension was not clear.

Discussion and Conclusion

As in many other studies the distinction between positive and negative emotion words was evident for both cultures in the multi-dimensional scaling as well as in the hierarchical cluster analysis. This points to cross-cultural similarity at the level of superordinate categorization. At the level of basic clusters, the five categories (love, joy, sadness, anger and fear) from the Shaver et al. study (1987) were replicated. There was one exceptional finding, namely a small guilt cluster in the Dutch data that disappeared after the removal of only two terms. Unfortunately, cluster analysis does not give very stable solutions and probably this is a case in point.

More shifts were found at the level of subclusters, but the implications are not clear. The fact that the shifts all concern a single cluster (love) is probably significant. However, it should be noted that the status of love as a basic emotion is not undebated. For example, Ekman does not include love in his listing of basic emotions. The numerous shifts for the separate translation equivalent emotion terms have to be left uninterpreted, since these shifts may refer to emotion blends. This means that an emotion word actually belongs in more than one basic cluster. Hierarchical cluster analysis does not allow for this possibility, revealing another shortcoming of the hierarchical cluster analysis as a tool to model the emotion domain.

As expected we found three dimensions in the Indonesian data, but only two in the Dutch data. According to Russell (1991) this is not an unusual finding, perhaps related to the sampling of words. In our study with extensive samples this does not seem a likely solution. More likely is that it has to do with the difference in mean size of the clusters in both samples, which could be a matter of how the task is understood and carried out by the subjects. We can not exclude that the component of activity is less prominent among the Dutch, but we did not find other support for this interpretation of the data.

In final evaluation, this preliminary analysis suggests extensive similarities in the cognitive structuring of emotions between Indonesian and Dutch students. We have been able to demonstrate this with indigenous lists of emotions terms, minimizing any danger of imposing indigenous structures of one culture on the other.

References

Ekman, P. (1972). *Universal and cultural differences in facial expressions of emotions.* In J.K. Cole (Ed.), Nebraska symposium on motivation, 1971 (pp. 207-283). Lincoln: University of Nebraska Press.

Ekman, P. (Ed.). (1982). *Emotion in the human face* (2nd ed.). Cambridge: Cambridge University Press.

Ekman, P., & Friesen, W.V. (1971). Constants across cultures in the face and emotion. *Journal of Personality and Social Psychology, 17,* 124-129.

Everitt, B. (1980). *Cluster analysis* (2nd ed.). New York: Halstead Press.

Fehr, B., & Russell, J.A. (1984). Concept of emotion viewed from a prototype perspective. *Journal of Experimental Psychology: General, 113,* 464-486.

Heelas, P. (1986). Emotion talk across cultures. In R.M. Harré (Ed.), *The social construction of emotions* (pp. 234-266). Oxford: Basil Blackwell.

Izard, C.E. (1977). *Human emotions.* London: Plenum Press.

Lutz, C. (1988). *Unnatural emotions: Everyday sentiments on a Micronesian atoll and their challenge to Western theory.* Chicago: University of Chicago Press.

Markam, S.S. (undated). *Lekseikon dantaksonomi emosi.* Report Jakarta, Universitas Indonesia.

Mesquita, B., & Frijda, N.H. (1992). Cultural variations in emotions: A review. *Psychological Bulletin, 112,* 179-204.

Osgood, C.E., May, W.H., & Miron, M.S. (1975). *Cross-cultural universals of affective meaning.* Urbana: University of Illinois Press.

Rosch, E. (1973). On the internal structure of perceptual and semantic categories. In T.E. Moore (Ed.), *Cognitive development and the acquisition of language* (pp. 111-114). New York: Academic Press.

Rosch, E. (1978). Principles of categorization. In E. Rosch & B.B. Lloyd (Eds.), *Cognition and categorization* (pp. 27-48). Hillsdale, New York: Erlbaum.

Rosch, E., Mervis, C.B., Gray, W.D., Johnson, D.M., & Boyes-Braem, P. (1976). Basic objects in natural categories. *Cognitive Psychology, 8,* 382-439.

Russell, J.A. (1991). Culture and the categorization of emotions. *Psychological Bulletin, 110,* 426-450.

Shaver, P., Schwartz, J., Kirson, D., & O'Connor, C. (1987). Emotion knowledge: Further exploration of a prototype approach. *Journal of Personality and Social Psychology, 52,* 1061-1086

Young, F.W., & Hamer, R.M. (Eds.). (1987). *Multidimensional scaling: History, theory, and applications.* Hillsdale, NJ: Erlbaum.

The Perception of Ability Scale for Students in Africa and New Zealand

Adebowale Akande
University of the Western Cape, Bellville, South Africa

> *Our life is what our thoughts make it*
> *Whatever is formed for long duration arrives*
> *slowly to its maturity*
> *The mind is an enchanting thing*
> *It is not enough for adults to understand*
> *children*
> *They must accord children the privilege*
> *of understanding them*
> *Peter Morakinyo of DTS, Oranyan*
> *1970*

Interest in self-report constructs has increased over the last three decades. In the 1990s, review of self-concept research (Burns, 1979, 1982; Shavelson, Hubner, & Stanton, 1976; Wylie, 1974, 1978) emphasized a lack of "theoretical models and the poor quality of measurement of instruments" (Marsh, 1972, p. 35). To this end, some of the credit for the considerable progress that has been made in this area of measurement is due to the lead provided by Shavelson et al. (Watkins & Akande, 1992). In their model, they formulated a hierarchical and multifaceted model divided into academic and nonacademic components of self-concept. Subsequent research, especially based on the set of the Perception of Ability for Students or PASS (Boersma & Chapman, 1992a) have supported the multi-dimensionality of self-concept and many aspects of the Shavelson et al. model (Akande, 1994a, 1994b; Byrne, 1984; Fleming & Courtney, 1984; Marsh, Byrne, & Shavelson, 1988; Mboya, 1989, 1993; Watkins & Akande, 1992).

The Perception of Ability Scale for students, formerly known as the student's Perception of Ability Scale (Boersma & Chapman 1992a) is a self-report instrument designed for the measurement of six of the most important self-concept dynamic factors, which have been identified by factor analysis among elementary school children (Chapman & Boersma, 1986). PASS is a 70 forced-choice "Yes-No" item-scale which purports to measure feelings and attitudes about academic ability and learning related achievements (Chapman & Boersma 1992). The six subscales in PASS measure five basic academic areas: reading, spelling, language, arts, arithmetic, printing/writing, and also school in general. The items are contained in six subscales, derived through factor analysis, and include Perception of General Ability, Perception of Arithmetic Ability, Perception of Penmanship and Neatness (each of which contains 12 items) and confidence in Academic Ability (10 items). Published psychometric data for PASS are reported in various sources (Boersma & Chapman, 1983, 1992a; Boersma, Chapman, & Maguire, 1979). Just like other measures of self-concept the developers of PASS (Boersma & Chapman, 1992a, 1992) were consistent with the Marsh/Shavelson model (Marsh, Byrne, & Shavelson, 1988). Therefore they provided procedures to estimate the internal validity of individual's responses to the scale in line with the need for two second-order academic factors as posited in the Marsh/Shavelson revision (Akande, 1993, 1994b).

There is considerable evidence to indicate that the PASS is meaningfully related to students, teachers or various school-related affective characteristics and contributes significantly in the prediction of school grade point average over a one year period. In addition, studies have shown that the scale is strongly differentiated from general self-concept, hence it enables investigators to usefully discriminate between gifted, average or normally achieving students, and learning disabled or intellectually handicapped students, and is sensitive to change after remedial intervention (Boyle, 1979, 1985, 1988a; Chapman & Boersma, 1986).

Empirical evidence that compares scores for gifted, average, and learning disabled students for the PASS in developed and less developed countries is lacking. Comparative studies are important theoretically and practically, now that education and social contact between developed and developing are increasing (Ezeilo, 1983; Olowu, 1990). It is also a matter of interest to examine the fundamental assumption of cross-cultural psychology that the components of cultural environment - beliefs, norms, values, customs and folkways - that the student brings to the school have a strong influence on school behavior. Therefore this study has two goals: (1) to provide a comparative descriptive account of the scores for gifted, average, and learning disabled students, and (2) to assess the applicability of theoretical analyses devised from studies of Western industrialized societies to the two samples. In particular, I investigated whether academic self-concept is associated with poor cognitive-motivational characteristics among children.

Accordingly, the present study utilized both a mixed sample of African children as well as sample of New Zealand children in order to increase the measurement variance on PASS indicators, in accord with sound analytic guidelines (Gorsuch, 1983). While the Boersma and Chapman (1992a) study attempted to elucidate the factor analysis in PASS, it is evident that the results obtained were rather tentative and required cross-validation (Boyle 1988a; Boyle, Start & Hall, 1988; Boyle & Start, 1989). The present paper reported the findings of such cross-validation investigations. The clinical usefulness of PASS is limited by inadequate literature measuring its reliability, validity, and normative data. Research directed toward generating these parameter is essential if the potential of the test for detecting six academic self-concept dynamic factors is to be realized, especially cross-culturally (Boyle, 1988b, 1988c; Boersma & Chapman, 1992b; Ezeilo, 1983; Olowu, 1990; Mwanwenda, 1991).

Method

Subjects

The sample was composed of 92 learning disabled students (43 boys, 49 girls); 74 normally-achieving students (41 boys, 33 girls); and 43 gifted students (21 boys, 22 girls). The African students were in form 2 (Grade 6) classroom in ten public schools in Nigeria and Zimbabwe. The sample ranged in age from 10 to 13 years with mean age 11.3 years ($s.d.$ = 4.8). All subjects participated voluntarily in the study and appeared motivated to answer the items in the respective instruments honestly and as accurately as possible. As the instrument was administered during the subjects' usual class periods the children appeared to accept the task without noticeable hesitation or concern. In selecting schools an attempt was made to ensure that the sample represented a wide range of socioeconomic and geographical influences in the African society. The schools selected included two city girls' high schools, three suburban high schools, three country high schools, and two boys' high schools. The children were all in the normal classes. No child with a major handicap, with major social-emotional problems, or difficulties with English as a second language was included in the sample (Akande, 1992, 1993).

Procedure

Disagreement over definitions of learning disabilities (Chapman, 1988; Lerner, 1985) pose difficulties for researchers identifying LD children in school systems that do not provide remedial programs for learning disabilities. In order to allow comparison with other studies in the LD area, the present researcher adopted the ability-achievement approach of Chapman (Chapman & Boersma, 1986; Chapman, 1985, 1988, 1992b) in identifying LD children. Although the approach is not without problems, this method has been noted as virtually the only common sampling characteristic in research on learning disabilities (Algozzine & Ysseldyke, 1986; Shepard, Smith, & Vojir, 1983). The sample was recruited from the cohort of 980 form 2 (Grade 6) children attending ten public schools in Nigeria and Zimbabwe.

Estimates of ability were obtained from scores on the Wechsler Intelligence Scale for Children - Revised (1974). Because intellectual assessments are not routinely administered to children in Africa, and because it was not feasible to administer the full WISC-R to a large number of children, a short form comprising four subscales was used. According to Chapman, (1988), though not an ideal procedure, it was considered an acceptable compromise for obtaining IQ estimates for research purposes. The subtests included were Information, Vocabulary, Picture Completion, and Block Design. They were chosen because of their high correlation (.94) with the full scale IQ (Sattler, 1974) and because of their relative ease of administration. Full scale IQ scores were prorated from summed scale scores, in accordance with the procedure recommended by Chapman, (1988) and Tellegen and Briggs, (1967).

The WISC-R was administered to children obtaining scores not greater than one standard deviation below the mean on the group administered Test of Scholastic Abilities (Reid, Jackson, Gilmore, & Croft, 1981). This test had been administered to the total form 2 cohort (N = 980) in regular classroom groupings, early in the 1992 academic year (February (in Nigeria) and August in Zimbabwe).

In accord with Chapman's (1988) method, achievement was assessed with four tests in the Progressive Achievement Test (PAT) series. These were Reading Comprehension (Elley & Reid, 1969), Reading Vocabulary (Elley & Reid, 1969), Listening Comprehension (Elley & Reid, 1971), and Mathematics (Reid & Hughes, 1974). These tests are the most frequently used standardized measures of achievement in schools in the developing world. The PAT data were collected after routine teacher-administered testing early in the 1992 school year. Selection of LD children was made on the basis of their having a prorated WISC-R IQ score of 80 or above and at least one PAT below the 16th percentile. The IQ cutoff point was selected because it is approximately equal to 76 plus one standard error of estimate ($SEy.x = 4.33$) for the four-subtest short form of the WISC-R (as explained more fully by Chapman, 1988; and Chapman, St George, & van Kraayenoord, 1984). Use of the upper bound of the $SEy.x$ confidence interval about the one standard deviation point of 76 was designed to increase the chances that children with a prorated IQ of 80 or above would be at least in the normal IQ range. The estimate is quite conservative (Boyle, 1989). In accordance with the recommendations of Chapman (1988) and Cone and Wilson (1981), all children met a regression method criterion for establishing an ability-achievement discrepancy; they had at least one PAT score that was below the lower bound of the standard error of estimate of confidence interval (at the 10% level) predicted from their IQ score. The selection of normal achievers (NLD) was made from those who had a prorated WISC-R IQ score of 80 or above and who had obtained scores at or above the 30th percentile on all four PATs. Most gifted children were achieving above the 85th percentile while learning disabled group were achieving below 16th percentile. Mean WISC-R Full scale IQs were 125.41 for gifted, 80 for normal achieving, and 79.6 for learning disabled groups.

The PASS scale was administered in regular class groups by the author, and two graduate assistants. The order of administration of the scale in schools was randomly counterbalanced. On each occasion, the children were informed that the purpose of the scale was to find out about the "thoughts and feelings" that they had about school, and they were assured confidentiality. Instructions and items were read aloud. No child was made aware that he or she was being assessed because of his or her particular achievement level (Boyle, 1988b, 1988d).

The internal consistency reliability *alphas* of the six subscales of PASS were found to vary from 0.48 - 0.63 (median 0.58) for the African subjects.

Results

Scores on the PASS and other scales were treated by two-way ANOVAS. Means and standard deviations for the African children are shown in Table 1. Significant main effects were found for all groups; even Penmanship/Neatness subscale yielded a significant effect. The finding is not consistent with results from studies of learning-disabled children. In addition, there were significant differences between the groups ($p < .001$). On each occasion individual comparison of means revealed that identifying learning disabled children is a problem because of disagreement over definitions of learning disability (Chapman, 1988). However this definition problem is more difficult in the African system because there is no meaningful systematic remedial provision for learning disabled children. Thus in accord with Chapman & Boersma (1986), "the learned children in this study were compared with peers who had similar ability but were performing in the average or above average range on standardized achievement tests" (p. 62).

Table 1. Means and Standard Deviations on PASS

Scale	Gifted		Average		Learning Disabled	
	M	SD	M	SD	M	SD
General Ability	11.01[a]	2.01	10.28[a]	3.04	8.11	2.13
Arithmetic	10.84[a]	2.48	9.14[a]	2.82	7.23	1.42
General School Satisfaction	7.11[a]	3.04	5.11[a]	3.21	3.43	2.23
Reading/Spelling	10.14[a]	1.74	8.52[a]	2.11	7.13	2.43
Pen/Neatness	8.11[a]	2.21	7.32[a]	2.03	6.14	3.18
Confidence	8.21[a]	2.10	5.04[a]	2.17	3.36	2.23
Full Scale	48.42[a]	6.82	39.43[a]	9.28	30.12	12.15

[a] Statistically significant differences between the African and New Zealand scores (N = 195 for New Zealand sample; see Chapman 1988, p. 359).

The main effect for gender indicated that girls obtained higher PASS scores ($M = 39.68$) than boys ($M = 32.89$). Once again, this finding is not unusual as it was similar to the New Zealand sample. That girls have a higher academic self-concept than boys is a trend consistent with other studies (Chapman, 1988). However the learning disabled (LD) boys had lower PASS scores ($M = 29.03$) overall than LD girls ($M = 37.92$). The LD girls also reported academic self-concept barely lower than those of normally-achieving students, which has been found in other studies (Boersma & Chapman, 1989; Chapman & Boersma, 1992b; Chapman, 1985, 1988).

For gifted students, boys had higher PASS scores ($M = 58.72$) than girls ($M = 49.82$) and the difference for the full scale was just under one raw standard deviation unit. For the normally-achieving and low-achieving student groups, the difference was about one standard deviation unit. The basic picture here is that although the mean scores of the African sample were statistically significant, the actual differences in mean scores were generally trivial in practical terms. Indeed of all the scores, only one - general ability - differed by more than one sten unit, with the mean African scores being lower than those for the New Zealand sample (Marsh, Smith & Barnes, 1985).

Discussion

Drawing together the findings of this study, the following picture emerges. Learning disabled children appear to be clearly different in affective characteristics than normally achieving children. These affective characteristic are associated with low self-perceptions of ability resulting in negative self-concept, along marked tendencies toward learned helplessness and lukewarm expectations for future success in schools. These findings of lower self-confidence and optimism about success for LD children are in line with those reported in other studies (e.g., Chapman, 1988; Chapman & Boersma, 1992).

In terms of practical significance as compared to statistical significance, we have no reason to reject the applicability of the New Zealand norms to the African context at the present moment.

Furthermore, the average correlation between the scales and the subscales primary scores for the African sample was .08, which was the same as the New Zealand sample. Hence the data gathered under African conditions closely matched those derived from the New Zealand sample (Chapman, 1988). These, then, are the first results from an analysis of this form in Africa. The Perception Ability Scale for Students does not appear to have a ceiling effect which prevents differentiation of normal or average and low-achieving from those of gifted ability. The findings do provide a point of departure for future explorations of other factors beyond those already described by Boersma, Chapman, & Maguire, (1979), and Boersma & Chapman, (1992a). The present findings do not however, suggest that any particular subscale and the PASS exhibit a discernible systematic pattern that is remarkably different between the African and New Zealand samples. Furthermore, it appears that the PASS is suitable for use with African high school students. In any event, what is now required is extensive research using large and varied sample with Boersma and Chapman's instrument.

Nevertheless, the PASS should be viewed not only as an instrument for measuring children's academic self-concept and the prediction of school grades but, more importantly, as an assessment and diagnostic tool. For example, knowledge of the comparative strength of general school satisfaction, perception of penmanship and neatness and confidence in academic ability, in regard to each dynamic trait, would show if a student's success and failure in school is also influenced by various affective and nonintellective variables. Or, whether prolonged failure will have an increasingly detrimental influence on a student's sense of self-worth as negative learning outcomes give rise to diminished self-concept, low achievement expectations and feelings of helplessness (Bandura, 1977; Chapman 1988; Licht & Kistner, 1986). Such advances in the assessment and understanding of children's school related affect and achievement should ultimately benefit the practice of early child health and care.

In summary, the present findings provided tentative support for the PASS as well as for the validity of its six dynamic factors themselves. Clearly the Boersma and Chapman scale requires refinement and further development. Research using PASS seems, however, to offer more promise of significant new knowledge within the self-concept domain, particularly in view of the multi-facet nature of previous academic self-concept measures (Boyle, 1979, 1985; Ezeilo, 1983). Clearly, more attention needs to be paid to the issue of academic self-concept in Africa. This is because empirical assessment of self-concept might influence learning task performance which might therefore provide an educationally useful focus of some future research project. In accord with Chapman, Lambourne and Silva (1990) "... academic self-concept would seem to be closely linked to actual achievement outcomes in school. Therefore any ongoing effects that home background may have on the reciprocal interplay between the achievement and academic concept probably relate to the nature of achievement-related feedback and encouragement in the home" (p. 150). Consequently it might be possible to teach specific strategies which could stimulate positive academic self-concept in African students, and thereby enhance school learning. Taken together, comparative measurement of self-concept in developed and developing nations (especially countries in Africa) should have considerable potential for new insights in both research and applied contexts.

References

Akande, A. (1992). Children's motivation analysis test (CMAT) - Normative data. *Early Child Development and Care, 87,* 105-110.

Akande, A. (1993). *Self-esteem: A familiar affair.* Unpublished manuscript.

Akande, A. (1994a). *IQ is irrelevant to the definition of apartheid education.* Submitted for publication.

Akande, A. (1994b). *African society and the adolescent self-image.* Unpublished manuscript.

Algozzine, B., & Ysseldyke, J.E. (1986). The future of the LD field: screening and diagnosis. *Journal of Learning Disabilities, 19,* 394-398.

Bandura, A. (1977). *Social learning theory.* Englewood, Cliffs, NJ: Prentice-Hall.

Boersma, F.J., & Chapman, J.W. (1983). *Manual for the Student's Perception of Ability Scale* (Rev. ed.). Edmonton, Canada: University of Alberta.

Boersma, F.J., & Chapman, J.W. (1992a). *Manual for the Perception of Ability Scale for Students.* Los Angeles, CA: Western Psychological Services.

Boersma, F.J., & Chapman, J.W. (1992b). *Perception of Ability Scale for Students.* Los Angeles, CA: Western Psychological Services.

Boersma, F.J., Chapman, J.W., & Maguire, T.O. (1979). The students' perception of ability scale: An instrument for measuring academic self-concept in elementary school children. *Educational and Psychological Measurement, 39,* 1035-1041.

Boyle, G.J. (1979). Behavioural management of hyperactive learning disabled children. *Australian Journal of Remedial Education, 11,* 6-10.

Boyle, G.J. (1985). The paramenstruum and negative moods in normal young women. *Personality and Individual Differences, 6,* 649 - 652.

Boyle, G.J. (1988a). Elucidation of motivation structure by dynamic calculus. In J.R. Nesselrode & R.B. Cattell (Eds), *Handbook of Multivariate Experimental Psychology* (2nd ed., pp. 737-787). New York: Plenum Press.

Boyle, G.J. (1988b). Contribution of Cattellian psychometrics to the elucidation of human intellectual structure. *Multivariate Experimental Clinical Research, 4,* 267- 273.

Boyle, G.J. (1988c). Exploratory factor analytic principles in motivation research. In Nesselrode, J.R. & Cattell R.B. (Eds.), *Handbook of Multivariate Experimental Psychology* (pp. 745 - 745). New York: Plenum Press.

Boyle, G.J. (1988d). *A Guide to the Military use of the 16PF in Personnel Selection.* Australian Army Psychology Research Unit, Canberra.

Boyle, G.J. (1989). Re-examination of the major personality-type factors in the Cattell, Comrey and Eysenck scales: Were the factor solutions by Noller et al. Optimal? *Personality and Individual Differences, 10,* 1289 - 1299.

Boyle, G.J. & Start, K.B. (1989). Comparison of higher-stratum motivational factors across sexes using the children's motivation analysis test. *Personality and Individual Differences, 10,* 483-548.

Boyle, G.J., Stanley, G.V., & Start, K.B. (1988). Canonical/redundancy analyses of the Sixteen Personality Factor Questionnaire, the Motivation Analysis Test, and the Eight State Questionnaire. *Multivariate Experimental Clinical Research, 7,* 113 - 132.

Boyle, G.J., Start, K.B., & Hall, E.J. (1988). Comparison of Australian elucidation of human intellectual structure. *Multivariate Experimental Clinical Research, 4*, 267-273.

Burns, R. (1979). *The self-concept: Theory, measurement, development and behaviour.* London: Longman.

Burns, R. (1982). *Self-concept, development and education.* London: Holt, Rinehart & Winston.

Byrne, B.M. (1984). The general/academic nomological network: A review of construct research. *Review of Educational Research, 54*, 427 - 456.

Chapman, J.W. (1985). *Self-perceptions of ability, learned helplessness and academic achievement expectations of children with learning disabilities: A comparative longitudinal study.* Palmerston North, NZ: Massey University.

Chapman, J.W. (1988). Cognitive-motivational characteristics and academic achievement of learning disabled children: A longitudinal study. *Journal of Educational Psychology, 80*, 357-365.

Chapman, J.W., & Boersma, F.J. (1986). Student's perception of ability scale: Comparison of scores for gifted, average, and learning disabled students. *Perceptual and Motor Skills, 63*, 57-78.

Chapman, J.W., & Boersma, F.J. (1992). Performance of students with learning disabilities on validity indexes of the perception of ability scale for students. *Perceptual and Motor Skills. 75,* 27-34.

Chapman, J.W., Lamboourne, R., & Silva, P.A. (1990). Some antecedents of academic self-concept: A longitudinal study. *British Journal of Educational Psychology, 60,* 142 - 152.

Chapman, J.W., St George, R., & van Kraayenoord, C.E. (1984). Identification of "learning disabled" pupils in a New Zealand form 1 sample, Australia and New Zealand. *Journal of Developmental Disabilities, 10,* 141-149.

Cone, T.E., & Wilson, W.R. (1981). Quantifying a severe discrepancy: A critical analysis. *Learning Disability Quarterly, 4,* 259-271.

Elley, W.B., & Reid, N.A. (1971). *Progressive Achievement Tests: Reading Comprehension and Reading Vocabulary.* Wellington, New Zealand: Council for Educational Research.

Ezeilo, B.N. (1983). Age, sex, self-concept in a Nigerian population. *International Journal of Behavioural Development, 6,* 497-507.

Fleming, J.S., & Courtney, B.E. (1984). The dimensionality of self-esteem II: Hierarchical facet model for revised measurement scales. *Journal of Personality and Social Psychology, 46*, 406-421.

Gorsuch, R.L. (1983). *Factor Analysis* (2nd ed.). Hillsdale, NJ: Erlbaum.

Hull, C.H., & Nie, N.H. (1984). *SPSS-X.* New York: McGraw-Hill.

Lerner, J.W. (1985). *Learning Disabilities: Theories Diagnosis, and Teaching Strategies* (4th ed). Boston: Houghton Mifflin.

Licht, B.G., & Kistner, J.A. (1986). Motivational problems of learning disabled children: individual differences and their implications of treatment. In J.K. Torgeson & B.Y.L. Wong (Eds.), *Psychological and Educational Perspectives on Learning Disabilities* (pp. 225-255). Orlando, FL: Academic Press.

Marsh, H.W. (1988). *Self-description Questionnaire I, SDQManual and Research Monograph.* San Antonio, TX: The Psychological Services.

Marsh, H.W. (1992). Content specificity of relations between academic achievement and academic self-concept. *Journal of Educational Psychology, 84*, 35 - 42.

Marsh, H.W., Byrne, B.M., & Shavelson, R.J. (1988). A multifaceted academic self-concept: Its hierarchical structure and its relation to academic achievement. *Journal of Educational Psychology, 80*, 366 - 380.

Marsh, H.W., Smith, I.D., & Barnes, J. (1985). Multidimensional self-concepts: Relations with sex and academic self-concept. *Journal of Educational Psychology, 77*, 581-596.

Mboya, M.M. (1989). The relative importance of global self-concept and self-concept of academic ability in predicting academic achievement. *Adolescence, 24*, 39 - 46.

Mboya, M.M. (1993). Development and construct validity of a self-description inventory for African adolescents. *Psychological Reports, 72*, 183 - 191

Mwamwenda, T.S. (1991). Sex diifferences in self-concept among African adolescents. *Perceptual and Motor Skills, 73*, 191 - 194

Olowu, A.A. (1990). *Contemporary Issues in Self-Concept Studies*. Ibadan, Nigeria: Sheneson

Piers, E.V. (1984). *Piers-Harris Children's Self-Concept Scale: Revised manual*. Los Angeles, CA: Western Psychological Services.

Reid, N.A., & Hughes, D.C. (1974). *Progressive Achievement Tests: Mathematics*. Wellington: New Zealand Council for Educational Research.

Reid, N.A., Jackson, P., Gilmore, A., & Croft, C. (1981). *Test of Scholastic Abilities*. Wellington: New Zealand Council for Educational Research.

Sattler, J.M. (1974). *Assessment of Children's Intelligence*. (Rev. ed). Philadelphia: W.B. Saunders.

Shavelson, R.J., Hubner, J.J., & Stanton, J.C. (1976). Self-concept validation of construct interpretations. *Review of Educational Research, 46*, 407 - 441. Services.

Shepard, L.A, Smith, M.L., & Vojir, C.P. (1983). Characteristics of pupils identified as learning disabled. *American Educational Research Journal, 20*, 309-311.

Tellegen, A., & Briggs, P.F. (1967). Old wine in new skins: Grouping Wechsler subtests into new scales. *Journal of Consulting Psychology, 31*, 499-506.

Watkins, D., & Akande, A., (1992). The internal structure of the self-description questionnaire: A Nigerian investigation. *British Journal of Educational Psychology, 62*, 120 - 125.

Wechsler, D. (1974). *The Wechsler Intelligence Scale for Children - Revised*. New York: Psychological Corporation.

Wylie, R.C. (1974). *The self-concept: A review of methodological considerations and instruments* (Vol. 1*)*. Lincoln, NE: University of Nebraska Press.

Wylie, R.C. (1978) *The self-concept* (Vol. 2*)*. Lincoln: University of Nebraska Press.

Note

This paper has benefited from comments of Dr. Bolanle Adetoun, and Professor Davis.

Urban and Rural People's Conceptions of Intelligence in Equatorial Guinea

Manuel Juan-Espinosa & Antonio Palacios
Universidad Autonoma de Madrid, Madrid, Spain

The study of intelligence in different environments may be approached either from the point of view of how people solve problems, or from the opinions they voice about intelligence as a concept in itself and/or as a property of people living in their environment. In the first case, scientists develop and expound theories through a network of hypotheses about intelligence, whether they be general or differentiated, in which they try to test using appropriate methods and techniques. This approach is called *explicit theories* or *formal theories*. Alternatively, in attempting to uncover the conceptual network underlying opinions about intelligence, it is said that one is using the *implicit theories* approach. While in the formal theories approach the concepts relating to intellectual functioning are useful scientific "inventions" or scientific constructs (Eysenck, 1979), implicit theories are people's constructions - whether scientists or lay people - that reside in the mind of these individuals, which scientists try to "discover" (Sternberg, Conway, Ketron, & Bernstein, 1981).

In psychometric or differential formal theories, most research has been devoted to the study of the structure of intelligence in different cultures. Administration of tests as procedure, test scores as the empirical database, and factorial analysis as the main method, have consituted the norm in this kind of research.

As far as testing procedures are concerned, many authors suggest caution. For example, Vernon (1969, 1979) presents a long list of the problems concerning measuring intelligence cross-culturally in that factors that affect test scores but may not reflect the intelligence of the subjects. This has resulted in concern with methods and in making cautious inferences. The main concern has been to ensure that the test item receives adequate representation in the the subject's mind, so that the cognitive processes that are assumed to be universal or particular components of mental life can operate upon it. The test score is seldom if ever, taken to mean what the test manual declares it to mean without extensive trials and checks. The inference of the source of variance in test scores, when tests are used cross-culturally, are often termed *X-scores* (Irvine & Berry, 1988).

Despite the above-mentioned problems and cautions, the evidence for the abilities of mankind from psychometric formal theories reveals a remarkable similarity in the constructs psychologists use to account for individual differences. No cross-cultural evidence exists to deny the validity of these broad divisions of capability that have been accessed by tests (Irvine & Berry, 1988).

Alternatively, when we study intelligence from the implicit theories approach, the goal is to discover the form and content of people's informal theories when they live in different habitats, and to contrast them in order to uncover shared and nonshared characteristics in the division of capabilities. The data of interest are people's communications (in whatever form) regarding their notions as to the nature of intelligence tied to their ecocultural contexts. In this sense, people's conceptions of intelligence have the aim of reducing the complexity and variability of one's own and others' problem-solving behaviors to a structure of limited concepts that allows them to categorize people through prototypes. Therefore, people make the implicit theories consisting of the structural and functional interlacing that underlies the mental ability concepts and which makes them useful as mechanisms of knowledge about the world and about ourselves (Juan-Espinosa, 1994).

The literature on implicit theories, although not abundant, is quite interesting. Some examples of them are the work done by Chen, Braithwaite and Jong (1982) with Taiwanese and Australian populations; Gill and Keats (1980) in Malasia; Keats (1982) in China, Klein, Freeman and Millet (1973) in Guatemala; Sternberg et al (1981) in the U.S.A.. This kind of research is very promising, not only with respect to what their meaning within these theories, but also because such studies allow for contrast with formal theories, and so provide a heuristic potential of significant theoretical, methodological and technological importance (Goodnow, 1984) to complement formal theories and increase our ability to understand these phenomena.

Most of the cross-cultural literature on intelligence highlights the idea that different habitats set different problems for man that must be solved. In terms of the *law of cultural differentiation* (Ferguson, 1956) this means that "cultural factors prescribe what shall be learned and at what age; consequently, different environments lead to the development of different patterns of ability" (p. 121). Our idea, in this sense, is that implicit theories of intelligence, as a mechanism of knowledge, may be of a quasi-universal nature, and that their attributes could be organized in a diffuse hierarchy where the continuum *globality-molecularity* of their components or properties would reflect the degree of immersion of groups and/or individuals in particular cultures, according to the law of cultural differentiation.

Berry (1987) proposed some theoretical mechanisms for the law of cultural differentiation. One of them, acculturation, focuses on the issue of how people's problem-solving behaviors are affected by influences stemming from culture contact in the sociopolitical context of one's group. It may be reflected by the presence of schooling and urbanization among other forms. Our interest in the present work is, precisely, to uncover how acculturation influences the structure and content of people's conception of intelligence.

Equatorial Guinea

Our research was conducted in the Spanish ex-colony Equatorial Guinea, an African country located between Cameroon and Gabon, where 387,000 inhabitants live in 28,051 square kilometers. Bioco is the main island and Malabo is the capital. The main ethnic groups in the continental area, where the research took place, are *Combes, Bujebas* and *Fang*. The sample was entirely composed of Fang people, a sub-ethnic Bantu group, who are the majority ethnic group in the continental area of Equatorial Guinea. Until very recently, they had no written language. The official langage is Spanish, although little spoken among the rural population. The Fang have a clan-based social structure. Marriages are exogamic and patrolinear.

In order to study the acculturation influence, two ecocultural contexts were selected: (a) A traditional/rural, represented by *Nsoc* (2,345 Km^2), an isolated northeast region near the Cameroon and Gabon border, and (b) an acculturated/urban, represented by Bata city, the economic capital of Equatorial Guinea.

Traditional/Rural Context

The Fang people live in villages consisting of a small group of houses around a rectangular plaza (*ufeng*) or alongside a central street (*atang*). The village is situated in an open clearing of the rain forest. Electric power or running water are not available, even in Nsoc-nsomo, the capital. Family plantations are located on spurs in the same clearing where they grow bananas, malangas, etc. The most important institution in the village is the *Aba* or House of Word, where lawsuits are disputed, a place for leisure and recreation, for meals, a place to receive foreign visitors and a centre for transmitting the oral traditions. In the oldest villages the *Ngun-Melan*, or House of Initiation still stands. Although one Catholic mission can be found, the cult of the ancestors, called *Melan,* is the main religious belief. The type of economy is mainly subsistence farming, and confined to the family plantation. The village is governed by a council of adults called *Be-Mda-mbor.*

Acculturated/Urban Context

Bata is located in the continental region, on the Atlantic coast. It is a small sprawling city, with 26,000 inhabitants, and with ministries of justice, education, tourism, culture, and economics. Some schools, a branch of the Spanish Open University, hospitals, some factories, markets, etc., are indicators of Spanish influence. Wood factories (privately owned by foreigners) and services are the main economic activities. Work in bars, shops in markets, such as *Mundoasi,* and casual jobs are the main sources of income among the Fang people.

The entire research project was carried out using the Fang language. It was undertaken in two stages plus a preliminary stage. In the preliminary stage, we determined whether quantitative terms were used to compare people, and if people's intelligence was assessed in any systematic way. The next stage was: (1) to discover the general properties of people's implicit theories of intelligence in each context; and (2) to obtain a master list of characteristics of an intelligent person. The last stage was to ask for a quantitative evaluation of these characteristics in order to uncover the structure and content of people's implicit theories in both contexts.

Preliminary Phase

The objectives of this stage were: (1) to determine if any Fang concept was employed to assess a person's intelligence, and which other concepts were related to it; (2) to determine whether the comparative classification through the use of Fang quantitative terms was culturally viable for Fang people; and (3) to prepare the materials for research and test them with a group of subjects.

This stage was undertaken in Bata city. In order to carry out all of these objectives we collaborated with a bilingual Fang/Spanish anthropologist, a native of Equatorial Guinea.

Method

Procedure

Since there is no reliable Spanish/Fang dictionary, we asked the native anthropologist about: (a) the main terms commonly used in the Fang language that were equivalent to "intelligent" and other terms related to it; and (b) which terms (if any) were used to compare people quantitatively.

The Fang language has one word to describe an intelligent persons: *nfefeg*. It is interesting to note that *nfefeg*, or "intelligent", is derived from the word *feg*, which can be translated as, "intelligence", "rule", or "vision". The last item is particularly interesting when related to the concept opposite to that of intelligence, *Ndjimam,* which comes from the word *Ndjibe,* meaning "darkness". In the Fang culture, the intelligent person or *nfefeg* is someone with *achi* or "profound thinking", the person who can see beyond the superficial appearance of things, while a *ndjimam* is a person who only sees outside appearances ("s/he is in the dark").

In addition to the word *Nfefeg* used to describe an intelligent person, *Mbomam* is used to describe a smart and skilled person who gets on well in everyday life, and *Eyemam* is used to describe a knowledgeable person or someone able to know and learn things. The opposite term to *Eyeman* is *Eyeyemam*. *Mbomam* does not have an opposite term.

With regard to qualifying people through the use of quantitative terms, we found that the comparison of one person with another was culturally viable. In order to build a classification continuum, our native anthropologist's advice was followed, and three quantitative terms were used: *ka djam* or "nothing", *avichang* or "little", and *abui* or "a lot".

In order to record answers quantitatively in each stage, a colored design showing a nine-step ladder, something that almost every village had, was employed. The first step was *avichang* or "little" and the ninth was *abui* or "a lot". The concept of nothing or *ka djam* was shown as a line on the ground. Although "tenth or ten" terms are commonly used in Spain meaning the highest point in any scale, we take the "ninth or nine" because these terms have the same cultural meaning for the Fang as "tenth or ten" in Spain. For example, a child is not a child until he or she reaches nine years old. Ten years old is considered the *beginning* of the process of becoming a man or a woman; and, s/he is not considered a man or a woman (or *mbot* in the Fang language) until s/he reaches nineteen years old. Twenty years old is the *beginning* of becoming an adult man or woman, or *Mda-mbot*, and so on.

Under the supervision of the anthropologist, two bilingual guides were trained for two days, carried out a small pilot study, to test the guides and the scales.

First Stage

The objectives were: (1) to uncover the general properties of the implicit theories concerning the intelligent, knowlegeable, and skilled person in the two different contexts, and (2) to obtain a master list of behaviors characteristics of the intelligent person.

Method

Subjects

This stage involved 218 subjects, 94 (50 males and 44 females) living in villages in the *Nsoc* area (traditional context) and 124 (72 males and 52 females) in Bata city (acculturated context). The mean age was 32, ranging from 19 to 54 years old.

Materials

Three interview protocols were prepared, all in the Fang language, to be read by the guides. The same question pattern, the characteristics of an ideal person, was asked in all of them; but only one of the three target concepts - *intelligent person, knowledgeable person,* and *skilled person* - was asked in each protocol. The answers were given orally and recorded on tape.

Procedure

People were approached by one of the two guides, accompanied by the first author, and asked what were the personal features of: an (a) intelligent, or (b) knowledgeable, or (c) skilled person. That is, each subject were asked about only one of the three atributes of a person.if they wanted to answer a simple question. Responses were tape-recorded.

The interviewers translated the responses into Spanish; the anthropologist translated them back to Fang, and then a contrast between the antropologist's translation and the tape-recorded responses was made. After this process, the master-lists of characteristics were contructed.

Results

Compilation of Master List of Behaviors

A total of 56 and 62 behaviors were obtained from the traditional and the acculturated contexts respectively, which were compiled into a single master list to be used in the final stage. Behaviors were included if they were listed only once, no matter in which context, although obvious redundancies were eliminated, and examples of the same kind of behavior were grouped in only one. As a result, the final list consisted of 46 behaviors.

As in previous research (Sternberg et al., 1981), since responses were summed over subjects, these data can be interpreted for both subgroups - the traditional and the acculturated - but not at the level of individual respondent. In this case, correlations were taken as an indicator of shared variance among intelligence concepts by the subjects within each context and between contexts.

Correlations of Frequencies of Listed Behaviors

Table 1 shows the correlations between the frequencies with which each of the 46 intelligent behaviors was listed by subjects in each context for each type of intelligence.

Table 1. Correlations between Frequencies of Listed Behavior

Traditional/Rural Context			
	Intelligent	Knowledgeable	
Knowledgeable	.4851**		
Skilled	.4461**	.3011*	
Acculturated/Urban Context			
	Intelligent	Knowledgeable	
Knowledgeable	.7486**		
Skilled	.2015	.2442	
Urban/Rural Contexts			
	Intelligent	Knowledgeable	Skilled
Urban/Rural	.5845**	.2956*	.3817**

Note: Correlations are based on frequencies for the 46 intelligent behaviors.
* $p < .05$ ** $p < .001$

In the traditional/urban context, frequencies of listed behaviors were significantly correlated for the three concepts under research, but they were greater for *intelligent and knowledgeable*, and *intelligent and skilled* than for *knowledgeable and skilled*. In the urban/acculturated context the results were different. Correlations between frequencies was very significant for *intelligent and knowledgeable*, but not for *intelligent and skilled* or *skilled and knowledgeable* concepts. In addition, intelligent and knowledgeable concepts were more closely related in the acculturated/urban context than in the traditional/rural. Finally, when the two contexts were correlated each of the three concepts, the highest correlation found was for *intelligent* person, next for *skilled*, while the smallest and least significant was for *knowledgeable*.

Conclusions

Judging by the frequencies of behaviors listed for each type of intelligent person in each of the contexts, we concluded that people in the traditional/rural context perceived the type of intelligent person as being substantially knowledgeable and skilled. In contrast, people in the urban/acculturated context perceived the type of intelligent person, not only as substantially more similar to the knowledgeable one than people in the traditional context, but as the only related concepts, since there were no correlations with or among the remaining concepts under research.

Taking correlation as an indicator of shared concepts between people living in each context, we concluded that the more shared concept was that of the intelligent person, the skilled and knowledgeable concepts being less shared.

Second Stage

In this stage we sought to uncover the conceptual structure underlying the idea of an intelligent person in each selected context.

Method

Subjects

With a mean age of 34 and a range from 17 to 56 years old, 135 (81 males and 54 females) and 148 (88 males and 60 females) adults (other than the subjects of the first stage) from the traditional and acculturated contexts were interviewed. The protocols were read by the interviewers, and the people were asked to point to the step they considered most appropriate in the colored design ladder.

Materials

Materials for this experiment consisted of: 1) a list of 46 behaviors compiled from the first stage. Two interview protocols were prepared in the Fang language. In each protocol the items were placed in random order to eliminate any effect due to order of presentation. Each subject answered only one of the protocols.

Procedure

Subjects were asked to think of a person (real or imaginary) who was very intelligent and to judge the importance of each of the items in describing that person. All items had to be answered on the above mentioned nine-step ladder on a scale from "nothing" or *ka djam* to "a lot" or *abui*. The respondents pointed with their finger to the step on the ladder they considered most appropriate.

Table 2. Behaviors Loading on each Factor in the Traditional and Rural Contexts

Nª FACTOR I: *Social competence*	Nª FACTOR II: *General reasoning*
To orient people properly. To do things for the good of others. To give good ideas. To solve the problems in the village. To give good advice to others. To solve conflicts among people.	To interrelate things among themselves To think well and methodically. To see problems before they appear. To reason well. To solve difficult problems
Nª FACTOR III: *Visual-Spatial ability*	Nº FACTOR IV: *Verbal ability*
To imagine how to do things before doing them To imagine things before doing them To know going in the jungle and not get lost. To know how to build houses and to build them.	To know what you are saying. To know what a person is going to say before s/he speaks. To be convincing when talking. To solve conflicts among people.
Nª FACTOR V: *Learning ability*	Nº FACTOR VI: *Traditional-verbal ability*
To be patient when thinking To learn things properly. To know how to bear suffering in order to get something. To know how to organize things.	To explain events through stories and sayings To have a good memory. To speak well. To speak in such a way as to be undertood.

Results and Discussion

Structure and Content of Intelligent Person Concept

In order to analyse the structure and underlying content of the conception of the Intelligent Person in the traditional/rural and acculturated/urban contexts, two factorial analyses were performed on the list of 46 characteristic behaviors, employing the Main Components method and Varimax rotation. Kaiser and Guttman's K1 rule (Kaiser, 1960) and Cattell Scree-Tests were used as criteria to determine the number of factors to extract. Comparability with the work done by Sternberg et al. (1981) with American samples, where orthogonal solution was adopted, was the rationale for using the varimax solution. Behaviors with loading of .50 and above were employed for interpretation purposes.

Traditional/Rural Context

Six interpretable factors emerged from the analysis of the 11 factor extracted. The six factors explained 55.5% of the variance. These factors were labelled *Social competence* (11%), *General reasoning* (10%), *Visual-Spatial ability* (9%), *Verbal ability* (8,6%), *Learning ability* (8,5%), and *Traditional-verbal ability* (7,7%). Table 2 shows the main behaviors loading each of the six factors.

Several points are worthy of note in these data, especially concerning the first, fifth and sixth factors. The first factor, *social competence*, may be seen as containing two types of social behaviors: a) *counselling and guidance*, which includes behaviors such as, "To orient people properly", etc. and b) *social problem-solving*, e.g., "To solve the problems in the village. The fifth factor, learning *ability*, can be characterized by persistence in thinking and learning, e.g., "To be patient when thinking.

The sixth factor, *traditional-verbal ability* can be seen as a specific Fang culture-bound factor. At sunset, when people meet in the House of Word, some old men and women in the villages tell stories, talk about the village ancestors, or recall sayings with reference to the ongoing events in the village. This ability seems to be captured by this factor through variables such as, "To explain events through stories and sayings", "To speak well", etc.

The second factor may be similar to *Gf* or Fluid Intelligence proposed by Cattell's theory (Cattell, 1971, 1987) but including an anticipation component (i.e., "To see problems before they appear"). The third and fourth are similar to the *Gv* or Visual-Spatial ability, and the fourth similar to Cattell's *Gc* or Crystallized Intelligence.

Acculturated/Urban Context

Six interpretable factors emerged from the analysis of the responses from the acculturated/urban context, which explained 69% of the variance. These factors were labelled: *Social competence* (22%), *Verbal-Educational ability* (11,6%), *General reasoning* (11%), *Memory and Learning ability* (8.7%), *Visual-Spatial ability* (7.7%), and the last one was labelled *Cautiousness* (7.5%). Table 3 shows the behaviors loading on each of the six factors.

While factors IV and V were easy to interpret as *Memory and Learning ability* and *Visual-Spatial ability*, factors I, II, III, and VI deserve some comment. As in the traditional context, the *Social competence* factor is the most important. With 22% of variance explained (compared to the 11% in the traditional context), it is a complex factor with many variables loading on it.

Table 3. Behaviors Loading on each Factor in the Acculturated Urban Contexts

Nª FACTOR I: *Social competence*	Nº FACTOR II: *Verbal-Educational ability*
34. To see problems before they appear. 15. To be convincing when talking 5. To know how to economize 37. To know how to negotiate with money. 21. To give good advice to others. 8. To solve conflicts among people. 26. To know how to organize things. 1. To orient people properly. 14. To treat others well. 41. To know the intentions of others. 42. To know how to distribute among others. 16. To solve difficult problems 38. To reason well. 9. To know how to observe things 10. To be patient when thinking	45. To know what a person is going to say before s/he speaks. 13. To know what you are saying. 23. To know writing. 46. To know how to play with numbers. 29. To know reading. 40. To know how to bear suffering in order to get something. 39. To speak well.
Nª FACTOR III: *General reasoning*	Nª FACTOR IV: *Memory and learning ability*
28. To think well and methodically. 3. To plan things 16. To solve difficult problems. 38. To reason well. 18. (Negative) To know how to plant and when to harvest. 11. (Negative) To explain events through stories and sayings	43. To have a good memory. 30. To learn things properly. 44. To learn quickly. 9. To know how to observe things
Nª FACTOR V: *Visual-Spatial ability*	Nº FACTOR VI: *Cautiousness*
7. To imagine how to do things before doing them 27. To know going in the jungle and not get lost. 4. To imagine things before doing them.	24. To be able to stop evil coming into the home. 35. To do things for the good of others. 10. To be patient when thinking

Although not as clearly characterizable as the one in the traditional context, one type of variable deserves to be mentioned: *money management*, which includes "To know how to economize", and "To know how to negotiate with money". Again *counselling and guidance* skills appears in this context but its social meaning is not as clear here as in the traditional context.

Money management skills appearing as part of the *Social competence* factor in the acculturated/urban context, deserves additional explanation. This type of skill was not present in the traditional context. In Bata, "To know how to economize", has the western cultural sense of the careful management of resources (a socially desirable skill), while in the traditional contexts this does not make much sense. The almost total absence of preservation methods together with high temperatures and humidity make products and meat perishable in a very short period of time. Thus, economy means the daily business of subsistence, something that all people have to do. As far as "To know how to negotiate with money" is concerned, we have the same interpretation problem. Given the relative isolation of the traditional context, goods for buying and selling are rare, and animals and field products are the main currency. Thus, money is not as important as in Bata. Although money exists in this area, its main value is for marriage purposes (dowry). Once a bride's family has this dowry, the money will be invested again in another marriage (for sons, or even father, where polygamy is the case), and seldom in buying goods. In this sense, money has the same meaning that *ekuele* (spear point) had in "the old days" - a warning and a guarantee of the bride's safety, and not a purchase. In the Fang tradition, the economy concept allows thing for thing and person for person exchange, but not thing for person (slavery never existed among the Fang people). Thus, the bride (person) cannot be bought with the dowry money (thing). In cities and other areas this meaning is changing towards that of purchase as a result of the acculturation process. As a consequence, "wealth" as an abundance of money has different meanings in the two contexts. While in Bata it means the possibility of buying things and developing economic skills, in the traditional area it means having an amount of money for marriage purposes and thus, "lots of personal problems". So, while being rich in Bata can be taken as an indicator of *Social competence*, this is not the case in the traditional context. Morover, though a person may be rich, he cannot be part of the *Be-Mda-mboro* or village council ("he has too many personal problems to think about those of others" was the repeated Fang explanation)

The second factor includes behaviors such as: "To know what a person is going to say before s/he speaks". It can be interpreted as the $V{:}ed.$ or Verbal-Educational factor in Vernon's theory, or Gc (Crystallized Intelligence) in Cattell's theory. Variables such as, "To know writing", etc. give this factor a shade of western culture's educational skills. In contrast to the traditional context, the existence of schools in Bata city could explain the importance of these variables and also be a sign of resocialization (Berry, Poortinga, Segall, & Dasen, 1992).

The third factor, although similar to the *General reasoning* found in the traditional context, includes two behaviors with negative loadings, "To know how to plant and when to harvest", and "To explain events through stories and sayings", and these made it difficult to interpret. A possible explanation can be found in the observed rejection of traditional life by some people, especially young people or those who have come to Bata to escape from the rural poverty and/or to look for a better life.

Factor six, *Cautiousness*, seems to represent the quality of being cautious, e.g. "Being able to stop evil coming into the home". As a result of acculturation, the *Melan* as religious reference and the council of adults or *Be-Mda-mboro* as authority figure disappeared in the cities. The *Mbueti* replaced *Melan* as religous reference (despite the influence of the Catholic Church); *Mbueti*'s "bishops" replaced *Be-Mda-mboro* as authority figure, *Mbueti*'s *ngui* societies (secret societies) being those who are responsible for administering punishments. Since this punishment often involves black magic and can reach all members in a family (no matter who "deserves" that punishment), cautiousness in activities and behaviors, to avoid *ngui* interventions, is considered a valuable quality in a person.

Conclusions

Judging by the obtained factors and their included characteristic behaviors in each context, we can infer some conclusions.

First, most of the factors obtained for *intelligent* prototype in both contexts overlapped considerably, as would be expected from the simple correlations made in the first stage between contexts, but they are different in some ways. The order of appearance of factors differs, probably reflecting their order of importance for differential adaptation purposes. Content in each factor also differs, probably according to how each ability takes form in relation to the ecological demands of skills in each context. Finally, the two contexts seem to demand two different specific culture-bound factors: *Traditional-verbal ability* in the rural and *Cautiousness* in the urban context.

Second, as far as factors in themselves are concerned, outwardly, some of them are quite similar to the main factors obtained in formal psychometric theories (Carroll, 1993; Cattell, 1971, 1987; Gustafsson, 1988; Vernon, 1969, 1979). *General reasoning, Verbal and Verbal-Educational ability, Memory and Learning ability,* and *Visual-Spatial ability* mirror the "g" and Gf factor, Gc and $V{:}ed$ factor, $2Y$ (General Memory and Learning in Carroll's theory), and Gv factor respectively. On the other hand, *Social competence*, though proposed as "Social Intelligence" by Thorndike (1920) and missing for explicit psychometric theories (Keating, 1978), can be found in some implicit teory research (Sternberg, 1990), and plays the main role in people's prototype of an intelligent person from both contexts in our research.

General Discussion

Fang people have well-developed implicit theories of intelligence and ability concepts employed for evaluation purposes according to the different contexts in which they live.

First, people appear to have organized general prototypes of intelligent behavior in, but their complexity differs according to context; in Bata the prototype is very similar to that found by Juan-Espinosa, Palacios & Garcia (1993) in Madrid, where Intelligent and Knowledgeable person prototypes were closely related. This difference between Bata and the traditional culture in the rural context. could be interpreted as an result of acculturation. However, people from the two contexts seem to share a general prototype of intelligent person. This prototype was the focus of the second stage of our research.

Second, the presence or absence of schools, urbanization, etc., indicators of acculturation, influence both the structure and content of intelligence prototypes. Structures are seen to be influenced in (a) the order of importance of abilities, and (b) in the presence of specific culture-bound factors. Traditional context demands more *General reasoning* and *Visual-Spatial ability* for adaptation and survival, and the urban context demands more *Verbal-Educational ability* and less *Visual-Spatial ability.* Rural context conditions demand a specific factor of *Traditional-verbal ability*, while the urban context demands *Cautiousness* people. *Social competence* is the most important ability in both contexts. It is in the characteristic problem-solving behaviors underlying the abilities where the influence of acculturation is most manifest. As discussed above, although *Social competence* is of great importance for both contexts, underlying skills differ. This factor was also identified by Sternberg et al. (1981), although different variables, such as "Accepts others for what they are", "Admit mistakes", "Displays interest in the world at large", etc., indicate cultural differences in the role played by this factor in both cultures. In Equatorial Guinea, while in the rural context this ability had skills oriented towards others; money management skills are very important in the urban context. *Verbal-Educational ability* includes academic skills which are very important in Bata, but not in the rural context and, therefore, does not appear in its *Verbal ability* factor. Even the *General reasoning* factor includes negative variables in Bata that can be interpreted as a rejection of traditional ways of life.

In general, our results with the Fang people show consistency with the law of cultural differentiation, and can be taken as an example of the acculturation mechanism proposed by Berry and of its influence on people's beliefs about intelligence.

References

Berry, J.W. (1987). The comparative study of cognitive abilities. In S. H. Irvine & S. E. Newstead (Eds.), *Intelligence and cognition: Comparative frames of reference.* Dordrech: Nijhoff.

Berry, J.W., Poortinga, Y.H., Segall, M.H., & Dasen, P.R. (1992). *Cross-cultural psychology: Research and applications.* Cambridge, N.Y.: Cambridge University Press.

Carroll, J.B. (1993). *Human cognitive abilities. A survey of factor-analytic studies.* Cambridge: Cambridge University Press.

Cattell, R.B. (1971). *Abilities: Their structure, growth and action.* Boston: Houghton-Miflin.

Cattell, R.B. (1987). *Intelligence: Its structure, growth and action.* Amsterdam: North-Holland.

Chen, M.J., Braithwaite, V., & Jong, T.H. (1982). Attributes of intelligent behavior: perceived relevance and difficulty by Australian and Chinese students. *Journal of Cross-Cultural Psychology, 13,* 139-156.

Eysenck, H.J. (1979). *The structure and measurement of intelligence.* Berlin-Heidelberg-New York: Springer Verlag.

Ferguson, G.A. (1956). On transfer and the abilities of man. *Canadian Journal of Psychology, 10,* 121-131.

Gill, R., & Keats, D. (1980). Elements of intellectual competence: Judgements by Australian and Malay university students. *Journal of Cross-Cultural Psychology, 11,* 233-243.

Goodnow, J.J. (1984). On being judged "Intelligent". *International Journal of Psychology, 19* (4-5), 391-406.

Gustafsson, J-E. (1988). Hierarchical models of individual differences in cognitive abilities. In R. J. Sternberg (Ed.), *Advances in the psychology of human intelligence* (Vol. 4). Hillsdale, N.J.: Erlbaum.

Irvine, S.H., & Berry, J.W. (1988). The abilities of mankind: A revaluation. In S. H. Irvine & J. W. Berry (Eds). *Human abilities in cultural contexts.* Cambridge N.Y.: Cambridge University Press.

Juan-Espinosa, M. (1994). *Estudios transculturales en Inteligencia [Cross-cultural studies on intelligence].* Unpublished report.

Juan-Espinosa, M., Palacios, A., & García, O. (1993). *Teorías implícitas de la inteligencia en contextos urbanos [Implicit theories of intelligence in urban contexts].* Unpublished report.

Kaiser, H.F. (1960). The application of electronic computers to factor analysis. *Educational and Psychological Measurement, 20,* 141-151.

Keating, D.P. (1978). A search of social intelligence. *Journal of Educational Psychology, 70,* 218-223.

Keats, D. (1982). *Cultural bases of concepts of intelligence: a Chinese versus Austrlian comparison.* Proceedings of the Second Asian Workshop on child and Adolescents Development, Bangkok.

Klein, R., Freeman, H., & Millet, R. (1973). Psychological tests performance and indigenous conception of intelligence. *Journal of Social Psychology, 84,* 219-222.

Sternberg, R.J., Conway, B.E., Ketron, J.L., & Bernstein, M. (1981). People's conception of intelligence. *Journal of Personality and Social Psychology, 41,* 37-55.
Sternberg, R.J. (1985). Implicit theories of intelligence, creativity, and wisdom. *Journal of Personality and Social Psychology, 49,* 607-627.
Sternberg, R.J. (1990). *Metaphors of mind: Conceptions of the nature of intelligence.* Cambridge: Cambridge University Press.
Thorndike, E.L. (1920). Intelligence and its uses. *Harper's Magazine, 140,* 227-235.
Vernon, P.E. (1969). *Intelligence and cultural environment.* London: Methuen.
Vernon, P.E. (1979). *Intelligence, heredity and environment.* New York: W.H. Freeman.

Part IV

Values

Characteristics of the Ideal Job among Students in Eight Countries

Geert Hofstede, Institute for Research on Intercultural Cooperation,
Maastricht / Tilburg, the Netherlands
Ludek Kolman, Agricultural University, Prague, Czechia
Ovidiu Nicolescu, Academia de Studii Economice, Bucarest, Romania
Indrek Pajumaa, EMOR Ltd., Tallinn, Estonia

Work Goals as a Topic for Cross-Cultural Research

The term *work goals* is used in social psychology to indicate a special area of human values operating in the working part of a person's life (Hofstede, 1977). Sociologists will rather speak of "work values". The U.S. or U.S.-inspired psychological literature contains many studies measuring work goals. They commonly use sets of questions in paper-and-pencil questionnaires asking respondents "how important" to them is each of a list of aspects (facets) of the work situation, such as high earnings, job challenge, or good physical working conditions. Instead of "important", other terms have occasionally been used, like "extent in ideal job" (Rosenberg, in Robinson, Athanasiou & Head, 1969, p. 233), "would like to have" (Hackman and Lawler, 1971, p. 269), or "attractive" (De Leo and Pritchard, 1974). Answers have either been given by ranking facets or, more often, by separately rating each facet on a scale with anything between two and twenty-one points, usually five.

The measurement of work goals has been carried out mostly in the context of surveys measuring aspects of morale and satisfaction as well. Work goals questions were first included in unsuccessful attempts to test theoretical models for predicting job satisfaction (Quinn and Mangione, 1973), and for showing management in survey feedback which aspects of the work situation were most important to subordinates (Youngberg, Hedberg & Baxter, 1962). Only later on were work goals studied for their own sake, as part of comparative research into respondents' values, as in Hofstede's (1980, 1991) IBM Hermes studies on differences in national cultures. The present paper continues this line of research.

Work goals are values "as the desired", rather than "as the desirable" (Levitin, 1973; Hofstede, 1980, p. 20). The *desired* is closer to *actual* behaviour than the *desirable* which explains the interest in work goals for values research. In the international IBM studies up to twenty-six work goals questions were included in successive survey rounds; fourteen of these were present in all survey versions. A factor analysis of the mean scores matrix for these fourteen work goals and for forty countries led to the identification of the culture dimensions of Individualism/Collectivism and Masculinity/Femininity. In the 14 x 40 matrix Ind/Col explained 24 % of the variance and Mas/Fem 22 % (Hofstede, 1980, p. 241).

Ind/Col and Mas/Fem are dimensions of *national cultures* and should not be confused with dimensions of personality. They can only be measured in studies comparing matched samples of respondents from two (or rather more) nations. Therefore, they are not meant to describe individuals, but dominant patterns of socialization ("mental programming") in nations; these dominant patterns will affect different individuals to different degrees, and some components of a national culture pattern may be found in one individual, while other, complementary components will be found in other individuals within the same society.

Individualism stands for a society in which the ties between individuals are loose: everyone is expected to look after himself or herself and his or her immediate family only. *Collectivism* stands for a society in which people from birth onwards are integrated into strong, cohesive ingroups, which throughout people's lifetime continue to protect them in exchange for unquestioning loyalty.

Masculinity pertains to societies in which social gender roles are clearly distinct: men are supposed to be assertive, tough, and focused on material success whereas women are supposed to be more modest, tender, and concerned with the quality of life. *Femininity* pertains to societies in which social gender roles overlap: both men and women are supposed to be modest, tender, and concerned with the quality of life (Hofstede, 1991, pp. 260-2).

From the various replications of the IBM studies the one by Hoppe (1990) is the most extensive. Hoppe surveyed elites from nineteen countries: politicians, employers, labor leaders, artists and intellectuals who were alumni of the Salzburg Seminar, a high-level international conference center in Salzburg, Austria. Eighteen of his, mostly European, countries overlapped with the IBM set.

Hoppe calculated country scores on Individualism and Masculinity for his population. His Individualism Index (IDV) scores were in the same range as for the IBM studies, and the two sets of scores showed a correlation of .69, significant at the .001-level. His Masculinity Index (MAS) scores were much lower than for the IBM population, and the two sets of scores showed a correlation of .36, not significant. This was mainly due to the deviant results for one country, Sweden; without Sweden, the correlation became .56, significant at the .05-level.

Method

Sample

The present study combines data from ten different samples of students, surveyed with the same questionnaire between 1989 and 1993. It is motivated by an interest in comparative data between Eastern and Western European population samples: four of the samples in the study are from Eastern Europe.

Hofstede and Vunderink (1994) collected scores on twenty-two work goals from female and male U.S. business administration students (n = 89) attending semester-long study programmes in Maastricht, Holland, in 1989 and 1990. These were compared to samples of female and male Dutch students. The questionnaire used in the present study was designed for their project. It refers to "the job you would like to get after graduation". For the Dutch students it was randomly administered in English ($n = 129$) and in a Dutch translation ($n = 100$). For reasons to be explained below, only the English language version results were used for the analysis.

Hofstede, van Twuyver, Kapp, de Vries, Faure, Claus and van der Wel (1993) used the same questionnaire to collected scores on samples of male and female business administration students from the neigboring universities of Aachen, Germany ($n = 100$), Liège, French-speaking Belgium ($n = 100$), Diepenbeek-Hasselt, Dutch-speaking Belgium ($n = 100$) and Maastricht, Holland ($n = 229$), in 1992. The questions were included in a survey which was part of a study of the police forces in the three countries. The students were surveyed as a control group. The questionnaire was administered in German, French and Dutch.

In 1989, before the disintegration of the Soviet Union, Indrek Pajumaa, a market researcher from Tallinn, Estonia, collected data from students with Estonian and Russian versions of the same work goal questionnaire. In Estonia the students were from Tartu University, Tallinn Technical University and the Teachers Training College ($n = 95$). In Russia the scores were collected at Ivanovo University ($n = 70$). These students were not matched with the U.S. and Western European groups on field of study; it is questionable whether in the Soviet Union a field of study equivalent to business administration could have been found.

In 1993 Dr Ludek Kolman from Prague, Czechia, collected data among business administration students at Prague University ($n = 101$), using a Czech translation of the same questionnaire.

Finally, in 1994 Professor Ovidiu Nicolescu from Bucarest, Romania, collected data from business administration students at Bucarest Economic University ($n = 99$), using a Romanian translation.

The ten country samples in the present study ($N = 983$) can be considered as more or less matched, but the match is not perfect. The Western samples were collected in two different studies two years apart, and the Eastern European samples in three different initiatives, partly before and partly after the demise of communism. In view of the sweeping changes having affected these countries this is a serious limitation which has to be kept in mind. However, we believe the data still merit a comparative analysis.

Procedure

For the Dutch students surveyed in 1990 only the English language half of the sample was kept in the analysis, in order to create a maximum difference with the sample surveyed in 1992, in Dutch only, which was also kept. Keeping both has offered insight into the reliability of the measures: the dependence of the scores on the language used, on the year of surveying, and on the groups surveyed.

The Dutch translation of the set of goals proceeded in steps described in Hofstede and Vunderink (1994, p. 344). In 1989 and 1990 Dutch students were surveyed in classroom sessions; questionnaires were distributed from a pile in which every second copy was in English, the other in a Dutch translation. In this way the language in which the questionnaire was answered by the Dutch students was randomized. For the 1989 surveys the ranking of the goals on the English and on the Dutch questionnaires showed a correlation of .84. On the basis of a comparison of the two language versions the Dutch translation was subsequently corrected. In the 1990 surveys the corrected version was used in the same random way. This time the two language versions showed a correlation of .92.

The goal rankings of the two Dutch samples in the present study show a correlation of .81. The 1992 sample and the 1990 sample surveyed in *Dutch* (not used in the present study but referred to in Hofstede and Vunderink, 1994) show a correlation of .86. The difference between the 1990 and the 1992 versions in the present study is therefore partly due to the language of the questionnaire but more to other factors: (1) changes in people's minds in the two years in between, and (2) unplanned differences between the student groups surveyed.

The 1992 students were on average younger and less advanced in their studies than the 1990 students. The biggest shifts in importance occurred for "use skills" (first rank in 1990, twelfth in 1992) and for "living area" (fourteenth rank in 1990, fifth in 1992). It is possible that these are influenced by the 1990 students being closer to graduation.

Results

The aim to obtain equal numbers of female and male students was not perfectly achieved. Therefore the total country scores were computed as $(F+M)/2$, in which F is the mean score for the women and M for the men. The small samples of Dutch (1990) and Russian females therefore carry the same weight as the larger samples of males.

"Importance" scales are notorious for their acquiescence (response set) component; different respondent groups use different parts of the scale, because it has no natural zero point. This is evident from Table 1 which for the twenty groups of respondents (2 genders, 10 countries) lists the overall mean scores across all twenty-two goals.

Table 1. Mean Raw Scores on Work Goals Questions across 22 Goals

Country	female	male
U.S.A.	1.94	2.09
Romania	2.14	2.24
Estonia	2.27	2.37
Holland (E)	2.31	2.34
Russia	2.53	2.17
Czechia	2.37	2.41
Belgium D	2.35	2.52
Holland (D)	2.39	2.51
Belgium F	2.43	2.50
Germany	2.55	2.55
Overall M	2.33	2.37

On the five-point scales used 1 = "of utmost importance", so a lower score indicated higher importance. U.S. respondents in Table 1 showed the strongest tendency to rate all goals as more important, and German respondents the weakest. The American response set can be interpreted as a manifestation of the preference for superlatives in the American culture (Indrek Pajumaa noticed that something which is "good" for a European is "fantastic" for an American). Dutch respondents rated goals more important when answering in English than when answering in Dutch; they differentiated less in the foreign language than in their own (this had also been found in comparing the two language versions in 1989 and 1990, see Hofstede and Vunderink, 1994, p. 344).

In Table 1 women scored lower than men in eight out of the ten cases (*Sign* test, $p = .05$). In earlier research, persons of lower status have been found to show more acquiescence (Hofstede, 1980, p. 79). Could female business administration students still on average have a lower status ?

Before the scores for the different work goals can be compared across countries, the acquiescence response set must be eliminated, by calculating for each respondent group the relative position of each goal in comparison to the other twenty-one goals. Importance is by its very nature a relative concept; behavior is affected by the relative importance of one outcome over another.

The relative position of a goal is computed by standardization (Hofstede, 1980, pp. 77-80). For every respondent group one deducts from every raw mean score the overall mean across the 22 goals, and one divides the outcome by the standard deviation across the 22 means. As a low score indicates greater importance, *a negative standard score* also means *a high relative importance*. The standard scores thus computed can be directly compared from country to country.

Table 2 presents the twenty-two goals in the order in which they were rated as relatively important across all ten student samples. So "cooperation" ("work with people who cooperate well with one another") was on average the most important goal and "serve country" the least important. The most preferred goals on average (top five) are cooperation and a good boss (social), use of skills, challenge and freedom (job content).

Table 2. Overall Order of Importance of the 22 Work Goals across 10 Country Samples

nr in work goal quest	mean of 10 standard scores	s.d. of 10 standard scores
8. cooperation	-1.00	.29
22. use skills	-.85	.32
2. challenge	-.85	.48
7. freedom	-.65	.42
5. good boss	-.65	.38
13. living area	-.64	.50
19. training	-.58	.44
11. earnings	-.41	.55
4. work conditions	-.41	.26
1. private time	-.39	.66
10. contribute	-.38	.46
14. advancement	-.26	.77
21. recognition	-.25	.40
15. variety	-.09	.97
6. security	.15	.62
20. benefits	.36	.28
9. be consulted	.70	.80
17. help others	.75	.40
18. clear job	.96	.69
3. stress-free	1.04	.97
16. prestige comp	1.14	.52
12. serve country	2.31	1.04

low scores mean high importance

The standard deviations across the ten country samples indicate for which goals the country differences were largest. For mathematical reasons the large standard deviations (over .70) are all found in the lower half of the table; the distribution of the means is skewed. When there is less consensus across countries, the means are also higher. The widest spread occurs for "serve country" which scores least unimportant in Russia (.52), Estonia (1.00) and Romania (1.03) and most unimportant for the Dutch speaking Belgians (3.48, which is very extreme). Other goals that vary widely in importance among the countries are "variety" (from -1.00 for the Dutch to 1.81 for the Russians);

"stress-free", from -.56 for the Estonians and -.43 for the Russians to 1.83 for the Americans; "be consulted" (from -.28 for the Dutch to 1.99 for the Russians; and "advancement" (from -1.27 for the Americans to 1.39 for the Russians).

Table 3. Results of a Cluster (factor) Analysis of the Scores for 10 Country Samples on 22 Work Goals

Cluster 1		Cluster 2	
loading	Country	loading	Country
.95	Holland (E)	.96	Russia
.94	Belgium D	.75	Estonia
.94	Holland (D)	.60	Czechia (2)
.92	U.S.A.	.50	Romania (2)
.91	Belgium F		
.85	Germany		
.64	Czechia (1)		
.50	Romania (1)		

(1) and (2) mean first and second loading. Goal scores have been standardized and based on F+M/2.

The 10 x 22 matrix was factor analysed, producing clusters of countries that are similar in terms of their goals. Two clusters explain 83% of the common variance. After orthogonal rotation, the clusters are as listed in Table 3.

There is one strong cluster including all Western country samples, and marginally also Czechia 1993 and Romania 1994. The second cluster combines the four former socialist countries and is headed by Russia, which was still under communism in 1989.

Table 4 shows the goals that make most difference between cluster 1 and cluster 2. The Western country samples attach much more importance to *challenge, advancement, recognition* and *being consulted*. The samples from countries once under a communist system attach much less importance to these four, but much more than the Western samples they value *using their skills, freedom to adopt their own approach to the job, little tension and stress on the job* and *serving their country*.

Table 4. Mean Standardized Work Goal Importance Scores Differentiating most between Western and Former Communist Countries

	Western	Former Communist
2. challenge	-1.16	-.39
14. advancement	-.67	.36
21. recognition	-.46	.08
9. be consulted	.15	1.52
22. use skills	-.64	-1.17
7. freedom	-.45	-.96
3. stress-free	1.64	.15
12. serve country	2.94	1.36

Based on F+M/2; lower scores mean higher importance

Table 5. Most Important Goals by Country, Absolute and Relative, based on (F+M)/2

Belgium (D)	Belgium (F)	Germany
1. cooperation	1. **challenge**	1. challenge
2. challenge	2. living area	2. **living area**
3. **good boss**	3. good boss	3. **training**
4. living area	4. **contribute**	4. cooperation
5. **recognition**	5. **use skills**	5. **security**
6. **advancement**	13. **be consulted**	
19. **prestige comp**		

Holland (D)	Holland (E)	U.S.A.
1. **challenge**	1. use skills	1. **advancement**
2. cooperation	2. **variety**	2. **good boss**
3. **variety**	3. cooperation	3. challenge
4. good boss	4. challenge	4. variety
5. living area	5. good boss	5. cooperation
	9. **recognition**	7. **work conditions**
	10. **be consulted**	14. **benefits**

Czechia	Estonia	Romania
1. **cooperation**	1. **living area**	1. **cooperation**
2. use skills	2. **use skills**	2. **contribute**
3. challenge	3. **training**	3. **freedom**
4. **private time**	4. **private time**	4. **use skills**
5. earnings	5. variety	5. **earnings**
13. **clear job**	7. **work conditions**	6. **security**
	10. **stress-free**	15. **prestige comp**
	13. **help others**	
	19. **serve country**	

	Russia	
	1. **freedom**	
	2. cooperation	
	3. **use skills**	
	4. **earnings**.	
	5. living area	
	9. **stress-free**	
	12. **benefits**	
	15. **clear job**	
	16. **help others**	
	17. **serve country**	

The top 5 goals are all listed; from the remaining goals only those scored **relatively** important compared to the other countries are listed (highest 2 among 10 country samples). Goals scored relatively important are printed in **bold**.

In Table 5 the five most important goals are shown by country sample; the remaining goals are only listed for a country if they are *relatively* important for this country compared to the others: if the country is one of the top two on this goal. When a country is among the top two, the goal is printed in bold.

Table 6. Comparison of Factor Analyses of the Scores on 14 Work Goals for Three Studies

IBM studies, 40 countries (Hofstede, 1980:241)

Factor 1 individual-collective		Factor 2 social-ego	
loading	goal	loading	goal
.86	private time	.69	good boss
.49	freedom	.69	cooperation
.46	challenge (2)	.59	living area
.35	living area (2)	.50	security
-.82	training	-.70	earnings
-.69	work conditions	-.59	recognition
-.63	use skills	-.56	advancement
-.40	benefits	-.54	challenge
-.37	cooperation (2)	-.40	use skills (2)

Elite study, 15 countries (Hoppe, 1990:149)

Factor 1		Factor 2	
loading	goal	loading	goal
.81	private time (1)	.90	contribute
.60	living area	.51	prestige comp (2)
.49	security (1)	.49	security (2)
.31	earnings	.42	freedom (1)
-.90	challenge	-.82	cooperation
-.82	advancement	-.72	work conditions
-.53	prestige comp	-.38	private time (2)
-.36	good boss		
-.35	freedom (2)		

Student studies, 10 country samples

Factor 1		Factor 2	
loading	goal	loading	goal
.94	advancement	.64	cooperation
.92	good boss	.59	security
.84	challenge		
.81	recognition		
.32	benefits		
-.81	use skills	-.88	private time
-.79	training	-.85	work conditions
-.77	freedom	-.67	living area
-.61	earnings		

(1) and (2) mean first and second loading

Data Analysis - Dimensional Structure

Hofstede (1980, p. 241) shows the factor structure in a factor analysis of mean scores for fourteen goals across IBM subsidiaries in 40 countries. All fourteen were used in the present study, (a fifteenth, "contribute", had been used in some of the IBM surveys). The same type of factor analysis with the same fourteen goals was repeated for the 10 country samples in the present study. Table 6 lists the loadings in the IBM study, the Hoppe replication and the present study. The Hoppe results cover only fifteen of his countries (in view of some missing data) and they include the goals "contribute" and "prestige company", but not "recognition", "use skills", "training" and "benefits".

Table 7. Scores on the Individualism and Masculinity Index by Country and Gender

IDV	IBM 1970	students around 1991		
		women	men	total
U.S.A.	91	23	36	30
Holland (D)	80	31	51	41
Holland (E)		(23)	(32)	(28)
Belgium D	78	47	58	53
Belgium F	72	73	53	63
Germany	67	69	70	70
Estonia		52	53	53
Czechia		34	56	45
Russia		28	50	39
Romania		13	33	23
total		39	49	44
MAS	IBM 1970	students around 1991		
		women	men	total
Germany	66	28	29	29
U.S.A.	62	54	69	62
Belgium F	60	18	54	36
Belgium D	43	15	58	37
Holland (D)	14	-4	19	8
Holland (E)		(-11)	(42)	(16)
Czechia		20	58	39
Romania		8	63	36
Russia		-6	65	30
Estonia		10	38	24
total		13	50	32

Approximation formulas: $IDV = -43m_1 + 76m_4 + 30m_8 - 27m_{13} - 29$
$MAS = 30m_6 + 60m_8 - 66m_{11} - 39m_{14} + 76$
in which m_1 is the raw mean score for question 1 (private time, see Appendix A), etc. Item m_8, cooperation, occurs in both formulas.

Factor 1 in the IBM studies (individual-collective) led to the identification of the national culture dimension of Individualism/Collectivism; factor 2 (social/ego) to Masculinity/Femininity.

In the first and the third part of Table 6 we find advancement, challenge and recognition together (ego); cooperation and security (social); use skills and training (collective); and private time and living area (individual). However, the "ego" goals for the students oppose the "collective" goals, not the "social" goals; and the "social" goals oppose the "individual" goals. Moreover, "good boss", "earnings", "benefits", "work conditions" and "freedom" are not part of the same clusters of goals in the student studies as they were in the IBM studies.

Hoppe's factors resemble more the student structure than the IBM structure. He finds "advancement", a "good boss" and "challenge" together on the same factor, and "earnings" on the same factor but with a negative sign. On a second factor he finds "security" (marginally) opposing "private time" and "work conditions".

In spite of the different factor structure in his data, Hoppe found the Individualism Indices (IDV) calculated on the basis of these data significantly correlated, and the Masculinity Indices (MAS) marginally correlated, with the IBM scores. He calculated the indices by the simplified formulas developed for this purpose (Hofstede, 1982; Bosland, 1985).

The IDV formula (see Table 7) uses the importance scores of "private time" and "living area" minus those of "work conditions" and "cooperation". In Hoppe's factor structure "private time" and "living area" appear on the same factor with high loadings; "work conditions" and "cooperation" also appear together with high loadings, but on the other factor. The IDV formula therefore combines his two factors.

The MAS formula uses the importance scores of "advancement" and "earnings" minus those of "security" and "cooperation". Unfortunately in Hoppe's factor structure "advancement" and "earnings" appear on opposite sides of the same factor (so the country scores are negatively correlated) and the same is true for "security" and "cooperation", on the other factor. In computing the MAS index for Hoppe's data one therefore adds scores that partly cancel each other out. It should be no surprise that the resulting MAS scores from Hoppe's study did not correlate as well with the IBM scores as the IDV scores did.

For the student goals related to Individualism, "private time" and "living area" do appear on the same side of the same factor (Table 6), and "cooperation" appears with an opposite sign on the same factor, but "work conditions" has the same sign as "private time". Unfortunately, "work conditions" in the IDV formula is precisely the goal that carries the most weight. The scores computed with this formula for the students are therefore unlikely to make much sense. This is evident from Table 8 which shows the results of the computation.

Between the IBM and Hoppe studies on the one side, and the student studies on the other, the order of the Western countries on IDV has been reversed: the Germans seem to score the most individualist and the Americans the least. This conclusion should however be taken with a grain of salt: the formula used is based on a selection of goals which for the students contains contradictory information.

For the student goals related to Masculinity, "cooperation" and "security" do appear on the same side of the same factor; but "earnings" and "advancement" appear on the other factor and with opposite signs; in the MAS formula, "earnings" carries the most weight. The formula again adds scores that are negatively correlated, but the effect will not be as strong as in the case of IDV.

For MAS, in the student studies, the U.S.A., Dutch speaking Belgium and Holland score at the same level as in IBM; French speaking Belgians seem less masculine, joining their Dutch speaking compatriots, and Germans even less masculine than the Belgians. Hoppe also found that on MAS, the U.S.A. and Holland continued to represent opposites, and that Belgium and Germany were in between, Germans less masculine than Belgians. But for the reasons exposed, these conclusions cannot be upheld.

Data Analysis - by Gender

According to Table 7 men score more "individualist" than women in nine out of the ten samples, but by a small margin (10 points on average). This can be explained by the influence of acquiescence in the formula, the tendency shown above for the women in our samples to score everything more important.

In Table 7 men score more masculine than women in all ten country samples by a much wider margin (37 points on average). Masculinity in the IBM studies was the only dimension to show a consistent gender effect (Hofstede, 1980, p. 279). The student scores do not follow the pattern of a widening gap between the sexes for the more masculine countries (Hofstede, 1991, Figure 4.1). In Hofstede and Vunderink (1994) this deviation from earlier results was also signalled. It is now clear that the formula used introduces too much error for a secondary effect like this to become visible.

Table 8 shows for each goal separately the difference between the standardized scores for men and for women, (M - F), and the number out of the ten samples for which M > F or F > M. The tendency for men or for women to score a goal as more important has been tested with the *sign* test across ten cases (country samples). For significance at the .05-level, 0 or 1 case may be scored in the opposite direction. By this standard, only "earnings" and "advancement" are rated more important by men, and "a good boss" by women. All three also showed significant gender effects in the IBM studies (Hofstede, 1980; pp. 272-3).

The other goals for which the gender effect in the IBM studies was significant, were: more important for men: training; for women: work conditions and cooperation, but these differences are no longer significant in the student studies. Goals with 8-2 votes in the student samples (approaching significance); for men: prestige company and serve country; for women: use skills, a clear job and a stress-free job. From these only "use skills" was also used in IBM, but it showed no clear gender effect there.

Table 8. Differences of Standardized Mean Scores on 22 Work Goals for Women and Men in the same Student Country Samples

Goal	Mean score	M>F	F>M
earnings	-.59	10*	0
advancement	-.48	10*	0
freedom	-.45	7	3
contribute	-.27	7	3
prestige comp	-.26	8	2
serve country	-.14	8	2
be consulted	-.13	6	4
challenge	-.10	6	3
recognition	-.07	5	5
living area	-.06	5	5
variety	-.06	5	5
training	-.02	5	5
benefits	.10	4	6
use skills	.16	2	8
help others	.17	4	6
work conditions	.26	4	6
cooperation	.33	3	7
good boss	.33	1	9*
security	.35	3	7
stress-free	.37	2	8
private time	.37	4	6
clear job	.47	2	8

* Gender effect significant at .05-level (sign test). Negative scores mean more important for men.

One can check to what extent the East-European countries account for the minority votes in Table 8, such as showing men rather than women to prefer a "good boss", or women rather than men to value "freedom". Could the communist experience have led to a different gender role socialization?

There is one case in Table 8 for which only East-European countries provide the minority votes: "prestige company" is a more important goal for women than for men in Estonia and Czechia. Three out of four minority votes are from Eastern Europe for "helping others"; this is more important for men in

Russia, Estonia and Romania. And two out of three for "freedom" (more for women in Russia and Estonia), "challenge" (more for women in Czechia and Romania), and "cooperation" (more for men in Estonia and Czechia). There is no overall reversal of gender roles in Eastern Europe, but there are a few exceptions to the Western norms.

Discussion

Country Differences in Work Goals

Based on the cluster analysis in Table 3 it makes sense to divide the countries into Western and East-European. The first cluster puts more value on challenging work, advancement, recognition and on being consulted by one's boss (Table 4). These are (so far) relatively less important in the former communist countries; according to Indrek Pajumaa from Estonia, the socialist system did not reward competition and individual success. Students in these countries put relatively more stress on using one's skills, freedom, little tension and stress on the job, and serving their country. The first three indicate things which under the communist system could not be taken for granted; often one could not change an undesirable job.

The stress in the Eastern European countries on "serving your country" (which is only relative, it remains an unimportant goal) indicates the emotional appeal of nationalism in this part of the world at this moment in history. In a nation-wide opinion poll in Estonia conducted in March 1991, after independence, Indrek Pajumaa surveyed samples of Estonian and non-Estonian (mainly Russian) inhabitants of Estonia with the Rokeach Value Survey questions; the results were compared with earlier data from Finland, Sweden and the U.S.A. The Estonians and Russians put much more value than the other groups on "national security" (Pajumaa, 1991).

For the Western countries the differences between Americans and Dutch were already analysed in Hofstede and Vunderink (1994). These differences are sustained in the new Dutch data. Americans stress earnings, advancement and benefits; Dutch stress freedom, being consulted and training. The importance to the Dutch of being consulted is one of the central themes in d'Iribarne (1989): Holland is a consensus society. The new data show Holland to lead in the stress on variety and adventure on the job.

The importance of consensus is also found in Dutch-speaking Belgium. All in all the difference between the two Belgiums is not that large. Both stress the importance of a good boss (Table 5) which they share with the U.S.A. This could be a sign of a larger Power Distance (Hofstede, 1991), but in the U.S.A. it was interpreted as a consequence of the boss' influence on one's promotion (Hofstede and Vunderink, 1994, p. 353). Belgians score very low on "serve country"; the Dutch speaking Belgians the lowest of all. The loyalty to their own nationality among Belgians is known to be very weak; they prefer to identify with their region. Germans stress training and security. On-the-job traineeships have a high status in Germany and Switzerland, higher than in any other country. The stress on security fits with a relatively large Uncertainty Avoidance.

Estonia shares with Germany a focus on training, and on the living area. Estonia and Czechia stress the individualistic goal of private time. Romania and Russia put the highest value on freedom and using their skills. Russia and Czechia attach most value to a clear job.

The work goals of students (and others) in the Eastern European countries are likely to be in the same state of flux as their countries' political and economic situation. In a few years they will probably resemble more the work goals in Western countries, and only differentiate themselves from other countries according to those traditional cultural elements that survive political systems. We saw already that the goals for Czechia and Romania, surveyed in 1993-94, are more like those in the Western countries than the goals for Estonia and Russia, surveyed in 1989 (Table 2).

Gender Differences in Work Goals

Our study allows an investigation of gender differences. Table 8 showed that only three of the twenty-two goals show a consistent gender effect across all ten (or nine out of ten) country samples: earnings and advancement are more important to male students and "a good boss" is more important to female students. Marginal cases (eight out of ten one way) are working in a prestigious, successful company and organization, and serving one's country, which are more important for males, versus fully using one's skills and abilities on the job, having little tension and stress, and working in a well-defined job situation where the requirements are clear, which are more important to females - or less important to males.

These differences confirm a universal dominant gender role distribution in all the countries surveyed: men are more ambitious and concerned with material gain - and also more nationalistic; women are more interested in interpersonal relations and the quality of the work situation - and less sure that their skills will be fully used. This is the dominant pattern that shows up in the statistical distribution of answers (of course it does not mean that every individual male or female student needs to correspond to this pattern). Not even the ex-communist countries deviate systematically from this dominant pattern. It represents a societal norm in the students' minds.

Against this background both the importance attached to "advancement" and the importance attached to "earnings" in a society can be used as indicators of Masculinity. Because of their different correlation patterns across countries (Table 6) a choice has to be made between these two. "Advancement" is clearly correlated with other ego-enhancing goals, and should therefore be preferred as a Masculinity indicator. The importance of "cooperation" is an undisputable indicator of Femininity.

Methodological Problems of Replicating IDV and MAS

The present study partially replicates the IBM studies conducted around 1970, but with several fundamental differences: (1) In IBM there were 40 well-matched country samples; in the student study only 10, and not so well matched. (2) The student sample covered only four Western countries also studied in IBM, next to four East-European ex-communist countries. The IBM studies did not cover communist countries. (3) The time lag between both studies is more than twenty years, and even in the Western countries the reference context has changed drastically in this period.

In the IBM studies the work goal scores were used to calculate the Individualism and Masculinity Indexes. The present study showed clearly that the formulas developed from the IBM data are not very suitable for computing credible IDV and MAS scores for the new populations. This problem existed already, at least partly in the Hoppe (1990) replication, especially for MAS. In the present student study the dimension most affected is IDV. The reason for the non-suitability of the formulas is that, for the new populations, the country means for the different goals are not associated in the same way as in the IBM populations.

The forty IBM subsidiary populations studied by Hofstede (1980) were exceptionally homogeneous as to work situation, which led to a shared perception of the various work goal items, even across countries. This perception, however, is not necessarily shared by new, non-IBM, populations. Let us take an example. In IBM the MAS score reflects among other things the combined importance of "advancement" and "earnings", because across the IBM subsidiaries these were positively correlated; in the IBM system the way to make more money was to be promoted. In Hoppe (1990, p. 149) we see that across fifteen countries the importance of "advancement" and the importance of "earnings" are negatively correlated. The same applies for the ten student samples in the present study. If respondents in a country attach more importance to "advancement", there are fewer respondents in that country who attach importance to "earnings". However, both "advancement" and "earnings" are rated more important by men than by women in all countries (Table 8). They are both more masculine; yet we cannot base a Masculinity score on "advancement" plus "earnings" like in IBM, because they are alternative ways of being masculine, cancelling each other out.

The same that applies for "earnings" applies to a "good boss" but in a reverse sense: this goal is more frequently rated important by women than by men, and in IBM it appears at the opposite pole from "advancement" and "earnings", but in Hoppe and in the student samples it appears at the same pole as "advancement".

Based upon research across the available replications, a new Values Survey Module has been designed (VSM 94, see Hofstede, 1994), in which only "private time" (positive) and "work conditions" (negative) have been maintained in the new IDV formula, and only "advancement" (positive) and "cooperation" (negative) in the new MAS formula; for either index two questions about other issues than work goals were added. In view of the correlations between country means in the student data (Table 8), the new MAS formula should work out well; the new IDV formula might produce problems. It did do well for the Hoppe replication and other replications, however, so one should await more information before passing judgment.

References

Bosland, N. (1985). *An evaluation of replication studies using the Values Survey Module.* Working Paper 85-2, Maastricht Neth.: IRIC.
De Leo, P.J., & Pritchard, R.P. (1974). An examination of some methodological problems in testing expectancy-valence models with survey techniques. *Organizational Behavior and Human Performance, 12,* 143-148.
d'Iribarne, P. (1989). *La logique de l' honneur : Gestion des entreprises et traditions nationales.* Paris: Seuil.
Hackman, J.R., & Lawler, E.E. (1971). Employee reactions to job characteristics. *Journal of Applied Psychology Monograph, 55,* 3, 259-286.
Hofstede, G. (1977). *Comparative measurements of work goal importance.* Working Paper 77-31, Brussels: European Institute for Advanced Studies in Management, December.
Hofstede, G. (1980). *Culture's consequences: International differences in work-related values.* Beverly Hills CA: Sage.
Hofstede, G. (1982). *Scoring guide for Values Survey Module 1982.* Maastricht Neth.: IRIC.
Hofstede, G. (1991). *Cultures and organizations: Software of the mind.* London UK: McGraw Hill.
Hofstede, G. (1994). *Values Survey Module 1994 - Manual.* Maastricht Neth.: IRIC.
Hofstede, G., & Vunderink, M. (1994). A case study in masculinity/femininity differences: American students in the Netherlands vs. local students. In A.M. Bouvy, F. van de Vijver, P. Boski, & P. Schmitz (Eds.), *Journeys into Cross-Cultural Psychology* (pp. 329-47). Lisse: Swets and Zeitlinger.
Hofstede, G., van Twuyver, M., Kapp, B., de Vries, H., Faure, M., Claus, F., & van der Wel, J. (1993). *Grensoverschrijdende politiesamenwerking tussen België, Duitsland en Nederland met speciale aandacht voor de Euregio Maas-Rijn.* Maastricht: Universitaire Pers Maastricht.
Hoppe, M.H. (1990). *A comparative study of country elites: International differences in work-related values and learning and their implications for management training and development.* Unpublished Ph.D. dissertation, University of North Carolina at Chapel Hill.
Levitin, T. (1973). Values. In J.P. Robinson & P.R. Shaver (Eds.), *Measures of social psychological attitudes.* Ann Arbor MI: Survey Research Center, Institute for Social Research, University of Michigan, 489-502.

Pajumaa, I. (1991). The Rokeach values in Estonia. *EMOR-Reports: Quarterly of Estonian Market and Opinion Research Centre Ltd., 1*, 1, 29-38.

Quinn, R.P., & Mangione, T.W. (1973). Evaluating weighted models of measuring job satisfaction: a Cinderella story. In R.P. Quinn & T.W. Mangione (Eds.), *The 1969-1970 Survey of working conditions* (pp. 85-114). Ann Arbor MI: Survey Research Center, Institute for Social Research, University of Michigan.

Robinson, J.P., Athanasiou, R., & Head, K.B. (Eds.). (1969). *Measures of occupational attitudes and occupational characteristics*. Ann Arbor MI: Survey Research Center, Institute for Social Research, University of Michigan.

Youngberg, C.F.X., Hedberg, R., & Baxter, B. (1962). Management action recommendations based on one versus two dimensions of a job satisfaction questionnaire. *Personnel Psychology, 15*, 2, 145-150.

National Differences in Value Consensus

Galit Sagie & Shalom H. Schwartz
The Hebrew University, Jerusalem, Israel

When examining values in cross-cultural perspective, two of the most frequent questions asked are, how national groups differ in the importance they assign to different values, and how these differences can be explained. In this paper, we discuss similar questions regarding *value consensus*; how do nations differ in the degree to which individual members of their society share the same value priorities, and what explains differences in societal value consensus?

Value consensus has been the subject of many theoretical discussions. Social theorists such as Comte, (as cited in Partridge, 1971), Durkheim, (1947), Parsons, (1968), and Shils, (1975) view value consensus as the basis of social order. According to these theorists, value consensus contributes to social stability by increasing coopeation, facilitating accomodations between conflicting interests and demands, and reducing the probability that violence will be used to resolve.

Despite a wealth of theoretical literature on value consensus, we found no sociological or social psychological research that systematically studied either value consensus or the factors that influence it. Our first step, therefore, was to develop a conceptual definition of value consensus and a method for measuring it empirically.

We adopted the definition of values proposed in Schwartz's theory of value sytems (Schwartz and Bilsky, 1987, 1990; Schwartz, 1992), according to which, human values are desirable goals, varying in importance, that serve as guiding principles in people's lives (see also Rokeach, 1973, Kluckholn, 1951). This theory identifies ten broad types of values distinguished by individuals in most cultures (Schwartz, 1992): benevolence, tradition, conformity, security, power, achievement, hedonism, stimulation, self-direction, universalism.

Value consensus is usually defined as concurrence among members of a society regarding their values (Comte, cited in Partridge, 1971; Cohen, 1968; Horowitz, 1962; Shils, 1975). Given our definition of values, this translates into agreement among the individuals in a society regarding the importance they attribute to different value types. The greater the agreement among individuals regarding the importance of a value type, the greater the value consensus regarding that value type. When individuals agree on the importance accorded to all ten value types, the overall level of value consensus in society is high.

A useful index of societal value consensus should provide scores that reflect the degree of agreement among societal members on particular values and on their overall value systems. A survey of the literature in psychology and sociology yielded no indicators of value consensus appropriate to these purposes. We therefore generated an index that follows directly from our definition of value consensus: the variance across the members of a sample in the importance they attribute to the different value types. Increasing variance reflects decreasing consensus.

This definition and operationalization of value consensus, provides a framework in which to compare the degree of consensus in different societies and to assess the relationships of value consensus to other variables. In this paper, we examine how societal characteristics might be related to cross-national differences in value consensus.

Our thinking was guided by the assumption that values are acquired and expressed in concrete social contexts such as, schools, families and jobs. The values that are transmitted in these settings depend upon how the major societal institutions operate - their goals, rules, demands, and so forth (Schwartz, 1993). We therefore considered social variables that might strongly influence the way value transmitting social institutions operate.

We focus on two societal characteristics that are likely to relate to value consensus: (1) *degree of modernization*, viewed in terms of socioeconomic development, and (2) *degree of democratization* of the political and social system.

Modernization

One of the major theories linking modernization to values is convergence theory. Convergence theory postulates that, regardless of their original structural diversity, modern industrial societies develop similar institutional features (Eisenstadt, 1973; Inkeles 1975; Levy, 1966; Meyer, Boli-Bennett & Chase-Dunn, 1975; Yang, 1988). Theorists who apply convergence theory directly to values propose that people become modern by incorporating the values implicit in modern institutions into their personal value systems (Inkeles & Smith, 1974; Kahl 1968). Convergence in the structure of modern societies is therefore thought to produce convergence in the value systems of people in different societies. That is, the values of modernized individuals *across* different industrialized societies are expected to become similar.

The reasoning underlying convergence theory can also be used to think about value convergence among individuals *within* modern societies. In the process of modernization, different segments of the population are brought under the influence of a single unified economic and political system. The mass of the population is integrated into the central institutional and value systems of society (Shils, 1975). Thus, both across and within societies, modernization promotes shared experiences in similar social institutions, so that people of differing socio-cultural backgrounds are likely to develop similar value systems. This interpretation of convergence theory leads us to hypothesize that: *the greater the socioeconomic modernization in a nation, the greater the overall level of consensus on values.*

Democratization

A second societal characteristic likely to relate to value consensus is democratization. In order to explain this relationship, we briefly describe aspects of democratic and totalitarian ideology that have implications for value consensus.

In democratic ideology the unique individual person is the basic unit of society. Democratic ideology emphasizes the natural rights, freedom, morality, rationality and equality of the individual (Christensen, Engel, Jacobs, Rejai & Waltzer, 1971; Graham, 1986; Groth, 1974). Democratic ideology has clear implications for the content of values likely to be inculcated. Equally important, however, is its opposition to coercion or manipulation of citizens to adopt particular values. Individuals should be encouraged to develop and express their own value priorities, and given opportunities to do so. Thus, to the extent that a democratic ideology prevails in a country, people are exposed to a wider range of viewpoints and lifestyles and given more freedom to choose among them. This is likely to foster societal variability in the value priorities of individuals.

In contrast, totalitarian regimes seek to control all realms of action and thought in society. Totalitarian ideology legitimizes a monopoly not only of power and authority, but also of wisdom and virtue. It defines groups that espouse opposing ideologies as inherently "evil", justifying their suppression by all means available (Barbu, 1956; LeFort, 1986; Christensen et al, 1971). To the extent that a totalitarian ideology prevails in a country, contact with the major institutions entails exposure to a monolithic value system. Individuals are indoctrinated to accept the values of the regime and to reject alternative value preferences. Of course, totalitarianism my sometimes elicit only public compliance to the values of the regime, with citizens hiding their divergent views. In either case, totalitarian states would be characterized by a high degree of reported value consensus.

Most nations are neither completely democratic nor completely totalitarian. We therefore phrase our hypothesis in terms of the closeness of the political systems of nations to the democratic versus totalitarian pole of the political continuum: *the greater the degree of democratization of the political system in a nation, the lower the overall levell of consensus on values.*

Note that the hypothesized effects on value consensus of democratization and of modernization are opposed. Value consensus is expected to decrease with democratization but to increase with modernization. Yet democratization and modernization tend to vary together (Portes, 1973; Yang, 1988). Most socioeconomically developed countries are also democratic. Hence, modernization and democratization are likely to suppress one another's effects on value consensus. For example, high modernization may expose all citizens to institutions that encourage autonomy and repudiate traditional values, thereby increasing consensus on these values. To the extent this society is democratic, however, subgroups that espouse traditionalism and reject autonomy will be free to socialize their members, thereby decreasing consensus. In order to discern the independent effects of modernization and of democratization on value consensus, we controlled each of these variables statistically while examining the other.

Method

Samples

Data were gathered from 36 samples of school teachers in 36 countries between 1988 and 1993. The countries and sample sizes are listed in the first two columns of Table 1. While no single occupational group can represent a culture, grade school teachers have a number of advantages: They play an explicit role in value socialization, they are presumably key carriers of culture, and they probably reflect the mid-range of prevailing value preferences in societies. Using teacher samples also controls fairly well for economic status and education, variables found to be related both to individual modernity (Inkeles, 1975) and to value priorities (Kohn & Schooler, 1983).

Because teachers are a more homogeneous group than the society at large, this study probably overestimates the degree of within-society value consensus. We assume, however, that the *order* of countries on value consensus, using matched teacher samples, is reasonably similar to the *order* one would obtain using other types of samples to represent these countries.

Measurement of Variables

Value Consensus

Values were measured using a survey instrument developed by Schwartz (1992); respondents were asked to rate each of 56 single values "as a guiding principle in my life" on a 9-pt scale ranging from 7 (of supreme importance) to -1 (opposed to my values). In order to measure value consensus in each national sample, we computed the variance in the ratings of importance that the individuals in the sample attributed to values. We computed the variances in the importance attributed to each of the ten value types separately. We then constructed an index of overall value consensus by averaging the variances for the ten value types.

Column three of Table 1 presents the index of overall value consensus for the sample from each country. The greater the variance, the less the consensus. Thus, East Germany has the highest level of overall value consensus relative to the other countries (.87) and Venezuela the lowest level of overall value consensus (1.50).

Table 1. Characteristics of the Samples Studied

Country	N	Overall Value Consensus*
East Germany	202	.87
Czech Republic	200	1.07
Taiwan	202	1.08
Singapore	149	1.09
Hong Kong	201	1.12
Israel	213	1.12
Slovenia	199	1.13
W. Germany	189	1.13
Australia	138	1.13
Canada	115	1.16
Japan	229	1.18
Finland	204	1.18
China	211	1.23
U.S.A.	261	1.24
Netherlands	187	1.24
Estonia	189	1.24
Hungary	141	1.25
Zimbabwe	185	1.25
Denmark	170	1.25
New Zealand	199	1.27
Malaysia	151	1.30
Sweden	211	1.30
Slovakia	189	1.32
Turkey	183	1.32
Poland	195	1.33
Switzerland	89	1.33
Greece	234	1.33
Spain	186	1.36
Italy	200	1.36
Brazil	160	1.36
Georgia	200	1.38
Portugal	192	1.43
Bulgaria	196	1.46
France	159	1.46
Bolivia	110	1.48
Venezuela	117	1.50

* Note: Smaller numbers signify *greater* consensus because the consensus index is the within-sample average variance rating of the importance of values.

Modernization

Modernization was measured by standardizing each of the following four indicators of socio-economic development, and then averaging them to form a single index: (a) Gross national product per capita in U.S. dollars, 1986; (b) Percent of economically active population not engaged in agriculture,

1985-1988; (c) Enrollment ratio in secondary education, 1986; (d) Number of telephones per 1000 population, 1987.

Democratization

To measure democratization we drew upon Gastil's (1987) comparative measures of freedom consisting of the political rights and civil liberties granted to citizens in a country. Political rights include the rights of citizens to participate in the determination of the nature of the law and its administration. Civil liberties include those freedoms that make possible the mobilization of new, alternative or non-official opinions, as well as religious freedom and freedom of residence. These two aspects of freedom in the political system were highly related ($r = .94$). Hence, we combined them to from a single, more reliable index that ranges from 14 (most democratic) to 2 (least democratic).

Results

As can be seen from Table 1, there are substantial differences between the samples in their degree of value consensus. We first examined whether these differences were due to differences between the samples in sex, age, education, or ethnic and religious heterogeneity. That is, we considered whether demographic heterogeneity in the *sample* we studied, rather than characteristics of the society, might have led to a lack of value consensus and accounted for the observed national differences in consensus. Statistical analyses indicated that greater *sample* heterogeneity on these variables did not produce less consensus. As expected, modernization and democratization were highly intercorrelated ($r = .60$). The correlations of these two variables with value consensus are presented in Table 2. The correlations are shown both with and without controlling for the effects of the other variable. Because variance reflects *dis*sensus, we reverse the signs of the correlations when reporting associations with consensus.

Table 2. Correlations of Modernization and Democratization with Overall Value Consensus (N = 36)

	MODERNIZATION		DEMOCRATIZATION	
	Zero Order	Controlled for Democratization	Zero Order	Controlled for Modernization
Overall value concensus	.28*	.49**	-.18	-.45**

* ($p < .10$), 1-tailed   ** ($p < .05$), 1-tailed.

As shown in Table 2, modernization correlated positively with overall value consensus ($r = .28$). Controlling for the effect of democratization substantially increased this correlation ($r = .49$). Democratization correlated slightly negatively with overall value consensus ($r = -.18$), and controlling modernization yielded a substantial negative correlation ($r = -.45$).

These results support the hypotheses that value consensus increases with socioeconomic development but decreases with political democratization. Because modernization and democratization tend to vary together, the effect of modernization on value consensus is partly masked by the opposing effect of democratization, and vice versa. The conflicting influences of modernization and democratization on value consensus imply that consensus is greatest in modern, non-democratic countries and dissensus is greatest in non-modern democratic countries.

In order to examine whether this is the case, we split the countries studied into four groups: (1) high on both modernization and democratization; (2) high on democratization and low on modernization; (3) low on democratization and high on modernization; (4) low on both modernization and democratization. Table 3 lists the countries in each of the four groups. The unequal distribution of countries into the four cells reflects the high correlation between modernization and democratization.

Table 3. Overall Value Consensus in Countries High and Low in Modernization and Democratization

	(1)	(2)	(3)	(4)
Modernization	HIGH	LOW	HIGH	LOW
Democratization	HIGH	HIGH	LOW	LOW
	(N = 15)	(N = 5)	(N = 3)	(N = 13)
	Australia	Bolivia	E. Germany	Bulgaria
	Canada	Brazil	Hong Kong	China
	Denmark	Greece	Singapore	Czech Rep.
	Finland	Portugal		Estonia
	France	Venezuela		Georgia
	W. Germany			Hungary
	Israel			Malaysia
	Italy			Poland
	Japan			Slovakia
	Netherlands			Slovenia
	New Zealand			Taiwan
	Spain			Turkey
	Sweden			Zimbabwe
	Switzerland			
	USA			
Mean Consensus*	x = 1.25	x = 1.42	x = 1.03	x = 1.26
Range of Consensus	1.12 - 1.46	1.33 - 1.50	.87 - 1.12	1.07 - 1.48

*Higher scores indicate lower consensus

As expected, more modernized countries with more totalitarian regimes (cell 3) are characterized by the highest degree of value consensus - that is, the variance in values is smallest on average for this set of countries (1.03). The lowest degree of consensus characterizes less modern but more democratic societies (cell 2 -1.42). The other combinations fall in between. An analysis of variance reveals these differences to be significant ($F(3, 32) = 8.78, p < .01$). Nontheless, there was a fair range of consensus scores in each cell. Thus, democratization and modernization are insufficient to explain *all* variation in consensus. Equally important, because most modernized societies are also democratic, neither of these variables taken alone is an accurate predictor of societal value consensus.

Discussion

This study was a first attempt to examine societal value consensus empirically. We have provided a conceptual definition of value consensus and a method for measuring it empirically. We have shown that two societal characteristics - modernization and democratization - relate to the degree of value consensus in a society. Moreover, they affect value consensus in opposite directions.

The subject of value consensus, however, remains largely unexplored. There are doubtless additional societal characteristics that relate to cross-national differences in value-consensus. In a study in progress, for example, we are examining how ethnic and religious heterogeneity influence value consensus. Other possible sources of influence on value consensus include the relations between a country and its neighbors, the degree of conflict in a country, and the dominant religion in a country.

Furthermore, in this study we treated value consensus as a global concept, that reflects the degree of agreement on all types of values. But consensus on some value types may be independent of consensus on others, given that different motivational goals underlie the different value types. Societal characteristics relate differently to the *importance* attributed to different value types (Schwartz, 1993). It is therefore possible that they also relate differently to *consensus* on the different value types. Initial studies of this possibility show, for example, that modernization relates most strongly to consensus on values that are associated with the syndrome of individual modernity, that is, to values that express openess to change. This is another avenue that merits further research.

References

Barbu, Z. (1956). *Democracy and dictatorship*. London: Routledge and Kegan Paul.
Christensen, R.M., Engel, A., S., Jacobs D. N., Rejai M., & Waltzer, H. (1971). *Ideologies and modern politics*. London: Nelson.
Cohen, P.S. (1968). *Modern social theory*. London: Heinemann.
Durkheim, E. (1947). *The division of labor in society*. Glencoe IL: Free Press.
Eisenstadt, S.N. (1973). *Tradition, change and modernity*. New York: John Wiley.

Gastil, R.D. (1987). *Freedom in the world*. Westport: Greenwood.
Graham, K. (1986). *The battle of democracy: Conflict, consensus and the individual*. Brighton: Wheatsheaf.
Groth, A. J. (1974). *Major ideologies*. New York: Wiley.
Horowitz, I.L. (1962). Consensus, conflict and cooperation: A sociological inventory. *Social Forces 41*, 177-188.
Inkeles, A. (1975). The emerging social structure of the world. *World Politics 27*, 467-495.
Inkeles, A., & Smith, D.H. (1974). *Becoming modern*. Cambridge, MA: Harvard University.
Kahl, J.A. (1968). *The measurement of modernism: A study of values in Brazil and Mexico*. Austin, TX: University of Texas.
Kluckhohn, C. (1951). Values and value-orientations. In T. Parsons & E.A. Shils, (Eds), *Toward a general theory of action* (pp. 388-433). New York: Harper and Row.
Kohn, M., & Schooler, C. (1983). *Work and personality*. Norwood, NJ: Ablex.
LeFort, C. (1986). *The political forms of modern society*. Cambridge, MA: The MIT Press.
Levy, M. J. Jr. (1966). *Modernization and the structure of societies*. Princeton, NJ: Princeton University.
Meyer, J. W., Boli-Bennett, J., & Chase-Dunn, C. (1975). Convergence and divergence in development. In A. Inkeles, J. Coleman, & N. Smelser(Eds.). *Annual Review of Sociology*, (pp. 223-246). Palo Alto, CA: Annual Reviews.
Parsons, T. (1968). *The structure of social action*. New York: Free Press.
Partridge, P.H. (1971). *Consent and consensus*. New York: Praeger.
Portes, A. (1973). Modernity and development: A critique. *Studies in International Development, 8*, 247-279.
Rokeach, M. (1973). *The nature of human values*. New York: Free Press.
Schwartz, S.H. (1992). Universals in the content and structure of values: Theoretical advances and empirical tests in 20 countries. In M. Zanna (Ed), *Advances in Experimental Social psychology* (Vol. 25, pp. 1-65). Orlando, FL: Academic Press.
Schwartz, S.H. (1993). *Toward explanations of national differences in value priorities*. Address delivered at the XXIV Congress of the Interamerican Society of Psychology, Santiago Chile.
Schwartz, S.H., & Bilsky, W. (1987). Toward a universal psychological structure of human values. *Journal of Personality and Social Psychology, 53*, 550-562.
Schwartz, S.H., & Bilsky W. (1990). Toward a theory of the universal content and structure of values: Extensions and cross-cultural replications. *Journal of Personality and Social Psychology, 58*, 878-891.
Shils, E. (1975). *Center and periphery: Essays in macrosociology*. Chicago, IL: University of Chicago.
Yang, K.S. (1988). Will societal modernization eventually eliminate cross-cultural psychological differences? In M.H. Bond (Ed), *The cross-cultural challenge to social psychology* (pp. 67-85). London: Sage.

Notes

1. The research reported here was supported by grant No. 187/92 from the Basic Research Foundation (Israel Academy of Sciences and Humanities), by grant No. 88-00085 from the United States-Israel Binational Science Foundation (BSF), Jerusalem, Israel, and facilitated by the Leon and Clara Sznajderman Chair of Psychology.
2. The contributions of the following persons in gathering data are gratefully acknowledged: Krassimira Baytchinska (Bulgaria), Gabriel Bianchi and Viera Rozova (Slovakia), Klaus Boehnke (East Germany), Michael Bond (Hong Kong), Bartolo Campos and Isabel Menezes (Portugal), Agnes Chang Shook Cheong (Singapore), Aake Daun (Sweden), Karen and Kenneth Dion (Canada), J.-P. Dupont and F. Gendre (Switzerland), Andrew Ellerman (Australia), Maggye Foster (Bolivia), James Georgas (Greece), Suzanne Grunert (Denmark), Sipke Huismans (Netherlands), Saburo Iwawaki (Japan), Maria Jarymowicz (Poland), Cidgem Kagitcibasi (Turkey), Leo Montada (West Germany), Kathleen Myambo and Patrick Chiroro (Zimbabwe), George Niharadze (Republic of Georgia), Toomas Niit (Estonia), Henri Paicheler and Genevieve Vinsonneau (France), Darja Piciga (Slovenia), Maria Ros and Hector Grad (Spain), Jose Miguel Salazar and Sharon Reimel de Carrasquel (Venezuela), Jan Srnec (Czech Republic), Alvaro Tamayo (Brazil), Giancarlo Tanucci (Italy), Harry Triandis (United States), Shripati Upadhyaya (Malaysia), Antti Uutela (Finland), Zsuzsa Vajda (Hungary), Colleen Ward (New Zealand), Louis Young (Taiwan), Zhi-Gang Wei and Pel-Guan Wu (China).
3. Sources for the indicators were, respectively: (a) World Bank (1990) *World Tables: 1989-1990 Edition*. (b) United Nations (1990) *1987 Statistical Yearbook*; Britannica (1989) *Book of the Year*; Labour Office (1990) *1989-90 Yearbook of Labour Statistics*. (c) UNESCO (1989) *Statistical Yearbook*. (d) United Nations (1990) *1987 Statistical Yearbook*; Britannica 1987 Statistical Yearbook; (1989) *Book of the Year*. Data for certain Eastern European countries that have only recently gained independence (Czech Republic, Estonia, Georgia, Slovenia, Slovakia) were gathered from official statistical sources by our associates in those countries.

Worries, Values, and Well-Being: A Comparison of East and West German, Nepalese, and Fijian Undergraduates

Klaus Boehnke, University of Technology Chemnitz-Zwickau, Chemnitz, Germany
Murari Prasad Regmi, Tribhuvan University, Katmandu, Nepal
Bert O. Richmond, University of Georgia, Athens, Georgia, U.S.A.
Subhas Chandra, Australian National University, Canberra, Australia
Claudia Stromberg, University of Technology Chemnitz-Zwickau, Chemnitz, Germany

The present paper has its starting point in puzzling findings of what is nowadays called *macrosocial stress research*. Since there was a widespread fear of nuclear destruction in the 1980s, the question has attracted the interest of psychologists, how such fears, or, as it was later termed, such feelings of macrosocial stress may affect mental health or, in a broader conceptualization, the psychosocial well-being of youth.

Contrary to early expectations by scientists from the medical domain (see, e.g., Beardslee & Mack, 1982) such research provided little evidence that feelings of macrosocial stress were in any simple way related to negative mental health (for an overview of pertinent studies, see Solantaus, 1991). There even were such counterintuitive results that high feelings of macrosocial stress were related to positive mental health (Meyer-Probst, Engel, & Teichmann, 1989; Boehnke, Macpherson, Meador, & Petri, 1989). The other puzzle of research in the macrosocial stress domain was that feelings of such type of stress (as expressed in response to pre-formulated items or in response to open questions) were not at all related to an "objective" stress potential. Macrosocial worries were as much expressed in Central Europe as they were in Australia and New Zealand, two regions of the world which in the 80's were most certainly affected in a differential manner by the military and non-military nuclear potential deployed in the respective areas. (Oliver, 1990; Boehnke & Macpherson, 1992). Today the same is true for macrosocial stress experienced vis á vis global environmental problems. There indeed are cross-cultural differences (as there had been for nuclear anxiety), but seemingly these differences do not covary with objective measures of environmental destruction. Boehnke, Frindte, Hamilton, Melnikov, Reddy, Singhal, Solantaus, & Unterbruner (1994) showed that worries about environmental destruction run low among US, Russian, and Indian undergraduates, whereas they are high among East and West German, Finnish, and Austrian students. From certain pollution indicators (Encyclopædia Britannica, 1995) one would, however, expect Russia, India, and East Germany in the high stress group and the other countries in the low stress group. This unpredicted finding brought into play the idea that feelings of macrosocial stress may not primarily be a reaction to strong macrosocial stressors, but that a moderator/mediator exists.

One suggestion is that values could be the moderator/mediator macrosocial stress research is looking for. It was shown that feelings of macrosocial stress were positively related to certain value priorities in Western societies (Boehnke, 1994). People who hold universalist value orientations usually tend to experience more macrosocial stress than do people who hold other value priorities. Universalist value orientations are described by Schwartz (1992, p. 12) as an orientation concerned with the "understanding, appreciation, tolerance and the protection for the welfare of *all* people and for nature". Up to now, the hypothesis that value orientations are strongly related to the experience of macrosocial stress could not be tested outside of Western cultures. A first attempt to do this, is undertaken in the present study. In seeking a cross-cultural test of the assumption that values and feelings of stress are related, it has to be taken into consideration that values are not only guiding principles an individual holds, but are at the same time something that is specific to a certain culture. Values are sometimes seen as a definitive descriptor of culture (Kluckhohn, 1951).

A comprehensive model of the relationship of "objective" macrosocial stressors, values, feelings of macrosocial stress and psychosocial well-being was formulated by Boehnke and Frindte (1994) in the way depicted in Figure 1. Boehnke and Frindte assume that "objective" macrosocial stressors, as, for example, global or local environmental problems, are perceived - so to speak - through the glasses of the values of a society (macrosystem), the values of societal subgroups (exosystems), the values of the interaction groups to which an individual belongs (meso- and microsystems), and the values of a particular individual. This means that it is assumed that values moderate the experience of macrosocial stress: Only individuals who hold certain value orientations will experience high feelings of macrosocial stress. Furthermore it is assumed that values also function as mediators in the relationship of macrosocial stress and psychosocial well-being. Due to the quality of macrosocial stressors as potentially chronic sources of strain, there are no immediate ways an *individual* can cope with macrosocial stress(ors). Therefore, the individual who experiences macrosocial stress will evaluate if his or her coping attempts are in accordance with his or her value orientations. Value orientations of hierarchically higher systems will most certainly also play an important role. For individuals who evaluate their coping attempts positively vis á vis their own values and the values of their social reference groups, a positive relation between feelings of macrosocial stress and psychosocial well-being should be found. Macrosocial stress gives these individuals a chance to engage in coping acts that are in accordance with their value priorities; for them macrosocial stress may even function as *eu*stress (Example: Greenpeace activists). For individuals who evaluate their coping attempts negatively vis á vis their own values and values of their social reference groups, a negative relation between feelings of macrosocial stress and psychosocial well-being should be found; for them macrosocial stress may function as *di*stress.

Figure 1. The Relationship of Macrosocial Stressors and Psychosocial Well-Being

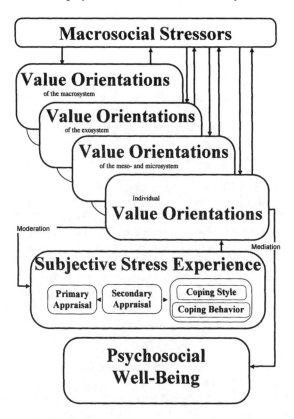

Of course, this model cannot be tested in a single study. The present research aims at shedding some light on the interrelation of values, feelings of personal/microsocial and macrosocial stress and psychosocial well-being in East and West Germany in contrast to two quite different cultures, namely the Asian countries of Nepal and Fiji.

Hypotheses are fairly global and it is too early to speak of a test of certain assumptions. First of all - without making this a formal hypothesis - we expect to find substantial differences in value priorities in the four samples/cultures included in the present study. Our formal hypotheses are that (1) feelings of macrosocial stress and feelings of personal/microsocial stress are related to different value priorities, and that (2) feelings of macrosocial stress do not covary negatively with mental health. A positive relationship would not come as a surprise. Such a relationship could be based on the experience that macrosocial stress and coping acts related to it are something participants see as being in accordance with their and their reference group's value orientations.

Method

Samples

Four samples were included in the present study: (1) A sample of undergraduate students of the teaching professions (i.e., students who intend to become teachers of various subjects in various school types) from Humboldt-University in East Berlin, formerly the capital of the German Democratic Republic[1]. Data from that sample were gathered in the early summer of 1991. (2) A sample of undergraduate education majors from the Free University in West Berlin; data gathered in summer 1992. (3) A sample of undergraduate students, predominantly psychology majors (but also business students), from Katmandu, Nepal; data gathered by the second author in mid 1992. (4) A sample of undergraduates from a Teacher's College affiliated to the University of the South Pacific in Suva, Fiji; data gathered by the third and the fourth author in late 1990. None of the samples were drawn at random from the population of undergraduates, but as all samples were taken from required courses in the respective departments, none of the samples were biased in an a priori manner. Sample sizes were: $N = 274$ for the East Germans, $N = 143$ for the West Germans, $N = 530$ for the Nepalese, and $N = 122$ for the Fijian sample. The average age at the time of surveying for the East German sample was 21.5 years ($sd = 3.0$), for the West German sample, 26.2 (6.4), and 25.9 (4.9) and 22.6 (5.9) for the Nepalese and Fijian samples respectively. The East German and the Fijian samples spent an average of 13.5/13.6 years in schools and universities, the West German and the Nepalese samples 14.8/14.9 years. This means that in spite of different average ages, all samples had approximately the same level of education. In the total sample there are equal proportions of men (49%) and women (51%) in the four subsamples, however, the proportions of men and women differ considerably: East Germany 62% women, 38% men; West Germany 75% women, 25% men; Nepal 36% women, 64% men; and Fiji 65% women and 35% men. These proportions are in accordance with the typical gender ratio among undergraduates from the respective departments in the countries included.

Instruments

The Schwartz Value Survey (Schwartz, 1992) was used to explore value orientations, the Goldenring-Doctor Scale of Existential Worries (Goldenring & Doctor, 1986) was used to gain information on feelings of micro- and macrosocial stress, whereas different scales were used to measure mental health. In the West German and the Nepalese samples the mental health subscale from Becker's (1989) Trier Personality Inventory was used. In the Fijian sample the Revised Children's Manifest Anxiety Scale (Reynolds & Richmond, 1978) was used for measuring psychosocial well-being. For the East German sample a very rudimentary measure of well-being was used: Subjects were asked, "Have the changes in Germany affected 'your chances to be successful' and 'your sense of control over your own life' positively to negatively". The items were be evaluated on a 5-point rating scale.

The Schwartz Value Survey is comprised of 56 value items, for each of which participants are asked to express the extent to which it is a "guiding principle" in their lives on a scale that ranges from -1 (opposed to my values) to 7 (of extreme importance). The list contains 30 terminal values, (end states of existence-nouns, Rokeach, 1973) and 26 instrumental values (adjectives, participles). The instrument measures 10 facets of value orientations (motivational types in Schwartz's terminology), universalism, benevolence, tradition, conformity, security, power, achievement, hedonism, stimulation, and self-direction. Schwartz assumes these ten facets to be related to each other in a circumplex manner with two underlying dimensions, namely the dimensions of self-transcendence (universalism, benevolence) vs. self-enhancement (power, achievement, hedonism) and of openness to change (self-direction, stimulation) vs. conservation of the status quo (tradition, conformity, security).

Cross-culturally valid associations of the 56 items with the ten facets have been developed by Schwartz (1992). The validity of the associations for the samples included in the present study was tested by calculating reliability analyses separately for all samples for all facets. Single items were excluded from further analyses if they did not reach a minimum item-total correlation of .10 for their a priori scales in one or more of the four samples. Two of the 56 items had to be excluded. Both stem from the Tradition value facet, the value *detachment* did not "work" in East Germany, the value *moderate* did not work either in Nepal or in Fiji. The *alpha* consistency coefficients for the ten value scales in the four samples were computed. Scale scores are average scores of the items belonging to the respective facets. For cross-cultural comparisons the difference between the sample grand mean for all 56 items and the answering scale's mathematical expected score, i.e., 3, was subtracted from the ten scale scores per country. If for example the grand mean in Nepal was 4.55, 4.55 - 3.00 = 1.55 was subtracted from each of the ten scale scores in that country.

Given the shortness of several of the subscales, the scale consistencies are acceptable, although some of them are low from a test theoretical point of view. It may be of interest to methodologists that consistency coefficients are significantly correlated with means ($r = .50$, $p < .01$, see below); if a value type is preferred in a sample, the measurement accuracy also tends to be higher.

The Goldenring-Doctor scale measures two types of worries, what has lately been called "personal-and-microsocial-stress" related and "macrosocial-stress" related worries. The entire scale is comprised of 20 items: five macrosocial items, e.g., "I am worried about pollution", and 15 personal and microsocial items, e.g., "I am worried about getting cancer". Scales were constructed as described above, i.e., a .10 item-total correlation of items with their respective scales was required. No item had to be excluded on the grounds of this rule. Consistency coefficients for the personal and microsocial worries scale ranged from .79 in both German samples to .86 in the Fijian sample. For the macrosocial worries scale, consistency coefficients varied from .60 in the West German to .69 in the Nepalese sample.

For the two mental health scales, consistencies were .64 (Nepal) and .81 (West Germany) for Becker's mental health scale. One item, "I have the feeling that everything is too much for me", had to be eliminated, because it did not suffice the .10 item-total correlation criterion in Nepal[2]. Kuder-Richardson's *alpha* for the dichotomous 28-item manifest anxiety scale used in Fiji was .82; no item had to be eliminated, showing that a manifest anxiety scale originally constructed for children was sufficiently valid for young adults as well. The two items used as a well-being index in East Germany were marginally correlated ($r = .11$). As this low correlation may stand for a questionable validity of the index, results pertaining to it should only be taken as very preliminary.

Neither the worry scores nor the mental health scores were adjusted for item grand mean differences. This has more or less "historical" reasons; all scales have been used in cross-cultural comparisons, and the scale authors have never proposed a mean adjustment.

Results

In a preliminary step, univariate analyses of variance were conducted for all dependent variables. Independent variables were gender, culture (European and Asian), and sample (East and West German, Nepalese, and Fijian) nested under culture. Detailed findings are omitted here. Their essence is that for gender significant differences ($p \leq .01$) were found for seven of the ten value types, substantial differences were, however, found for Benevolence only ($\eta^2 = 9.2\%$): Preference of benevolence values was much higher among women than it was among men in European as well as Asian cultures. Mean differences between European and Asian cultures - regardless of gender - were more substantial. Differences were nonsignificant only for achievement values. More than five percent of the variance in value preferences were accounted for by culture for Benevolence, Hedonism, Stimulation and Self-Direction (higher preference in Europe), Tradition, Conformity and, Power (higher preference in Asia). For Universalism (higher in Europe) and Security (higher in Asia) effects were smaller, but still significant at the 1% level. Sample differences, that is, differences between the two European samples and between the two Asian samples, were significant at the 1% level for eight out of ten value orientation (all except Achievement and Stimulation), but only in the case of Tradition was more than 5% of the variance accounted for by sample differences. The Fijian sample was more tradition oriented than the Nepalese sample, while the West German sample was a bit less tradition minded than the East German sample.

With regard to Worries, substantial gender differences were found for Personal and Microsocial Worries. They were higher for women than they were for men in both cultures ($\eta^2 = 4.0\%$). The same was true for Macrosocial Worries, but to a lesser extent ($\eta^2 = 1.1\%$). Women and men did not differ in mental health scores. Culture and sample differences were of even lesser importance than gender differences. The Personal and Microsocial Worries scores were higher in the Asian than in the European cultures ($\eta^2 = 2.3\%$). Scores in West Germany were particularly low, scores in East Germany and Nepal were in the middle, and Fijian scores were particularly high. For Macrosocial Worries neither culture nor sample differences were found. For Mental Health scores no differences between West Germany and Nepal were found[3].

Table 1. Partial Correlation Coefficients for Value Facets and Personal / Microsocial and Macrosocial Worries

Scale	East Germany		West Germany		Nepal		Fiji	
	Micro[a]	Macro[b]	Micro	Macro	Micro	Macro	Micro	Macro
Universalism	-.17	.64	-.17	.38	.11	.17	.03	.22
Benevolence	-.02	.38	.18	.17	.17	.06	-.25	.28
Tradition	.06	-.10	-.04	.12	-.01	.02	.09	.36
Conformity	.16	-.01	.06	.05	.18	.05	-.06	.28
Security	.11	.28	.22	-.00	.23	-.06	.12	.21
Power	.13	-.23	-.01	.09	.20	-.10	.18	-.07
Achievement	.06	.32	.16	-.01	.09	.05	.03	.22
Hedonism	.03	-.02	.08	-.19	.18	-.05	.40	.23
Stimulation	-.15	.36	.03	.11	.11	-.01	.09	.10
Self-Direction	-.14	.47	-.09	.04	.07	.06	.26	.09

[a] Micro = Personal and Microsocial Worries
[b] Macro = Macrosocial Worries

In the remainder of the results section the focus changes from descriptive information to relational information. Results on the interrelation of values and worries and of worries and well-being are presented. To explore the interrelation of values and worries, partial correlation analyses were calculated. The ten value types were correlated with one type of worries, the other type of worries was partialed out. This approach was taken to ensure that the two *intercorrelated* worry-types are related to values only with regard to that proportion of variance that is *not* connected with the other type of worries. Table 1 gives an overview of the partial correlation coefficients for values in relation to personal and microsocial worries as well as for values in relation to macrosocial worries.

The table shows that in all four samples, Security, Achievement, and Hedonism values were positively related to Personal and Microsocial Worries (with Security having the highest coefficient on average). All other value types were, at first glance, related differently to personal and microsocial worries in the four samples. The table also shows that in all four samples Universalism, Benevolence, and Self-Direction were positively related to Macrosocial Worries (with coefficient size in the above order).

These results show that there is a certain, but not overly impressive interrelation between individual values and worries. The adjusted R^2s from multiple regression analyses varied from 3% (East Berlin) to 27% (Fiji) for values predicting personal and microsocial worries, with 8% common variance in the total sample. For macrosocial worries the respective scores varied between 7% (Nepal) and 41% (East Berlin), with 16% in the total sample. The picture of a "mediocre" relationship improves, however, if one does not relate individual values per se to worries, but relates culture-specific standard z-scores and worries to each other. In that case it is appropriate to use the total sample as a data base. Table 2 reports partial correlation coefficients for the two worry types and the ten (z-standardized) value facets for the grand sample.

Table 2. Partial Correlation Coefficients for Intra-Culturally Standardized Value Priorities and Worries in the Grand Sample

Worry-Type	Universalism	Benevolence	Tradition	Conformity	Security	Power	Achievement	Hedonism	Stimulation	Self-Direction
Microsocial	-.01	.09	.01	.15	.18	.15	.08	.14	.03	-.00
Macrosocial	.34	.18	.02	.05	.06	-.11	.13	-.04	.12	.18

The table shows that value preferences were more closely related to macrosocial than to microsocial worries. For microsocial worries it shows that the three highest positive coefficients were found for Conformity, Security, and Power values with the only two negative (though certainly insignificant) coefficients found for Universalism and Self-Direction. For macrosocial worries the three highest positive correlations were found for Universalism, Self-Direction, and Benevolence with the only negative coefficients found for Power and Hedonism.

The improvement in clarity of results (comparing Tables 1 and 2) suggests that culture-specific means play a certain role when the relationship of values and worries is under scrutiny. In a final exploratory analysis we therefore probed into the question whether the average absolute discrepancy of an individual's value orientations from the means of his or her culture's value orientations predicts worries. In order to do this, we transformed the standard z-scores into absolute scores and calculated an individual mean of these absolute scores over all ten value facets. The partial correlation with microsocial worries was insignificant ($r = .02, p = .484$). The partial correlation with macrosocial worries was, however, significant ($r = -.12, p < .001$). If an individual's value preferences differed from the modal values of his or her cultural reference group, he or she was not very prone to express macrosocial worries.

The question remains, how worries (and values, respectively) are related to psychosocial well-being. For the West Berlin and the Nepal sample, mental health scores served as the criterion to be predicted in multiple regression by the two worry-types. In Fiji low manifest anxiety scores served as the criterion. In East Germany high index scores did. Results again were homogeneous across cultures. In all four samples the general relationship was comparable. Personal and microsocial worries were negatively related to mental health and to low manifest anxiety. High personal and microsocial worries covaried with bad mental health scores and high manifest anxiety ($\beta = -.165, p = .011$ in East Germany; $\beta = -.419, p < .001$ in West Germany; $\beta = -.217, p < .001$ in Nepal; and $\beta = -.142, p = .365$ in Fiji. If one combined the four probabilities according to Fisher's (1932) formula, the common χ^2-score was highly significant, $p < .001$. For macrosocial worries the opposite was true: Strong macrosocial worries covaried with good mental health scores and a low manifest anxiety ($\beta = .179, p = .006$ in East Germany; $\beta = .138, p = .103$ in West Germany; $\beta = .127, p = .009$ in Nepal; and $\beta = .329, p = .040$ in Fiji). The combined χ^2-score also was significant, $p < .01$). The amount of variance of the well-being scores explained by the two worry-types varied from 3.1% in East Germany to 14.1% (adjusted R^2s) in West Germany. This amount was considerably higher than the amount of common variance of values and mental health scores. The latter varied between less than 1% in Fiji and 7.6% in East Germany. In Nepal, the largest sample, we found one significant correlation coefficient, high achievement value preferences covary positively with the mental health scores ($r = .16, p < .001$) The same is the case in East Germany, the second largest sample. There, the preference of achievement values also covaried positively with the well-being index ($r = .20$,

$p = .007$). No other single correlation coefficient was significant in any of the samples.

Discussion

The current study dealt with the interrelation of values, worries, and well-being in a cross-cultural comparison. It tried to probe into a conceptual model of the interrelation of values and well-being that may help to integrate puzzling and partially contradictory earlier findings. Preliminary analyses of variance showed that mean differences in value preferences between European and Asian cultures are quite substantial for all except achievement values. This difference is most pronounced with regard to variation on the change vs. status quo dimension. European value preferences, in general, are more change-oriented, Asian value preferences more status-quo-oriented. To a lesser extent Asian and European students also differ on the self-transcendence vs. self-enhancement dimension of value priorities. Asians were less prone to put emphasis on self-transcendence values than are Europeans. Substantial differences between samples from a similar cultural background were found only with regard to Tradition; among the two Asian cultures, both strongly influenced by Hinduism, the Fijian sample showed substantially higher preference for tradition values than the Nepalese sample. These descriptive findings make it obvious that indeed value preferences differ strongly between samples from different cultures, as was assumed. Such a constellation makes cross-cultural commonalties even more informative with regard to theory-based hypotheses.

Our first formal hypothesis postulated that feelings of personal/microsocial stress and feelings of macrosocial stress would be related to different value preferences. With regard to this hypothesis results are somewhat more complex than had been expected. Partial correlation analyses show that microsocial worries are best predicted by security, power, and conformity values, although the amount of common variance of values and microsocial worries (8%) is not overwhelming. Macrosocial worries, on the other hand, have much more common variance with value preferences (16%). Their best predictors are universalism, self-direction, and benevolence values. Obviously, however, individual values are not the whole story. It is shown that the average discrepancy of an individual from his or her culture's modal values is negatively related to macrosocial stress. This finding suggests that the relationship of values and worries may not only be a matter of psychological functioning, but has sociological implications as well. If an individual holds value orientations that are in accordance with the values typically preferred by his or her reference group, this constellation increases the probability that the person will experience macrosocial stress. One might speculate that only the sense that one lives in accordance with the typical values of one's culture gives an individual the "psychic energy" to worry about the socio-political domain.

All in all, one can say that Hypothesis 1 was confirmed. Feelings of personal/microsocial and macrosocial stress are more or less closely related to different value preferences, the former to conservation and self-enhancement value preferences, the latter to openness and self-transcendence value preferences. Additionally, the study can serve as a starting point for research into the importance of person-society value discrepancies. The relationship of individual values and value preferences of the reference group had some importance in determining worries of individuals: Congruence of individual and group value priorities gave way to feelings of macrosocial stress.

Hypothesis 2 stated that feelings of macrosocial stress would not be related negatively to mental health. This hypothesis was confirmed in a convincing manner. In all samples macrosocial stress and psychosocial well-being covaried *positively*, although significance was not reached in all the samples. As would be expected on common sense grounds, feelings of personal/ microsocial stress were related to low mental health scores. This relationship was significant in the two German samples and in Nepal. Values showed no substantial direct relationship to well-being scores.

These results give room to a number of speculations. The first is that we now have some evidence that macrosocial stress is influenced by individual *and* reference group value preferences. In spite of the fact that the "objective" strength of macrosocial stressors was not measured in the current study, a look at easily available indicators of environmental life quality (see, Encyclopædia Britannica, 1994) makes it obvious that there is no linear relationship between the quantitative extent of, e.g., environmental problems in a society and the amount of stress these problems induce in individuals. Feelings of macrosocial stress depend much more on an individual's integration into the value preferences of his/her reference group. And, to make a last point, feelings of macrosocial stress should not be feared as having negative consequences for mental health.

Of course, the above paragraph is highly speculative. All data are correlational; causal inferences drawn from the study are tentative. Certainly worries can also influence values (see Boehnke, 1994) and not only vice versa as proposed in the present paper. Furthermore, it has to be pointed out that findings relate only to haphazard samples of college and university undergraduates. But, after all, cultures included in the present study are so distinct and different, that that very fact may be seen as adding validity to the reported findings.

References

Beardslee, W., & Mack, J. (1982). The impact on children and adolescents of nuclear developments. In R. Rogers (Ed.), *Psychological aspects of nuclear developments*. Washington, DC: American Psychiatric Association (Task Force Report # 20).

Becker, P. (1989). *Trierer Persönlichkeitsfragebogen* [Trier Personality Inventory]. Göttingen: Hogrefe.

Boehnke, K. (1994, August). *Macrosocial stress in young adulthood: A cross-cultural research project on the influence of politically-caused threats on psychosocial well-being*. Paper presented at the Third International Symposium on the Contributions of Psychology to Peace, Ashland, VA.

Boehnke, K., & Frindte, W. (1994). *Auswirkungen globaler Umweltveränderungen auf die psychosoziale Befindlichkeit von Jugendlichen: Zur Bedeutung von Werthaltungen* [Consequences of global environmental changes for the psychosocial well-being of adolescents: On the importance of value orientations]. Chemnitz: Proposal to the Deutsche Forschungsgemeinschaft.

Boehnke, K., Frindte, W., Hamilton, S.B., Melnikov, A.V., Reddy, N.Y., Singhal, S., Solantaus, T., & Unterbruner, U. (1994). Makrosoziale Besorgnisse und ethnisch-kulturelle Stereotype im Kulturvergleich. In J. Mansel (Ed.), *Reaktionen Jugendlicher auf gesellschaftliche Bedrohungen* (pp. 80-92). Weinheim: Juventa.

Boehnke, K., & Macpherson, M.J. (1992). Coping with macro-social stress: A comparison of adolescents in West Germany and Australia across time. In S. Iwawaki, Y. Kashima, & K. Leung (Eds.), *Innovations in cross-cultural psychology* (pp. 371-383). Lisse: Swets & Zeitlinger.

Boehnke, K., Macpherson, M.J., Meador, M., & Petri, H. (1989). How West German adolescents experience the nuclear threat. *Political Psychology, 10*, 419-443.

Encyclopædia Britannica (1995). *Book of the Year 1994*. Chicago: Encyclopædia Britannica Inc.

Fisher, R.A. (1932). On a property connecting χ^2 measure of discrepancy with the method of maximum likelihood. *Atti Congresso Internationale Matematica, Bologna, 1928, 6*, 94-100.

Goldenring, J., & Doctor, R. (1986). Teenage worry about nuclear war: North-American and European questionnaire studies. *International Journal of Mental Health, 15*, 72-92.

Kluckhohn, C. (1951). Values and value-orientations in the theory of action: An exploration in definition and classification. In T. Parsons & E. Shils (Eds.), *Toward a general theory of action* (pp. 388-433). Cambridge, MA: Harvard University Press.

Loch, A. (1995, May). *Liebe, Macht und Angst vorm Tod - ethnopsychologische Reflexionen über eine Eigenarten in Nepal* [Love, power, and fear of death--ethnopsychological reflection on some peculiarities of Nepal]. Paper presented in the colloquium Sozialwissenschaftliche Jugendforschung. Chemnitz

Meyer-Probst, B., Engel, H., & Teichmann, H. (1989). Wünsche und Befürchtungen 14jähriger Jugendlicher: Phänomenologie und Abhängigkeitsbeziehungen [Hopes and fears of 14-year-old adolescents: Phenomenology and interdependencies]. *Pro Pace Mundi, 5,* 36-54.

Oliver, P. (1990). Nuclear freedom and students' sense of efficacy about prevention of nuclear war. *American Journal of Orthopsychiatry, 60,* 611-621.

Reynolds, C.R., & Richmond, B.O. (1978). What I think and feel: A revised measure of children's manifest anxiety. *Journal of Abnormal Child Psychology, 6,* 271-280.

Rokeach, M. (1973). *The nature of human values.* New York: The Free Press.

Schwartz, S.H. (1992). Universals in the content and structure of values: Theoretical advances and empirical tests in 20 countries. *Advances in Experimental Social Psychology, 25,* 1-65.

Solantaus, T. (1991). Young people and the threat of nuclear war. *Medicine and War, 7,* Supplement 1.

Notes

1. For their help in facilitating the data gathering at Humboldt-University the authors owe sincere thanks to Professor Harry Dettenborn and Professor Karla Horstmann-Hegel.

2. After intensive interviewing in Katmandu, Loch (1995) found that this seems to be owed to a peculiarity in the understanding of English in Nepal, where the phrase that something is "too much" for a person bears the connotation that he or she does not deserve it. This, of course, is not the connotation intended by the item in the scale.

3. This comparison is the only one that possible, because in East Germany and Fiji different indicators were used.

Personal Values and Acculturation

Remko H. van den Berg & Nico Bleichrodt
Vrije University of Amsterdam, Amsterdam, The Netherlands

Migration and acculturation have always been important subjects for cross-cultural research. In Western-Europe and especially in the Netherlands, migration is also becoming a more important issue politically. More than 50,000 immigrants from different cultural backgrounds enter the Netherlands each year. More than one million people belong to ethnic minority groups, almost 7% percent of the total population. These immigrants live mostly in the suburbs of the four big cities: Amsterdam, Rotterdam, The Hague and Utrecht. It is estimated that around the year 2000 almost 50 percent of the population in Amsterdam will consist of first or second generation immigrants. These immigrants originate from different countries and can be broadly divided into three groups. Firstly, a large group originates from the former Dutch colonies of Indonesia, Surinam and the Netherlands Antilles. Secondly, a substantial group of migrant workers from the Islamic countries Turkey and Morocco, recruited during the sixties to work in the Netherlands. Thirdly, a group of refugees from different countries such as, for example Iran, Iraq and former Yugoslavia.

These ethnic groups and the growing numbers of refugees have resulted in enormously increased intercultural contact. The Netherlands can definitely be called a multi-cultural society. These cultural differences also cause problems. Housing problems, unemployment, criminality and lack of education are all examples of the problems facing ethnic minorities, refugees and society in general. For example, unemployment among ethnic minorities is more then four times higher than among the Dutch. For the immigrants from Morocco the unemployment rate is almost 40% and the average duration of unemployment is around five years. In addition, intolerance and racism in the Netherlands are also becoming more serious problems.

Acculturation

Cross-cultural psychology has two broad domains of interest: first, the search for psychological differences and similarities across broad ranges of cultures and second, the psychological changes that occur when individuals move from one culture to another. This process is also known as *acculturation*, in which we are interested analyzing individual behavior for continuities and changes that are related to the experience of two cultures.

Redfield, Linton and Herskovitz (1936) defined acculturation as phenomena which result when groups of individuals having different cultures come into continuous first-hand contact, with subsequent changes in the original culture patterns of either or both groups. The Social Science Research Council (1954) defined acculturation as culture change that is initiated by the conjunction of two or more autonomous cultural systems. In these definitions the group or population level is the most important. Graves (1967) has introduced the term *psychological acculturation*; the changes that an individual experiences as a result of coming into contact with other cultures and participating in the process of acculturation that one's cultural or ethnic group is undergoing. This distinction between the population level and the individual level is important because the phenomena differ at both levels and because not every acculturating individual participates in the collective changes that are underway in one's group in the same way (Berry, 1990).

At the population level of acculturation we can see physical changes (urbanization), biological changes (new illnesses/diseases), political changes (being under the control of a different cultural group), economic changes (employment), cultural changes (language and religion) and changes in social relationships. At the individual level we can see changes in behaviour, motives, values, identity, and health (i.e., acculturative stress).

Marin (1992) differentiates three levels on which culture learning as part of an acculturation process can occur. The first, and most superficial level, is the learning (and forgetting) of the facts that are part of one's cultural history or tradition. The second level involves the more fundamental behavior of an acculturating individual. Use of language and language preference, ethnicity of friends and neighbors, preference for ethnic media are examples of this intermediate level. The third, and most important, is the level at which changes can take place in terms of values and norms, the constructs that prescribe people's worldviews and interaction patterns. Changes at this level can be expected to be more permanent.

Most instruments for measuring acculturation focus on the first and second level. These levels are more superficial and, therefore, less reliable than the third level. There is some evidence that more central values such as familialism, e.g., sense of obligation and the power of the family as a behavioral referent, in Hispanics change as a function of acculturation (Sabogal, Marin, Otero Sabogal, Marin & Perez Stable, 1987). A problem is that changes in values and norms are not as easy to measure as language or media preference.

In addition, the *emic-etic* dilemma in cross-cultural psychology is also important for the cross-cultural study of values. The *etic* (culture-comparable) constructs are needed for cross-cultural comparisons, but their use may distort the meaning of concepts. An *emic* strategy that identifies culture-specific constructs, provides a more precise and thorough description of constructs within each culture, but makes cross-cultural comparisons more difficult.

When studying acculturation, we are interested in the changes of values, and especially changes due to the influence of the values held by the majority of individuals of the dominant culture. In this study we focus on the influence of the dominant interpersonal values of the host country. We do not investigate the cultural identity or the culture specific values of the migrant as such, nor do we attempt a comprehensive study of values. In this study we are interested in acculturation at the individual level and especially in the differences and changes in values when people are engaged in an acculturation process.

Considerable attention has been given to the cross-cultural study of values in the past decade. The important research conducted by Hofstede (1980, 1982, 1991), in which he identified four independent cultural dimensions has led to more research on this topic. Hofstede identified four dimensions: (1) Power Distance, referring to Hierarchical relationships between boss and subordinate, autocratic decision making versus consultative decision making; (2) Uncertainty Avoidance, which refers to avoiding uncertainties in life, rules should not be broken; (3) Individualism, referring to the relative importance of freedom and challenge, being independent; and (4) Masculinity, defined as the relative importance of recognition, earnings, advancement and challenge, relative unimportance of cooperation and relationship with the manager.

Both on the cultural level and on the individual level, researchers have tried to identify etic value-dimensions (Bond 1988; Feather & Peay, 1975; Hofstede & Bond, 1984; Ng, Akhtar, Ball, Bond, Hayashi, Lim, O'Driscoll, Sinha, and Yang, 1982; Rokeach, 1973; Schwartz, 1992, 1994; Schwartz & Bilsky, 1987). These approaches have used different instruments and have yielded different dimensions for which different interpretations have been given. The factors or dimensions found are sometimes represented by very few items, for example, Hofstede's dimensions Power Distance and Uncertainty Avoidance consist of only three items.

An instrument often used in cross-cultural research is the Rokeach Value Survey (Rokeach, 1967). The RVS consists of 36 items, each measuring a different value, which have to be rank-ordered in two sets of 18 items. Factor analyses across different cultures often result in different factors. Ng, Akhtar, Ball, Bond, Hayashi, Lim, O'Driscoll, Sinha, & Yang, (1982) found two dimensions: (1) Social-Oriented values, which emphasize submission to a hierarchical society, vs. Self-Oriented values (Dionysian) and (2) Self-Oriented values, which emphasize inner strength, vs. Materialistic-Oriented values, both factors explaining 50 % of the variance. Bond (1988) found the following dimensions for the Chinese Value Survey: Social Integration vs. Cultural Inwardness and Reputation vs. Social Morality, in total explaining 13.8%. For the RVS he found: Competence vs. Security, Personal Morality vs. Success, Social Reliability vs. Beauty, Political Harmony vs. Personal Sociability explaining 25.2 % of the variance. Schwartz and Bilsky (1987) found four dimensional domains using Smallest Space Analysis: Self-Direction vs. Restrictive Conformity, Achievement vs. Security, Achievement vs. Prosocial, and Enjoyment vs. Prosocial. They recommend using indices of the importance of value domains rather than using single values, because they are likely to be more reliable and to reflect a more comprehensive meaning. A few years later Schwartz (1994) identified three dimensions (on the cultural level) using data from more then 40 countries: Conservatism vs. Autonomy, Hierarchy vs. Egalitarian Commitment and Mastery vs. Harmony. On the individual level he identified 10 dimensions (value types) on the basis of which individuals can be compared: Power, Achievement, Hedonism, Stimulation, Self direction, Universalism, Benevolence, Tradition, Conformity and Security. In Schwartz's research, the dependence on the SSA technique needs addressing. This technique has its drawbacks and it is important that Schwartz's findings are replicated using, for example, factor analysis.

In general, convergence between the different value dimensions found by different researchers is needed. Therefore replication studies using different instruments and different samples should be conducted and the use of instruments with more reliable scales consisting of more items would be appropriate.

In our research we have chosen to use an reliable existing instrument that is frequently used in the Netherlands for selection and career counseling. We used Gordon's Survey of Interpersonal values (1960) which was adapted and translated for use in the Netherlands.

The factors (each scale consists of 15 items) measured by the SIV are: (1) Support: being treated with kindness and understanding; (2) Conformity: doing what is socially correct, following regulations closely; (3) Recognition; being looked up to and admired; (4) Independence; having the right to do whatever one wants to do; (5) Benevolence: Doing things for other people, sharing with others; (6) Leadership: Being in charge of other people, having authority over others. Unfortunately no research has been reported on correlations between scores on the SIV and Schwartz's dimensions or value types.

Hofstede (1980) compared the results of a study on the SIV and the country mean scores he found. The significant correlations for 17 countries are presented in Table 1.

High scores on Support were correlated with low Power Distance (consultative decision making) and high Individualism (importance of freedom and independence). High scores on Conformity were correlated with high Power Distance (autocratic decision making), low Individualism (freedom is less important) and high Masculinity (importance of earnings and recognition). High scores on recognition were correlated with high scores on Masculinity (importance of earnings and recognition). High scores on Independence were correlated with low Power Distance (consultative decision making), high Individuality (importance of freedom and independence) and low Masculinity (importance of cooperation with manager, low importance of earnings and challenge). High scores on Benevolence were correlated with low Uncertainty Avoidance (not avoiding uncertainties, rules can be broken) and low Masculinity (importance of cooperation with manager, low importance of earnings and challenge). Leadership had no significant correlations with any of the Hofstede dimensions.

Table 1. Significant Correlations between the SIV and the Hofstede Dimensions

SIV	Hofstede dimensions			
	Power distance	Uncertainty avoidance	Individualism	Masculinity
Support	-.70		.68	
Conformity	.80			.44
Recognition			-.76	.50
Independence	-.79		.41	-.54
Benevolence		-.49		-.59
Leadership				

Hypotheses

The first hypothesis is that the values of ethnic minorities/immigrants will resemble the national cultural values of their native countries. The second hypothesis is that individuals from western and non-western cultures will show differences, comparable to the differences Hofstede found. The third hypothesis is that length of residence will have an important influence on personal values. The longer the residence the more personal values will resemble the national cultural values of the dominant culture.

Method

Subjects

The SIV was given to 439 Dutch born and 577 immigrants from different countries. The average age of the subjects was 30 years. The mean length of residence for the immigrants was 8.7 years and ranged from 7 years for the Turkey/Morocco group to 12 years for the Surinam/Antilles group.

All subjects completed the SIV as part of a selection procedure for re-education programmes for unemployed. The re-education programmes were for various professions like computer programmer, computer-network organizer, nurse, security officer, police officer, bank employee and insurance company employee.

Instruments

Gordon's Survey of Interpersonal Values (1960) was used to measure personal values. The SIV was adapted and translated for use in the Netherlands. The instructions and most of the items were rewritten to make them more applicable to candidates from different cultural backgrounds. Typical Dutch sayings and expressions were removed and the language used was simplified to make it more easily understandable for immigrants with less knowledge of the Dutch language. The scale is in ipsative form. Three items are presented and the subject has to choose the most preferred and the least preferred value-statement. Items were paired on comparable social desirability. An example of an item: (a) To be free to do as I choose; (b). To have others agree with me; (c) To do my duty. The ipsative form prevents response-sets and forces subjects to choose between value-statements; a relative disadvantage is that the scores on the subscales are not totally independent.

Procedure

The subjects were all interested in re-education courses for the unemployed. Candidates were recruited through advertisements in local and national newspapers and by different employment agencies. In the selection procedure a psychological test consisting of an aptitudes test battery and one or more personality or values questionnaires was used. The subjects were told that the values questionnaires weren't used for the selection of candidates but only for research purposes. Candidates who passed the aptitudes test were invited for a selection interview.

Results

Table 2 presents the mean scores on the SIV for the different cultural groups. The results of the different analyses presented in Table 3, show substantial differences between the Dutch group and the immigrant groups. The results of some of the immigrant groups are combined because of the similarities of the cultural backgrounds of these groups.

The differences between the Dutch and total immigrant group were significant ($p < .001$, except Recognition $p < .05$) for all the scales. The Dutch scores were higher for Support, Independence and Leadership, and lower for Conformity, Recognition and Benevolence. These differences were found for all the immigrant groups, although the European immigrants scored more in the direction of the Dutch group (except for Leadership).

When we compare Hofstede's 1980 data with our results, we can make the following observations. In Hofstede's study the Netherlands scored lower then average on Power Distance and Uncertainty Avoidance, very high on Individualism, and extremely low on Masculinity. Turkey scored high on

Power Distance and Uncertainty Avoidance, low on Individualism, and average on Masculinity.

Table 2. Mean Scores on the Scale of Interpersonal Values and Number of Subjects per Country of Origin

	N	S	C	R	I	B	L
Total	1016	15.2	16.0	8.8	17.0	16.2	16.0
Standard deviation		4.8	5.0	4.1	5.8	5.0	6.0
Dutch	439	17.4	13.5	8.5	18.7	14.7	17.1
Immigrants	577	13.6	17.8	9.0	15.7	17.4	15.3
Turkey/Morocco	114	13.4	18.0	9.8	15.3	17.7	14.5
Surinam/Antilles	112	13.3	18.1	8.2	16.5	17.4	15.4
South-East Asia	84	13.6	16.7	9.1	16.4	17.2	14.8
Eastern/Southern Europe	23	15.3	16.4	7.9	17.3	16.8	15.1
Latin America	26	15.2	15.9	9.3	16.1	18.4	14.0
Africa	58	12.7	18.7	9.9	14.3	17.0	16.0

Taking into account the correlations reported in Table 1, we would predict the following for Turkish immigrants, as compared to the Dutch: (a) higher scores on the SIV scale Conformity and lower scores on Independence and Support (based on the correlations with Power Distance); (b) lower scores on Benevolence (based on the correlations with Uncertainty Avoidance); (c) higher scores on Conformity, lower scores on Support and Benevolence (based on the correlations with Individualism); (d) higher scores on Conformity and Recognition, lower scores on Independence and Benevolence (based on the correlations with Masculinity). All these predictions were confirmed by our data, except for Benevolence. A possible explanation may be that an unknown interacting variable correlated with immigration plays a role in the scores on Benevolence.

In Hofstede's research, Thailand scored lower than Turkey in all dimensions and higher than the Netherlands except for Individualism where Thailand scored much lower than the Netherlands. When we compare this to our data, most of the SIV scores show a similar pattern: South-East Asia scores between Turkey and the Netherlands. The only difference is that in the Hofstede data, Thailand scores lower on Individualism than Turkey, while our results don't show any evidence for that.

To find more general factors, we analysed the factor pattern of the subscales of the SIV. To deculture the data (see Bond, 1988) we standardized the subscale scores within each culture separately. The correlation between any two variables within a culture is thereby not affected, but the cultural confound (i.e., the cross-cultural correlation) is removed because the average score for the two variables in each of the pooled cultural samples is zero. Using Principal Component Analysis with varimax rotation we found three factors explaining 75.7 % of the variance. Because the subscale scores within each instrument are not independent due to the ipsative item format the factor loadings are higher than normal.

Table 3. Means for the Total Group (39 countries), the Netherlands, Turkey and Thailand on the Hofstede Dimensions

Country	Hofstede dimensions			
	Power distance	Uncertainty avoidance	Individualism	Masculinity
Total group	51 (20)	64 (24)	51 (25)	51 (20)
Netherlands	38	53	80	14
Turkey	66	85	37	45
Thailand	64	64	20	34

standard deviations in brackets

The first factor can be described as collectivism versus individualism (SIV: A .78, C .74, R -.41, I -.37, L -.32) and is comparable with Hofstede's dimension Individualism and perhaps more with Schwartz's Conservatism vs. Autonomy dimension (cultural level). The second factor can be described as Social Integration vs. Social Power (SIV: S .82, L -.85) and is comparable with both Hofstede's Masculinity dimension and Schwartz's Hierarchy vs. Egalitarian commitment dimension (cultural level). The third factor can be described as Independence vs recognition (social status) (SIV: I -.87, R .77) and is more or less comparable with the Power distance dimension of Hofstede and, to a lesser extent, to Schwartz's Dimension Mastery versus Harmony (cultural level). Hofstede's dimension Uncertainty Avoidance was not found in our analysis. It appears that this factor is included in the factor Individualism vs. Collectivism.

To analyse the relationships between different background variables of the immigrants and the scores on the SIV scales, correlations were calculated. Female immigrants scored lower on Leadership values (r (566) = -.11, p = .05), older immigrants found altruistic values less important (r (564) = -.11, p = .05), and the higher the educational level, the more they valued conformity (r (574) = .11, p = .05) The longer immigrants stayed in the Netherlands, the more they valued support (r (427) = .12, p = .01) and independence (p (427) = -.14, p = .01) and the less they valued recognition (r (427) = .14, p = .01).

Multivariate Analysis of Variance for the immigrant group for the SIV with age, gender, level of education and length of residence as independent variables showed a significant multivariate effect for length of residence (F = 3.61, df 6, p = .002); the other main effects and interaction effects were not significant. Univariate Analysis of Variance showed that length of residence was significant for Support ($F = 7.03, p = .008$), Recognition ($F = 9.59, p = .002$) and Independence ($F = 6.02, p = .015$).

Discussion

Our first hypothesis, that values of immigrants will resemble the national cultural values of their native countries was supported by the results of this study. Our results corresponded with the results reported by Hofstede for different countries. We found remarkable similarities between our results and the Hofstede data for the Dutch and the Turkish/Moroccan immigrant group. Furthermore, there was a significant difference between the Dutch and non-European countries. Most of these differences corresponded with the differences Hofstede found. The Dutch were more individualistic and less altruistic, more oriented towards social integration, less conformist and less oriented towards recognition (status) than the non-European countries.

Within the non-European groups we see that the two largest groups, the Surinam/Antilles group and the Turkey/Morocco group differed in some important aspects. The Turkey/Morocco group valued recognition (status) more and independence less than the Surinam/Antilles group. These differences confirm the views of different authors in the Netherlands (Eppink 1981, Pinto 1990).

Although the results are promising, some shortcomings need addressing. The different cultural groups were not evenly divided in our sample and the number of subjects was sometimes small. Secondly, if we try to identify similarities and differences in values of immigrants and compare these to the values of their native countries, a problem is that individuals who emigrate do not form a representative sample of their cultural group. Social, political, psychological and economic circumstances play an important role in the decision to emigrate and can cause differences between the values of immigrants and the values of the citizens who didn't emigrate. This means that comparison of the values of immigrants with mean country scores (Hofstede) will be difficult. In addition, the differences in values between immigrants and the values of their native countries are influenced by the acculturation process of the immigrants. The amount of contact with the dominant culture, the ease of learning a new language, change of social relationships and economic circumstances (unemployment) are examples of factors that play a role in this process.

This acculturation process was the object of our third hypothesis; length of residence will have an important influence on the personal values of immigrants: the longer the residence the more personal values will resemble the national cultural values of the dominant Dutch culture. This hypothesis was partly supported by the results of our study. We found a significant correlation with length of residence for the values: social support (social integration), recognition (social status), and independence (individualism).

Although small, the changes were all in the direction of the Dutch values; the longer immigrants lived in the Netherlands the more these values corresponded to the Dutch values. This finding was important because the mean length of residence was only 8.7 years and age, gender and level of education were of no influence. This results showed that the acculturation process influences the values that people hold.

Three values, Conformity, Benevolence and Leadership, as measured by the SIV, did not correlate with length of residence in the Netherlands. Why some values change and others do not is still unclear. Are some values perhaps more central to the immigrant's cultural identity? Or are some values more easily influenced by contact with another dominant culture? Perhaps there are other variables, as ethnicity of friends or language knowledge and preference, that influence the relationship between the duration of residence and change of values. Future research will have to clarify these relationships.

The current research has confirmed the existence of broad value dimensions comparable to the dimensions Hofstede found, and has also shown that the values held by immigrants resemble the values of citizens of their native countries. Furthermore it has shown that some values of immigrants change in the direction of values held by the dominant culture, although why some values change and others don't, remains unclear.

It is important that future research investigates the variables that influence this change of values. For example it would be useful to use acculturation lists (Mendoza, 1989) in combination with value surveys to study this relationship more closely. In this way it is possible to identify factors that facilitate the integration process of future immigrants.

References

Berry, J. W. (1990). Psychology of acculturation: Understanding individuals moving between cultures. In R.W. Brislin (Ed.), *Applied cross-cultural psychology*. Newbury Park, CA: Sage Publications.

Bond, M.H. (1988). Finding universal dimensions of individual variation in multicultural studies of values: The Rokeach and Chinese value surveys. *Journal of Personality and Social Psychology, 6,* 1009-1015.

Eppink, A. (1981). *Cultuurverschillen en communicatie.* Problemen bij hulpverlening aan migranten in Nederland. Alpen aan de Rijn: Samson.

Feather, N.T., & Peay, E.R. (1975). The structure of terminal and instrumental values: Dimensions and clusters. *Australian Journal of Psychology, 27,* 157-164.

Hofstede, G. (1980). *Culture's consequences: International differences in work-related values.* Beverly Hills Calif.: Sage Publications.

Hofstede, G. (1982), Dimensions of national cultures. In H.S. Rath, D. Asthana, & J.B.P. Sinha (Eds), *Diversity and unity in cross-cultural psychology*. Lisse: Swets and Zeitlinger.

Hofstede, G., & Bond, M.H. (1984), Hofstede's culture dimensions: An independent validation using Rokeach's value survey. *Journal of Cross-cultural Psychology, 15,* 417-433.

Hofstede, G. (1991). *Allemaal andersdenkenden: omgaan met cultuurverschillen.* Amsterdam: Contact.

Gordon, L. (1960). *Survey of interpersonal values.* Chicago: Science Research Associates.

Graves, T.D. (1967). Psychological acculturation in a tri-ethnic community. *Southwestern Journal of Anthropology, 23,* 337-350.

Marin, G. (1992). Issues in the measurement of acculturation among Hispanics. In K.F. Geisinger (Ed.), *Psychological testing of Hispanics.* Washington, DC: APA.

Mendoza, R.H. (1989). An empirical scale to measure type and degree of acculturation in Mexican-American adolescents and adults. *Journal of Cross-cultural Psychology, 20,* 372-385.

Ng, S.H., Akhtar, A.B.M., Ball, P., Bond, M.H., Hayashi, K., Lim, S.P., O'Driscoll, M.P., Sinha, D., & Yang, K.S. (1982). Human values in nine countries. In H.S. Rath, D. Asthana & J.B.P. Sinha (Eds.), *Diversity and unity in cross-cultural psychology.* Lisse: Swets and Zeitlinger.

Pinto, D. (1990). *Interculturele communicatie. Drie stappen methode voor het doeltreffend overbruggen en managen van cultuurverschillen.* Houten/Antwerpen: Bohn Stafleu Van Loghum.

Redfield, R., Linton, R., & Herskovits, M.J. (1936). Memorandum on the study of acculturation. *American Anthropologist, 38,* 149-152.

Rokeach, M. (1967). *Value survey.* Sunnyvale, California: Hallgren Press.

Rokeach, M. (1973). *The nature of human values.* New York: Free Press.

Sabogal, R., Marin, G., Otero Sabogal, R., Marin, B.V., & Perez Stable, E.J. (1987). Hispanic familialism and acculturation: What changes and what doesn't? *Hispanic Journal of Behavioral Sciences, 9,* 397-412.

Schwartz, S.H. (1994). *Values.* Paper presented at the IAAP congress, Madrid.

Schwartz, S.H. (1992). Universals in the content and structure of values: Theoretical advances and empirical test in 20 countries. In M.P. Zanna (Ed.), *Advances in experimental social psychology* (Vol. 25, pp. 1-65). San Diego: Academic Press,

Schwartz, S.H., & Bilsky, W. (1987). Toward a universal psychological structure of human values. *Journal of Personality and Social Psychology, 53,* 550-562.

Social Science Research Council (1954). Acculturation: An exploratory formulation. *American Anthropologist, 56,* 973-1002.

Values in Indian Villages

Meera Varma
University of Allahabad, Allahabad, India

Cross-cultural research in the area of value orientation reflects a wide variety of interests of researchers encompassing ethnic and cross-cultural comparisons (Feather, 1975; Furnham, 1984; Rokeach, 1973); value acculturation (Georgas, 1991, 1992); analysis of universal content and structure of values (Schwartz & Bilsky, 1990) and other significant dimensions. These studies have provided direction to new enquiries. Betancourt, Hardin and Manzi (1992) observed that within culture observation of dimensions known to vary cross-culturally as they relate to behavior could contribute to the understanding of the culture in social psychological phenomena. The present study employs this perspective in attempting to examine the value preferences of Indian villagers and delineate the relevance of some of the variables which have a significant bearing on their value orientation.

India is a country where the majority of the population resides in villages. Their main source of subsistence is agriculture. They do not have access to modern technology for agriculture. Illiteracy and superstition still are dominant features of their lives. Their attitude, thinking span and level of aspiration have not gone beyond their limited sphere of experience. Studies on Indian villages have revealed the hierarchical character of village society. The hierarchy is hereditary and is maintained everywhere by groups who own and control the use of land and power. The *dominant caste* wields economic and political power over all minor castes and acts as guardians of the social and ethical code of the entire village (Srinivas, 1960). Despite living under conditions of deprivation and social disadvantage, they have in no way been prevented from deriving satisfaction through spiritual, social and specific work related activities. Structural features of the village groups have influenced and reinforced their social norms and value patterns. In theoretical and empirical research programs, values have been defined as broad, trans-situational goals, varying in importance that serve as guiding principles in life (Schwartz, 1994). Schwartz and Bilsky (1987) rightly contend that, "Values are cognitive representations of three types of universal requirements: biologically based needs of the organism, social interactional requirements for interpersonal coordination, and social institutional demands for group welfare and survival" (p. 551).

The values considered in the present study were elicited from the interviews conducted with villagers. The categories of values generated represented the cognitive framework and behavioral patterns of villagers. These values were: Religiosity, Helping, Economic, Social Equality and Collectivism. The five values have been briefly described to understand the perspective within which they have been conceptualized.

Description of Values

Religiosity
Religion is a way of life in India. Religious devotion is still one of the strongest motives in our rural population. It is believed that religion provides a source of inspiration for the betterment of life in this world and after death. God fearing actions and behavior dominate the pattern and life style of people in villages. Religion in the form of prayer and worship of Gods and Godesses determine their behavior and belief system. Villagers also worship natural resources such as trees, river, land, rains because they are directly responsible for vegetation and agricultural yield. Religion is in the form of faith manifested in worshipping deities, natural resources and observing rituals.To be religious means being pure, pious and god-fearing. That God will reward good deeds and punish wrongdoing, is ardently believed by the villagers.

Helping
The concept of helping is based on the assumption that human beings are basically good or have potential to be good. Man cannot live alone as s/he has to depend on others for the fulfillment of his/her needs. Right from birth till death, it is others who provide care and assistance, show *Sahnubhuti* (Sympathy), *Day* (Mercy) and extend their help during the hour of need and distress. In India, particularly among rural masses, words like *Paropakra*, *Dna* and *Sahyat* are quite prevalent and carry special relevance in their day to day life. Here, helping implies serving others and virtuous acts which bring happiness in return.

Economic
It is argued that lower needs must be adequately satisfied before the higher needs can emerge in the development of individual. This phenomenon can be witnessed more in rural India. It appears that deficiency needs are so characteristic of the daily life of rural people that growth motives have not yet had a chance to play their part. Thus, earning money to meet the bare necessities of their lives predominates their thinking, efforts and social interactions. In the present study economic values refer to the preference for monetary gain.

Social Equality
Individuals' performance of rewards, costs, and needs vary due to situational and cultural variations. Norms that determine resource allocation are cultural and therefore the definition of fairness and just distribution might well vary from culture to culture (Singh & Pandey, 1994). Traditional Indian society, it has been said, was based on the premise of social inequality

and hierarchical values permeating every sphere of village life. In any society, people have strong beliefs in the equality of rights, benefits, opportunities and privileges, though in reality, it is uncommon. Social power, status, caste, religion and property ownership have led to discrimination among the people in society. Frustration, distress and dissatisfaction prevail in the deprived groups on account of this inequality. Contrary to this exploitation, misuse of power and power exist among the privileged groups. The Preamble of the Constitution of India underlines the resolve of the people to secure for all citizens, social, economic and political justice and equality of opportunity and status. In the present study, social equality denotes a belief in equal distribution of benefits to all without any discrimination as all are born equal.

Collectivism

Collectivism reflects the belief that the ingroup is of utmost importance. The ingroup may consist of family, friends, neighbors, relatives and others. When interdependence is deemed essential for survival, collectivism is preferred but when individuals are able to survive without much help from others, individualism may predominate. In collectivism the interest is directed towards the members of the extended family whereas in individualism one tries to look after his or her own interest or the interest of the immediate family members. Village society is more oriented towards collectivism than the urban counterpart. Tradition and custom-bound societies believe in collectivism. Feelings of oneness, participation and cooperation are some of the indices of collectivism. Here collectivism refers to importance given to collective efforts and also incorporates a cooperative spirit.

The purpose of this study was to examine the values of villagers with reference to certain salient factors such as caste, gender, occupation, size of land holdings and development. It was expected that the value patterns and preferences in villagers would vary as a result of these factors.

Method

Sample

The sample consisted of 456 respondents, 264 males and 192 females drawn from six villages of Allahabad district, U.P., India. Two heads of the family (one male and one female) were included in the sample. The independent variables were as follows. Caste was separated into high, middle and low. Occupation consisted of farming, farming and service, service and business. Size of Land Holding was separated into four categories: no land, small, medium, and large land holders. Education was separated into two categories: literate and illiterate. Villages were classified into developed and underdeveloped on the basis of literacy and employment. Other criteria for the assessment were: quality of land, road, transportation, health care, civic and educational facilities.

Measures

A value measure was constructed to assess the value preferences of villagers with respect to Religiosity, Helping, Economic, Collectivism and Social Equality. Each item in the measure delineated two different opinions followed by two choices. Each choice designated a value category. The respondents were required to give their relative preference between the two values. Only those combinations of values were presented which made sense to the villagers. Items were written in the Hindi dialect used in the region. In some cases Hindi aphorisms were used. An example of the items used is given for Sample Item 5: Some say , "If you do good to others, it will come back to you," whereas others say, "Money is above everything." Which of these two statements do you believe is truer, A or B?.

If the respondents selected A to be truer, it was coded as preference for the value category Helping. On the contrary, if the respondents selected B, it was coded as the value category Economic.

All preferences expressed by the respondents were coded under different value categories. The total number of possible preferences for different value categories were: Religiosity 3, Helping 2, Economic 2, Collectivism 2, Social Equality 2. The total score under each category was calculated by adding the number of preferences expressed by a respondent. None of the preferential categories overlapped with the other. One value preference was dropped at the time of coding as it was a single item on Power. Thus, Religiosity scores ranged from 0-3, whereas Helping, Economic, Collectivism, Social Equality ranged between 0-2.

Procedure

The individual interview method was used to obtain the responses on the value measure. In most of the cases one male and one female head of the family were contacted for the interview. The social arrangement in the villages is such that women stay in the interior of the house most of the time, whereas, men usually are in the exterior of the house. Therefore, men were interviewed by male investigators in the exterior portion of the house, while women in the interior of the house by females. As it was extremely difficult to use any scale measure with them, each item of the value measure was presented to the respondents in a certain order. First, the item was explained and then two categories of values were presented. They were then asked to compare and select one.

Results

Inter-Correlations between Values

The inter-correlations between the five selected values was determined with the *Pearson product-moment correlation* (Table 1).

Table 1. *Intercorrelations between the Values*

	Religiosity	Helping	Economic	Social Equality
Helping	-.41**			
Economic	-.47**	-.53**		
Social Equality	-.13*	.19**	-.03	
Collectivism	.20**	-.08	-.14*	-.03

*p<.01   **p<.001

The correlations between Social Equality and Helping, Religiosity and Collectivism show the highest relationships. High negative relationships were found between the values Helping and Economic, Religiosity and Economic, and Religiosity and Helping. Contrary to this, negative moderate correlations were found between Economic values and Collectivism. The values without significant correlations were: Helping and Collectivism, Economic and Social Equality, and Collectivism and Social Equality.

Value Preferences with Respect to Antecedent Variables

Mean scores on the five values with respect to six antecedent variables were computed. The results were analyzed with one way ANOVA and Newman-Keuls tests (Table 2).

Religiosity

The High Caste group was significantly higher than the low group on Religiosity. Gender difference did not appear. The Farming group was higher than farming and service, and service and business groups. With respect to the Size of Land holding, the No Land group was higher than the Small Land group. Education and Development did not have effects on values.

Helping

In case of Helping, only Education had an effect on values. The literate group was higher than the illiterate.

Economic

Gender and Development had no effect on Economic values. The Low Caste group was higher than the High and Middle Caste groups. The Service and Business work group and Farming and Service group were higher than the Farming group. The No Land group was higher than groups, Small, Medium and Large. Also, the Illiterate group showed greater preference for Economic values than the Literate.

Table 2. Mean Values with Respect to Six Antecedent Variables

Antecedent Variables	Religiosity	Helping	Economic	Social Equality	Collectivism
Caste					
High	2.28a	1.34a	.26a	1.08a	1.47a
Middle	2.22ab	1.44a	.21a	1.15a	1.36a
Low	2.04b	1.37a	.46b	1.21a	1.45a
Gender					
Male	2.00a	1.56a	.27a	1.33a	1.25a
Female	2.35a	1.14a	.43a	.92a	1.70a
Work					
Farming	2.34a	1.40a	.19a	1.13a	1.35a
Farming Service	2.06b	1.41a	.37b	1.18a	1.51b
Service & Business	2.01b	1.33a	.50b	1.20a	1.48b
Size of Landholding					
No Land	2.01a	1.34a	.49a	1.19a	1.46a
Small	2.23b	1.39a	.28b	1.19a	1.44a
Medium	2.15b	1.51a	.20b	1.06a	1.42a
Large	2.31b	1.37a	.19b	1.04a	1.27a
Education					
Literate	2.10a	1.59a	.19a	1.33a	1.27a
Illiterate	2.19a	1.28b	.40b	1.09b	1.52b
Development					
Underdeveloped	2.16a	1.34a	.35a	1.07a	1.47a
Developed	2.14a	1.43a	.32a	1.25b	1.41a

Common subscripts do not differ.
Means with noncommon subscripts differ significantly at $p < .05$.

Social Equality
 Caste, Gender, Work and Size of Land Holding did not effect Social Equality. The Literate group had higher values as compared to Iliterate group. Respondents of Developed villages preferred Social Equality more than those of Underdeveloped.

Collectivism
 Except for Work and Education, no differences were found on the other four variables for Collectivism. Farming and Service, Service and Business groups were higher than the Farming group. The Illiterate group had higher preference for Collectivism than the Literate group.

Discussion

Pursuing a certain kind of occupation, living a certain kind of life style, having some education and belonging to a certain group or class are all examples of behaviors that go far beyond a single specific act. The values, therefore, must be functionally related to all kinds of molar as well as molecular behaviors. It has been observed that attitudes, cognitive makeup and behavioral indices are value based. According to Schwartz, values are goals, and they must thus represent the interests of some person or group (Schwartz & Bilsky, 1987). The findings of the present study in the form of interrelatedness among the values show either parallel or opposite trends. Social Equality and Helping were given equal priority and importance by the villagers. It can be contended that these two values have larger bearing on the promotion of Social Welfare and Justice. Those who care for the interest of others are presumably intolerant of social discrimination, self-interest, possessiveness and injustice. Generosity and Helping may be regarded as important accompaniments of the belief in Social Equality.

Two more values, Religiosity and Collectivism, also demonstrated homogeneous trends in the responses of villagers. Village norms, social milieu and traditional roots are very strongly linked to interdependence and cooperative spirit. The villagers believe that, "a feeling of oneness is a strength in itself", and togetherness and brotherhood lessen their miseries and strengthen their efforts. The phenomenon of togetherness is not only linked to social needs but also to religious beliefs. Religion is a strong force which guides their cooperative and collective behaviors. Many of their rituals, customs and festivities are basically governed by canons of the Gods and must be performed jointly by all the members of the community. The Hindu concept of *Dharma* calls for meeting one's obligation to the family as well as to the larger community (Tripathi, 1988).

An interesting feature of the results could be seen in case of some of the values which presented opposite trends. It is interesting as it reflects conflict, indecisiveness and diverse thinking which seem to predominate the life style of villagers. Helping and Economic values were found to be in the opposite direction. Similarly, Religiosity and Economic values, and again, Religiosity and Helping, indicated the same reverse trends. The economic condition of Indian villagers is still not satisfactory. The domination of basic needs and their satisfaction mainly govern their life goals and aspirations. Helping can be perceived at different levels. Villagers intend to help, wish to cooperate, and have concern for others. But when it comes to monetary or material help, it is likely that their inability to cope with it prevents them from doing so. It has also been observed that in performing any kind of customary or norm-related religious activity, expenses are high. For such activities villagers go to the extent of securing loans and subletting their property, jewellery and other household articles. This seems to be a valid reason for the opposite trend which has been demonstrated between Religiosity and Economic value. Religiosity and Helping presented an inverse relationship against the conviction that Religiosity implies caring and helping others. Kirkpatrick (1949) concluded that intensity of religious belief was negatively correlated with humanitarianism as assessed by attitudes towards criminals, blacks and foreigners.

Inequality underlying the caste system in India is the application of moral and evaluative standards in the hierarchical categorization of castes. The religious basis of castes was found in the work of Hocart, who viewed castes as a system organized for the performance of sacrifical rituals. The king in particular bore the expenses for the sacrifical offerings. The *Brahmins* performed the rituals and the *Vaishyas*, as farmers, were required to support the king through their lands and cattle. These three Varnas, by virtue of being vehicles of the immortal Gods, had to avoid coming into contact with death and that which caused death in the form of decay and disease. The handling of death and decay was assigned to the *Shudras* whose specialization in "polluting tasks" excluded them from involvement in the sacrifical rites (Bteille, 1983).

Antecedent Factors in Value Preferences

The results show that the low caste group, the Schedule caste, evaluated monetary gain as the most important, in comparison to the higher caste groups, most likely because for them, the satisfaction of the fundamental needs have distinct priorities over social needs. Similarly, villagers with no land, in comparison to those who have land, must depend completely on earning their livelihood through other means. They work as laborers, carpenters, potters, or engage in other small jobs or businesses. Andre Bteille (1983) argues that in the Indian rural context where land is scarce and where a major population is dependent on land only for its livelihood, those who own land are privileged not because it is a valued commodity but because it can be used as an instrument for controlling the landless.

Education played an important role in determining preferences for Economic values. Perhaps the illiterate, as compared to the literate, feel more insecure due to their lack of qualifications for finding a better job and their limited scope for progress.

Out of the three work groups, the service and business groups had sufficient ground to opt for Economic values because in the villages, this group is supposed to be better off economically as compared to the rest of the groups. The business group has always surpassed others in gains and benefits, saving and monetary multiplication.

Contrary to the emphasis on Economic values, the Service and Business group scored lowest on religious values. Here the farming group surpassed all other groups. Fear of God, and faith in work, most probably, determined their religious outlook and benefits. Farmers take natural resources to be a blessing of God. Most of the Indian religious festivals are related to the harvesting season. Rain is believed to be the blessing of *Indra,* the God of Rain. Small land holders, as compared to those who have no land, have tendencies similar to the Farming groups since dependency on land because of agriculture influences their belief in religion.

Farming and Service, and Service and Business groups faired higher on Collectivism as compared to the Farming group. The multiplicity of work will always require collective efforts to achieve the desired goal. Greater prevalence of the joint family system could be witnessed in these groups because more hands mean more accomplishment. Another explanation of this result could be related to their social obligations, interdependence and mutual interests. Therefore, Collectivism may be considered closer to power, strength and satisfaction.

Collectivism was given a prominent place by Illiterates. Illiteracy is strongly associated with ignorance and backwardness. Social realities are beyond their comprehension and hope. Therefore, traditions, conventions and age old patterns dominate their life style and behavior. They depend on other's help and ingroup support for achieving their goals. They have the fear of social norms and duties defined by the ingroup and individual pleasure, beliefs and values are shared by the members of ingroup. There is a readiness to cooperate rather than compete with the group members (Triandis, 1988). Sinha (1984) pointed out that, "the logical corollary for man to grow with the social groups and collectivities rather than strive for personal excellence which might alienate him from the person around him".

The Literate group showed greater preference for Helping as compared to the Illiterate. Their knowledge that cooperation and mutual help are needed to live harmoniously with others has a rational base. This view implies that one's satisfaction of needs and desires depends on his willingness to contribute to the welfare of others and that a person has potential for developing optimal social characteristics.

When all the antecedent factors are compared, keeping in mind the five values, there are no gender differences. In the present study 86% of the sample included couples. In villages, the majority of women were illiterate; as a result they become dependent on the male members. Over the period of time women tend to adopt and accept the ideas, attitudes and beliefs of their male counterpart. It appears to be a reasonable explanation for having no gender differences in the value patterns.

The overall results of the present study demonstrate the relevance of structural factors and development contributing to the value pattern of rural masses. The findings reported here are based on the first phase of the Omnibus Survey which is a longitudinal endeavor undertaken by the Department of Psychology of the University of Allahabad.

References

Betancourt, H., Hardin, C., & Manzi, J. (1992). Beliefs, value orientation, and culture in attribution processes and helping behaviour. *Journal of Cross-Cultural Psychology, 23*, 179-195.
Bteille, A. (Ed.) (1983). *Equality and inequality.* Delhi: Oxford University Press.
Feather, N.T. (1975). *Values in education and society.* New York: Free Press.
Furnham, A.F. (1984). Value systems and anomie in three cultures. *International Journal of Psychology, 19*, 565-579.
Georgas, J. (1991). Intra-family acculturation of values in Greece. *Journal of Cross-Cultural Psychology, 22*, 445-457.
Georgas, J., & Kalantzi-Azizi, A. (1992). Value acculturation and response tendencies of Biethnic adolescents. *Journal of Cross-Cultural Psychology, 23*, 228-239.
Kirkpatrick, C. (1949). Religion and Humanitarianism: A study of institutional implications. *Psychological Monographs, 63*, (9, Serial No. 304).
RoKeach, M. (1973). *The nature of human values,* New York: Free Press.
Schwartz, S.H., & Bilsky, W. (1987). Toward a universal psychological structure of human values. *Journal of Personality and Social Psychology, 53*, 550-562.
Schwartz, S.H., & Bilsky, W. (1990). Toward a theory of the universal content and structure of values: Extensions and cross-cultural replications. *Journal of Personality and Social Psychology, 58*, 878-891.
Schwartz, S.H. (1994). Studying human values. In A.-M. Bouvy, Fons J.R. van de Vijver, P. Boski, & P. Schmitz (Eds.). *Journeys into cross-cultural psychology.* Lisse: Swets and Zeitlinger.
Singh, P., & Pandey, J. (1994). Distributive decisions as a function of recipients' need performance variations and caste of allocator. In A.-M. Bouvy, Fons J.R. van de Vijver, P. Boski, & P. Schmitz (Eds.). *Journeys into cross-cultural psychology.* Lisse: Swets and Zeitlinger.
Sinha, J.B.P. (1984). Towards partnership for relevant research in the third world. *International Journal of Psychology, 19,* 169-177
Srinivas, M.N. (Ed.) (1960). *India's villages.* New Delhi: Asia Publishing House.
Triandis, H.C. (1988). Collectivism and development. In D. Sinha & Henry S.R.K. (Eds.), *Social values and development: Asian perspective.* New Delhi: Sage.
Tripathi, R.C. (1988). Aligning development to value in India. In D. Sinha & Henry S.R. Kao (Eds.), *Social values and development: Asian perspective.* New Delhi: Sage.

Part V

Social Psychology

The Perception of Interpersonal Action: Culture - General and Culture - Specific Components

John Adamopoulos, Grand Valley State University, Allendale, Michigan, U.S.A.
Ariana Stogiannidou, Aristotelian University of Thessaloniki, Thessaloniki, Greece

The identification of action (i.e., the definition of what constitutes a psychologically meaningful "unit" of behavior) has been one of the most significant and persistent problems in social-psychological theory. Many of the early pioneers of modern social psychology were clearly and consistently concerned with this problem, and offered important theoretical insights that led to recent developments in attribution and attitude theory (e.g., Heider, 1959; Lewin, 1951). However, the problem's complexity has been so vexing that social psychologists abandoned its systematic exploration until recently, when developments in methodological and analytic techniques permitted much more sophisticated experimentation.

A number of very different questions, and approaches to answering them, can be formulated regarding the perception of action. The core question asked in the present approach is: "What type of information is essential for the identification of any given interpersonal behavior (i.e., behavior involving at least two interacting individuals)?" Put differently, the present approach will explore the elements or perceptual attributes and structures that *constitute* interpersonal behavior.

As stated above, the focus of this research is on the specification of the kinds of information necessary for action identification. For example, following some of the most influential approaches in the philosophy of action (e.g., Davidson, 1980; Goldman, 1970), one could argue that the presence of an "actor" ("agent") is necessary in the labeling of the action. While the theoretical framework proposed later in this paper indeed focuses on such issues, the core question will always be framed in terms of the psychological meaning structures involved in the perception of the social environment. Thus, the identification of action will be explored as *both* a context-free set of abstract psychological structures, and as a context-dependent process, where cultural meanings play a crucial role in determining how people make sense of, and recall, a given configuration of components of interpersonal behavior.

Recent perspectives emerging in this area were developed along somewhat different lines, but may indicate some theoretical convergence. In the 1970s, the imaginative and original work of Newtson and his colleagues (e.g., Newtson, 1973; Newtson & Engquist, 1976; Newtson, Engquist, & Bois, 1977) showed that action units in ongoing behavior sequences are identifiable by distinctive changes in behavior. This research presented the possibility that certain configurations of components of action may exist that are critical in its identification (i.e., recognition and labeling). However, the research never specified these configurations, probably because it followed a heavily empirical approach, with little guidance from a cohesive theoretical framework.

More recently, this problem was revisited by Vallacher and Wegner (1985), who attempted a more systematic theoretical analysis. However, their approach differed radically from Newtson's in that Vallacher and Wegner asked the question of action identification from the perspective of the actor, focusing on an actor's interpretation of his/her action. Newtson's work, like the framework outlined in the present paper, takes the more general perspective of the observer of action. This is critical, because, clearly, the actor has much more information than is available in the general culture and in the perceptual field with regard to the action.

In addition to these developments in social psychology, work by cognitive psychologists has shown that, in any action sequence, the meaning of particular events, and the overall plan or goal of the action, are more important than the temporal order of specific acts in determining the cognitive activity surrounding the conceptualization, understanding, and recall of the action (e.g., Lichtenstein & Brewer, 1980; Nottenburg & Shoben, 1980).

Until now, these diverse research programs had not been integrated. In fact, they had been construed as completely independent pieces of research, bearing little significance for each other. However, we can reach a fundamental conclusion on the basis of all this work: structural configurations of behavior are important in action identification, but the psychological function of these configurations may be mediated by the intent, goal, or meaning of the action (which is a culturally-shaped construct). Thus, any theory of interpersonal action identification must incorporate both a structural and a meaning (semantic) analysis of behavior. This conclusion forms the basis for the framework described below.

Concept Definition

At this point it becomes essential to define a few of the theoretical constructs that will be explored in the *proposed* research. Figure 1 presents the main elements of the framework.

Action can be understood both in terms of universal (i.e., culturally invariant) structures (frame domain), and in terms of culturally variant phenomena that give meaning to these structures (semantic domain). For the purposes of the present analysis, the frame domain consists of 4 basic elements that may be useful in defining an act: the "actor," the "target," the "act," and the "environment" in which the interaction takes place. Combinations of these four elements generate various "frames" of action. Thus, frame 1 in Figure 1 consists of <ACTOR, TARGET>, and frame 2 consists of <ACT, TARGET, ENVIRONMENT>.

Obviously, in the present context there can be frames involving two, three, or all four elements.

It is important to recognize here that a number of other components have been identified and their significance has been debated in philosophical action theories. For example, the effect(s) of an act can be considered a separate component involving a goal or even a reason for the action. The present analysis focuses on a subset of possible components. Thus, "actor" refers to the person performing the act, "act" refers to the observable behavior, "target" refers to the recipient (person) of the behavior, and "environment" refers to the location of an event. Our approach does not include directly the concept of intentionality - so central to the philosophical debates on action (e.g., Davidson, 1980) - because it attempts to describe the identification of action by *observers*, who may make inferences about the actor's intentions but have no direct access to them.

Figure 1. A Framework for the Identification of Action

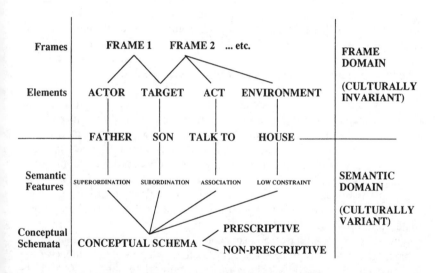

It should be clear from this discussion that our approach takes a more "molecular" perspective to the analysis of action than do many other action-theoretic frameworks (c.f., Oppenheimer & Valsiner, 1991). This limits the usefulness of the approach in discussions of the ontogenesis of action, morality, and so on - all issues of great significance in many action-theoretic approaches. On the other hand, the present framework is compatible with and complementary to the "molar" approaches, which generally also take a componential perspective.

Figure 1 also presents an example of a simple interpersonal behavior. Imagine that a father talks to his son in their house. Frame 1 would consist of:

<The father is with the son>, whereas frame 2 would consist of something like: <The son was talked to in the house>.

These frames are abstract structures that are assumed to be of perceptual significance in action identification, but are not thought to have any culturally-defined meaning. In other words, it is assumed that all humans believe that actions stem from specific individuals and are directed toward others in particular social contexts or environments.

As Figure 1 indicates, there is another level at which action is understood - the semantic level. "The father talked to his son in the house" can also be understood (at least in North America) as, "A superordinate person associated with a subordinate in a low-constraint environment."

There has been a great deal of research on the problem of identifying the dimensions (features) of meaning along which people perceive social interactions (Adamopoulos, 1982a, 1982b, 1984; Triandis, 1977). A number of these dimensions from all major domains of social stimuli involved in the present context (i.e., actor-target relationships [roles], behaviors, and environments) are thought to be universal. For example, the behavior domain is defined by at least three culture-common dimensions: (a) association-dissociation (affiliation); (b) intimacy; and (c) superordination-subordination (dominance). The role domain includes dimensions such as: (a) conflict; (b) status and power differences; and (c) intimacy. Finally, the environment domain, while less well understood, seems to involve at least the dimension of constraint [1].

These dimensions are thought to provide the features used to encode social information in memory, and to interpret observed actions. In addition, attempts have been made to account for the emergence of these dimensions in differing cultural contexts by assuming that all interpersonal interaction can be understood as a resource exchange process (Adamopoulos & Bontempo 1986; Adamopoulos 1988; 1991). One of the implications of this work is that certain combinations of semantic features may define *prescriptive* (normative) conceptual structures (schemata) for general patterns of behavior that are appropriate in different social situations. The term "prescriptive" is used in order to emphasize the notion that certain conceptual schemata of interpersonal action summarize long-term experiences and customs shared by members of a particular culture, and thus form norms for appropriate behavior (e.g., Kahneman and Miller, 1986). Other combinations of semantic features may define concepts with no normative implications, and thus may have little significance in understanding the social environment. We call such structures *non-prescriptive* schemata. The important question here concerns the role of these "conceptual schemata." This paper presents an initial effort to address this question.

Prescriptive conceptual schemata may be thought of as abstract rules of interpersonal behavior in various types of contexts or social environments. By definition, such schemata are culturally-defined ideas. As cognitive rules, they can be expected to facilitate the processing of relevant information about particular interaction episodes. Thus, prescriptive schemata can be expected to modify the role that various frames, defined above, play in the identification of action.

Cultural variables play a very significant role in this framework. At a minimum, it is anticipated that prescriptive schemata - which are components of the semantic domain - vary from culture to culture, even though they are constructed from culture-common features of meaning. Even more important, however, cultural variables may affect the cognitive functioning of frames in a direct fashion. For example, we can predict that in "tight" or even in collectivist cultures (i.e., cultures with highly interdependent interpersonal relations) people may utilize action identification frames that involve the actor-target relationship much more frequently than in "loose" and generally individualistic cultures, where actors are often seen as acting independently of the social context. These are novel predictions that are not easily generated by existing theories in the area of psychology and culture. Although long suspected, there is little direct evidence of the role of cultural "syndromes" like individualism-collectivism in the utilization of different processing strategies at such basic levels of intellectual functioning.

In order to investigate these issues, a research strategy was devised and is outlined below, along with the results of some initial investigations. The strategy focuses first on identifying prescriptive schemata in two cultures - one where interpersonal relations are "tight" (Greece), and one where they are "loose" (U.S.A.). In a second study, a methodology adopted from cognitive psychology was tested in a preliminary monocultural exploration of which of several possible frames are important in action identification, and of how the function of these frames is modified by the presence of prescriptive schemata.

Study 1: The Derivation of Conceptual Schemata

The basic methodology for the description of prescriptive and non-prescriptive schemata has been outlined in Adamopoulos (1982b). Subjects estimate the likelihood that particular individuals within certain roles (e.g., "father") will engage in a number of behavioral acts (e.g., "talk with") with other persons (e.g., "son") in various environments (e.g., "house"). Multivariate analyses (e.g., three-mode factor analysis) can then be used to derive the conceptual dimensions for each of the three stimulus domains (i.e., role-pairs, behaviors, and environments), and to describe the strength of the relationships among all combinations of the derived dimensions from the various stimulus domains. Table 1 presents, in a simplified manner, such empirical relationships obtained from ratings provided by U.S. and Greek respondents. Specifically, strong positive relationships (implying high likelihood that a particular type of behavior will occur within certain social and environmental contexts) are indicated by a "+" symbol, strong negative relationships with a "-" symbol, and weak relationships are represented with a "0." Within the present framework, strong positive and negative relationships are thought to reflect *prescriptive* schemata, whereas weak relationships are thought to reflect *non-prescriptive* schemata.

As mentioned earlier, the U.S. schemata in this case were obtained in an earlier investigation (Adamopoulos, 1982b). The Greek schemata reported here were derived from preliminary analyses of an investigation using somewhat different stimuli. Greek college students judged the likelihood of occurrence of each of 20 behaviors in the context of 12 role-pairs and 12 environments. Mean ratings for each of the 2880 stimulus combinations were analyzed using a variety of multidimensional scaling procedures, including 3-way INDSCAL.

Table 1. Relationships among Social Interaction Dimensions in Two Cultures

Roles	Culture	Social Environments					
		Low Constraint			High Constraint		
		Behavior					
		ASSOC	INTIM	SUPER	ASSOC	INTIM	SUPER
DPC	US	0^3	-	$+^2$	0	-	0
	Greece	0	-	+	-	-	-
INT	US	+	0	0	$+^1$	-	0^4
	Greece	0	0	0	+	-	-

Note. Roles: DPC = Differential Power and Conflict; INT = Intimacy
Behaviors: ASSOC = Association; INTIM = Intimacy; SUPER = Superordination
[1] Prescriptive Schema 1 for Study 2 (US)
[2] Prescriptive Schema 2 for Study 2 (US)
[3] Non-Prescriptive Schema 1 for Study 2 (US)
[4] Non-Prescriptive Schema 2 for Study 2 (US)

As expected, most of the usual culture-common dimensions described earlier were obtained, but analyses suggested that the prescriptive schemata that emerged in the U.S. sample were not identical to those of the Greek sample. For example, as can be seen in Table 1, superordination seems to be much less likely among Greek than among U.S. respondents in the context of differential power/conflict roles and high constraint environments. This may mean, for instance, that Greeks are less likely to assert authority toward people of lesser status in public places where behavior is highly regulated. In general, it appears that constraint is experienced across a wider range of behaviors by Greeks. Considering the more normative nature of Greek social life, such schemata are to be expected. Even though a degree of caution is necessary at this point because of the preliminary nature of the analyses, these findings are not surprising considering known differences in individualism - collectivism between the two cultures.

Cultural differences like the ones described above can be used eventually to test the culture-dependent aspect of the proposed framework, and to elaborate the manner in which culture interacts with culturally invariant conceptual structures to determine social perception processes.

Once the schemata have been defined, the second phase of the investigation can explore the differential influence of frames, and their interaction with prescriptive schemata, on action identification. In this case a study with U.S. respondents explored the viability of this approach. Specifically, a procedure developed by Bransford and Franks (1971) was modified for the present context. This procedure involves the presentation of partial information about a complex sentence to respondents, who then have to engage in a recognition memory task regarding the sentence. In this case, test sentences described an act performed by a role person toward another role person in a particular environment. Two of the sentences represented instantiations of prescriptive schemata, and two represented non-prescriptive schemata. Thus, each test sentence contained four components: actor, target, act, and environment. During the acquisition phase, respondents were exposed to selected combinations of two or three components. During the recognition phase, they were asked to recognize "old" sentences from a list that included both "old" and "new" sentences with two, three, or all four components.

The major hypotheses tested in the study were: (1) "Old" instantiations of prescriptive schemata should receive higher recognition scores than "old" instantiations of non-prescriptive schemata, because of the existing cognitive structures that facilitate the encoding of information. (2) "New" instantiations of prescriptive schemata with all four components should receive higher false recognition scores than those with three or two, because the existing cognitive structures are expected to interfere with the accurate recall of presented information. Finally, comparisons of false recognition scores among sentences with different components should provide information about the importance of various frames in action identification.

Study 2: Action Frames and Recognition Memory

Method

Subjects

Forty-three U.S. college students volunteered for this study. They all received academic credit for their participation.

Materials and Procedure

As indicated in Table 1, two prescriptive and two non-prescriptive conceptual schemata were selected, and a particular instantiation was generated for each of them. For prescriptive schema 1, the sentence was: "The student helped his roommate with the class assignment in the library" (Intimate roles, associative behavior, high constraint environment). For prescriptive schema 2: "The father yelled at his son about his low school grades in the cafeteria" (Differential power roles, superordinate behavior, low constraint environment). For non-prescriptive schema 1: "The student greeted the professor in the cafeteria" (Differential power roles, associative behavior, low constraint environment). For non-prescriptive schema 2: "The woman gave her husband some advice in the church" (Intimate roles, superordinate behavior, high constraint environment).

Four partial sentences were generated from each of the complete instantiations - two with two components, and two with three components: (1) <Actor, Target>, or AxT (e.g., "the student was with his roommate"); (2) <Target, Behavior> or TxB (e.g., "the roommate received some help with the class assignment") ; (3) <Actor, Target, Environment> or AxTxE (e.g., "the student and his roommate were in the library"); and (4) <Target, Environment, Behavior> or TxExB (e.g., "the roommate received some help with the class assignment in the library") . Respondents listened to each of these sixteen sentences once, and then had to respond to a single question for each sentence (e.g., "who was?" or ""what was done?") so that they could hold the idea in memory for a few seconds. At the end of the acquisition phase respondents were given a five-minute break.

The recognition list consisted of twenty-eight sentences - seven from each schema instantiation: (1) the four-component sentence (AxTxExB), a "new" sentence (i.e., a sentence that was not in the acquisition list); (2) two three-component sentences: (AxTxB) and (AxExB), which were both "new"; and (3) three two-component sentences: (AxE), (TxE), and (AxB), which were "new."[2] In addition, two "old" three-component sentences (TxExB for prescriptive schema 2 and AxTxE for non-prescriptive schema 2), and two "old" two-component sentences (AxT from prescriptive schema 1 and TxB from non-prescriptive schema 1) were included in the recognition list. For each of the 28 items respondents indicated whether or not they recalled it ("yes-no" response), and rated their confidence in their recollection on a 5-point scale ranging from "not at all confident" to "extremely confident."

Results

Dependent Variables

Recognition responses for each item were combined in the following manner: (a) "yes" was coded as "1" and "no" as "-1"; (b) responses on the confidence scale were coded from "1" ("not at all confident") to "5" (extremely confident"). Responses to (a) were then multiplied with responses to "b", creating a single recognition score for each item ranging from -5 (no recognition, low confidence) to +5 (recognition, high confidence).

"Old" Items

As hypothesized, the mean recognition score of the two "old" sentences from prescriptive schemata ($M = 2.16$) was significantly greater than the mean recognition score ($M = 0.93$) of the two "old" sentences from non-prescriptive schemata [$t(42) = 2.65, p < .05$].

"New" Items

Most of the analyses focused on "new" items only because the main purpose of the experiment was to investigate the manner in which actions acquire meaning. It should be remembered that in this section higher scores imply false recognition, and, consequently, the influence of cognitive constructions of meaning.

Mean recognition scores were computed for all two-component sentences stemming from prescriptive schemata (6 items), three-component sentences from prescriptive schemata (4 items), four-component instantiations of prescriptive schemata (2 items), and their parallel structures stemming from non-prescriptive schemata. The data were then analyzed in a 2 (prescriptive vs. non-prescriptive schemata) X 3 (4, 3, and 2-component sentence types) ANOVA. Mean false recognition scores appear in Table 2.

Table 2. Comparisons (t-tests) of False Recognition Scores for Sentence Type

Components	Mean Recognition Score	Sentence Type				
		3-Component		2-Component		
		AxTxB	AxExB	AxE	TxE	AxB
AxTxExB (4)	1.80	1.57	9.08*	5.29*	4.45*	10.12*
AxTxB (3)	1.10		8.27*	4.76*	4.09*	10.92*
AxExB (3)	-2.00			-2.37	-4.00*	2.36
AxE (2)	-1.04				-1.01*	4.41*
TxE (2)	-0.61					5.21*
AxB (2)	-2.58					

*$p < .003$.

As predicted, respondents were more confused by sentences generated from prescriptive schemata ($M = 0.30$) than non-prescriptive schemata ($M = -0.38$), $F(1, 42) = 4.90, p < .05$. In addition, the type of sentence had a significant effect on recognition scores, $F(2, 84) = 41.40, p < .01$. Specifically, respondents had the highest false recognition scores for the four-component sentences ($M = 1.81$), as predicted on the basis of Bransford and Frank's (1971) studies, and had lower scores for the three-component ($M = -0.45$) and two-component items ($M = -1.48$). The linear trend was highly significant, $F(1, 42) = 52.72, p < .001$.

The interaction of schema and sentence type was also significant, $F(2, 84) = 11.55, p < .01$. As Figure 2 indicates, false recognition appeared to increase consistently with number of components for sentences derived from prescriptive schemata, but not necessarily so for other sentences. Analysis of simple effects indicated the following: (1) There was less false recognition for two-component sentences from prescriptive schemata than for sentences of the same type from non-prescriptive schemata, $F(1, 42) = 5.10, p < .05$. (2) There was more false recognition for three-component sentences from prescriptive schemata than for three-component sentences from non-prescriptive schemata, $F(1, 42) = 93.63, p < .05$. (3) The difference between the two kinds of four-component sentences did not reach significance, $F(1, 42) = 2.04, ns$.

Figure 2. Mean False Recognition Scores for Sentences Representing Conceptual Schemata with Different Numbers of Components

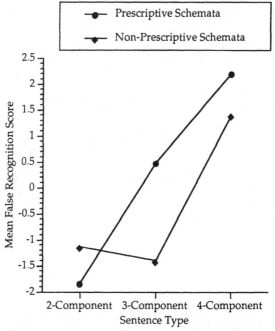

Action Frame Comparisons. The methodology outlined in this paper makes it possible to investigate the relative significance of particular frames in identifying action correctly. The limitations of the present design do not permit an exhaustive analysis of this problem, but certain suggestions are possible. Table 2 also presents the results of t-tests comparing false recognition scores for the various frames involved in the study. While a clear pattern does not emerge from these limited data, it appears that the presence of the components "actor" and "target" together is associated with higher levels of false recognition. On the other hand, the frame <Actor, Act (Behavior)> leads to significantly less false recognition than any other frame employed in the study.

General Discussion

The findings of this investigation provide support for all the basic hypotheses outlined earlier, and also validate the use of the Bransford and Franks (1971) methodology for the study of the identification of action.

Overall, it appears that conceptual schemata play an important role in determining how individuals understand social episodes by providing an interpretive framework for such episodes. This finding is quite consistent with previous research on the encoding and recall of information. However, the present study reaches beyond the conclusions of the initial investigation of Bransford and Franks (1971) by qualifying the operation of conceptual schemata. Specifically, it is primarily schemata that seem to prescribe behavior in a given cultural context that provide an interpretive framework for interpersonal behavior. Descriptions of ordinary events that are not generated from prescriptive schemata may be remembered individually, rather than within the broader context of social rules. However, this phenomenon appears to be less pronounced with both very simple and very complex events. The specific processes involved in these cases are not clear at this point.

An alternative way of expressing the findings of this investigation is that the more the information we receive from the environment approximates what we already know about interpersonal interaction in the social environment, the more we tend to confuse actual events with generalized expectations.

Results from the pair-wise comparisons of frames are not conclusive, though certainly instructive. For example, whereas it is clear that, at least for prescriptive schemata, the addition of components tends to make action identification more difficult because of the overwhelming tendency to interpret events on the basis of existing knowledge structures, the operation of specific components seems difficult to gauge. However, in general, it appears that the <Actor, Target> frame may involve conceptual properties that interfere with the accurate identification of action - perhaps as a result of the central function of role-pairs in generating an interpretive context for action. On the other hand, the <Actor, Act> frame appears to be the least likely to lead to false recognition, and thus may reflect a basic unit employed in the encoding of action in the present investigation. Clearly, however, these conclusions must be validated by investigations involving more extensive designs that will allow the direct comparison of many frames within the context of multiple instantiations of several different conceptual schemata (e.g., Logas and Adamopoulos, 1995).

It seems obvious that cross-cultural psychology should be able to make a significant contribution to the analysis of action identification. The methodology outlined in this paper appears to hold some promise for facilitating our understanding of this core psychological problem.

References

Adamopoulos. J. (1982a). Analysis of interpersonal structures in literary works of three historical periods. *Journal of Cross-Cultural Psychology, 13,* 157-168.

Adamopoulos, J. (1982b). The perception of interpersonal behavior: Dimensionality and importance of the social environment. *Environment and Behavior, 14,* 29-44.

Adamopoulos, J. (1984). The differentiation of social behavior: Toward an explanation of universal interpersonal structures. *Journal of Cross-Cultural Psychology, 15,* 487-508.

Adamopoulos, J. (1988). Interpersonal behavior: Cross-cultural and historical perspectives. In M. H. Bond (Ed.), *The cross-cultural challenge to social psychology* (pp. 196-207). Newbury Park, Cal.: Sage.

Adamopoulos, J. (1991). The emergence of interpersonal behavior: Diachronic and cross-cultural processes in the evolution of intimacy. In S. Ting-Toomey & F. Korzenny (Eds.), *International and intercultural communication annual*: Vol. 15. Cross-cultural interpersonal communication (pp. 155-170). Newbury Park, Cal.: Sage.

Adamopoulos, J., & Bontempo, R. N. (1986). Diachronic universals in interpersonal structures: Evidence from literary sources. *Journal of Cross-Cultural Psychology, 17,* 169-189.

Adamopoulos, J., Smith, C. M., Shilling, C. J., & Stogiannidou, A. (1993). *Cross-cultural invariance in the perception of social environments: A rule-theoretic approach to situational classification*. Unpublished manuscript.

Bransford, J. D., & Franks, J. J. (1971). The abstraction of linguistic ideas. *Cognitive Psychology, 2,* 331-350.

Davidson, D. (1980). *Essays on actions and events*. Oxford: Clarendon Press.

Goldman, A. (1970). *A theory of human action*. Englewood Cliffs, N.J.: Prentice Hall.

Heider, F. (1959). On perception, event structure, and psychological environment. *Psychological Issues, 1,* 1-123.

Kahneman, D., & Miller, D. T. (1986). Norm theory: Comparing reality to its alternatives. *Psychological Review, 93,* 136-153.

Lewin, K. (1951). *Field theory in social science: Selected theoretical papers*. New York: Harper and Brothers.

Lichtenstein, E. H., & Brewer, W. F. (1980). Memory for goal-directed events. *Cognitive Psychology, 12,* 412-445.

Logas, M., & Adamopoulos, J. (1995). *The perception of interpersonal episodes*. Paper presented at the meeting of the Midwestern Psychological Association, Chicago, Illinois, U.S.A.

Newtson, D. (1973). Attribution and the unit of perception of ongoing behavior. *Journal of Personality and Social Psychology, 28,* 28-38.

Newtson, D., & Engquist, G. (1976). The perceptual organization of ongoing behavior. *Journal of Experimental Social Psychology, 12,* 436-450.

Newtson, D., Engquist, G., & Bois, J. (1977). The objective basis of behavior units. *Journal of Personality and Social Psychology, 35,* 847-862.

Nottenburg, G., & Shoben, E. J. (1980). Scripts as linear orders. *Journal of Experimental Social Psychology, 16,* 329-347.

Oppenheimer, L., & Valsiner, J. (Eds.) (1991). *The origins of action: Interdisciplinary and international perspectives*. New York: Springer-Verlag.

Triandis, H. C. (1977). *Interpersonal behavior*. Monterey, Calif.: Brooks/Cole.
Vallacher, R. R., & Wegner, D. M. (1985). *A theory of action identification*. Hillsdale, N.J.: Lawrence Erlbaum.

Notes

1. Each of the three domains described here can be defined by a number of additional culture-general and culture-specific dimensions (e.g., Adamopoulos, 1988; Adamopoulos, Smith, Shilling, & Stogiannidou, 1993). For expediency, the present discussion focuses only on a partial list of dimensions from each domain.
2. One of these sentences, (AxE) for prescriptive schema 2, was inadvertently left out of the recognition list. However, since analyses were based in this case on the mean of five items, it is expected that this problem is of little, if any, significance in the overall pattern of results.
3. We would like to thank Jayne Edmundson and Heather Crandall for their help in the studies reported here, and Christine Smith for her many invaluable suggestions.
4. This research was supported by a research stipend and a research grant from Grand Valley State University to the first author.

The Development of Concepts of Fairness in Rewards in Chinese and Australian Children

Daphne M. Keats, University of Newcastle, Newcastle, Australia
Fang Fu-Xi, Institute of Psychology, Beijing, China

The aim of this study was to explore how Chinese and Australian children understand the concept of fairness in the giving of rewards in a situation of daily life. Whereas following Piaget's (1932) work on the development of moral judgement in the child there has been much written on fairness in meting out punishment for wrongdoing, much less has been written on giving rewards to children. Yet this is a situation very familiar to children and a sense of injustice when a reward is unfairly given or given to the wrong child can be very distressful.

Perhaps the closest is the work on distributive justice by Damon (Damon, 1977, 1981; Darley & Schultz, 1990). Damon's method used dilemmas based on Kohlberg's (1981) theory of moral development, but more suitable for children. In these studies children had to divide up a reward among themselves. The extent to which children's views of the justice or otherwise of their solutions to the problems would coincide with those of adults has not been examined. Also, although situational factors are known to affect fairness of allocation rules (Darley & Schultz, 1990), the role of culture has not been taken into account.

According to Damon's theory, children under five disregard differences in input, allocating to the ones who want the items most. This is followed by a stage in which a distribution is seen as fair if it corresponds to a characteristic of the recipient, for example, age or size. Next comes the use of the equality rule, whereby all receive the same reward regardless of other factors. This is followed by a proportional equity rule, which takes account of input, and finally "a distributive morality rule emerges in which distributions are considered fair if they correlate with recipients' assessed needs" (Darley & Schultz, 1990). In this developmental process the child's cognitive abilities to differentiate and integrate information are also developing (Anderson, 1980), so that older children take more factors into consideration in their judgements.

The issues are related to moral development and the development of social perspective taking, both of which are aspects of the more general development of values. Previous studies of social perspective taking in Chinese and Australian children (Fang & Keats, 1989; Keats & Fang, 1992) found an earlier development of this concept in the Chinese than the Australian children, but a similar developmental trend to that proposed by Selman (1976). In view of this finding it might be expected that Chinese children might also differ from Australian children on their ideas of fairness in the giving of rewards but that a similar developmental trend might be found.

According to a naive theory of how adults would treat the concept, various factors or rules would be weighed and combined in making a decision as to how to distribute rewards fairly. If this is so, then three important questions are raised. First, is there a model of adults' distributive practices which is a product of the culture and may thus be specific to a particular culture, or is there a more general model which transcends cultural differences? Second, if there is an adult model, whether culturally specific or more general, how do children progressively come to reach the mature model typical of the adults? Third, how and to what degree does the development of the concepts of fairness in the allocation of rewards in daily practices depend on the processes of socialisation involved in acquiring social values and the ability to integrate information?

In this study we were particularly interested in three factors: 1) rewarding on the basis of the outcome or product; 2) rewarding on the basis of the effort put in; and 3) rewarding on the basis of ability, i.e., the ability of the child to carry out the task for which the reward is given. Two further considerations were whether girls should be treated differently from boys, and whether the very young should be more favourably treated than older children. The latter is a traditional Chinese attitude derived from Confucian precepts of care and responsibility.

The two cultures involved in the study, China and Australia, present great cultural contrasts. China has a long history of child rearing principles steeped in Confucian morality. Australia's predominant heritage is in the western cultural tradition, and is egalitarian and individualistic in orientation. In contrast, the traditional Confucian Chinese cultural orientation was hierarchical and collectivistic. However, in modern China the dominant ideology of the socialist principle of distribution according to the contribution to the society - or production - should be expected to influence the values of adults and hence the development of values in children. The traditional value of mutual hierarchical responsibility remains in the custom of older persons showing indulgence to the very young.

On the basis of these cultural differences, the work of Damon discussed above and results from our previous studies on the development of social perspective taking, several predictions can be made. First, because adults would make decisions for the allocation of rewards in terms of the criteria or rules of fairness which should be strongly influenced by the social values with which they identify, there would be differences among differing sociocultural groups. However, adults within the same socio-cultural group and living at the same time would have relatively consistent patterns of allocation because they have the same cultural tradition and are influenced by the dominant ideology of the society.

Second, there is a universal process by which the development of children's cognitions of fairness in the allocation of rewards goes through a process in which the criteria of rules of fairness are differentiated and integrated. This process is constrained by both the development of socialisation by which children acquire social values and the development of information processing abilities. We expect that Chinese children would be more mature in the use of differentiated rules in decision making than Australian children because in China there is prevailing propaganda for the socialist principle according to the contribution or production. Furthermore, the strong competition for students to gain entry into the "key schools" or top universities leads teachers and parents to study hard, valuing effort highly and punishing the lazy.

Method

Sample

The subjects were 90 children in each culture, with 30 in each of three age groups, 7 years, 9 years and 12 years, with equal numbers of boys and girls in each group, drawn from middle range schools in Beijing and Newcastle. In addition there was a sample of 30 adults from each culture, drawn from a wide range of ages and socioeconomic backgrounds. The mean ages were 41 years for the Chinese and 37 years for the Australians. There were equal numbers of males and females in each group.

Design

Materials
To elicit the children's responses a story was created in which four children help a farmer to pick his apples when the farmer becomes ill. The story was brightly illustrated, with appealing child characters given currently popular names. The story was first created in the Chinese version, then a matching version with Australian children was prepared keeping the wording and illustrations as close as possible to the original version. The only changes were in the eye and hair coloring of the Australian children to include blue eyes and blond and brown hair and the inclusion of cicadas as distractions of the lazy boy.

Testers were trained in China and Australia and pilot runs conducted to ensure crosscultural comparability and tester reliability.

The Apples Story
This is a story about a poor farmer who had two lovely apple trees in his garden. This year the trees were laden with apples.

But when the apples were ripe and ready to pick, the farmer became sick and had to stay in bed. He looked sadly out of the window but could do nothing to pick the apples. He was worried that a storm would come and his apple crop would be ruined.

Now there were three children who often used to come and play in his garden. There was a little girl, whose name was Li Li/Kirsty, a boy, Xiao Qiang/Jason, a little bigger than Li Li/Kirsty, and a bigger boy, whose name was Xiao Tao/Adam.

When they came to play that day they found that the apples were ready to pick but the farmer was sick in bed and worried that a storm might come and ruin his apple crop. So they decided to help the farmer pick the apples.

Xiao Qiang/Jason worked very hard and soon had a large basket heaped full with apples.

Li Li/Kirsty worked very hard too and soon she also had a basket full of apples, but not as many as Xiao Qiang/Jason.

But Xiao Tao/Adam was lazy. Instead of working hard he chased butterflies and cicadas and only collected half a basketful.

While the children were picking the apples along came little Ming Ming/Kim. Ming Ming/Kim wanted to join in and help pick the apples too. Ming Ming/Kim was quite small so couldn't reach up very high. So the others said that Ming Ming/Kim could pick up those that fell on the ground. Ming Ming/Kim worked very hard and soon had some in the basket, although not as many as the others.

When they had finished, they had four baskets of apples. Xiao Qiang/Jason had the most, Li Li/Kirsty had the next most, Xiao Tao/Adam had the next, and Ming Ming/Kim had the smallest amount.

The farmer was so pleased that he felt much better and wanted to give the children a reward for their kindness. So he took a basket of eight apples to divide out among the children.

How should the farmer divide out the eight apples so it will be fair?"

Procedure

A basket of eight "apples" was then presented to the child who was asked to distribute them to the children in the story in such a way as to be fair. The "apples" were actually red plastic pencil sharpeners in the shape of apples. Using the final illustration showing the four children each with their baskets of apples, the child then gave out the "apples" to the children in the story according to his or her judgements of a fair allocation of the reward. When the child was satisfied with the allocation, the reasons for the judgement were sought.

After allocating the eight apples, the child was asked whether they could allocate them fairly with more apples. The basket of twelve apples was then presented. to the child and the reasons sought as before.

Then a series of 7 possible allocations was presented and the child was asked whether each one would be fair. Again the reasons were sought after each response. The patterns of these hypothetical allocations were chosen to represent combinations in which the key variables of product, effort and ability were differentially weighted and in interaction. It was not possible to isolate single variables, nor would that have been a true representation of an everyday situation.

Patterns

A. (4211): The child who picked the most apples receives the largest reward.
B. (2231): The biggest boy, most able to pick the apples, receives the largest reward irrespective of the effort put in by himself or the other children.
C. (3212): Both effort and product are rewarded. Lack of effort in the biggest child is penalised, effort by the other three is rewarded and an extra apple is given to the child who picked the most.
D. (3311): Effort and product are rewarded in the first two children but the smallest receives fewer and the lack of effort in the biggest child is penalised.
E. (2222): All children are treated in the same way.
F. (2213): The youngest child is given the most on the basis of being the youngest rather than for effort or product.
G. (3302): The biggest boy is punished for his laziness, effort and product are rewarded.

The various patterns were presented in random order. The adult subjects went through the same procedure as the children, with appropriately worded variation in the instructions to explain the purpose of the study and the inclusion of the adult group. In addition they were asked to rank the 7 hypothetical arrangements in order of their acceptability.

Results

A formula for representing the three factors of effort, product and ability was devised as set out in Tables 1 and 2.

Table 1. Ranking for Relevant Factors Attributed to Protagonists of the Story.

	Effort (E)	Product (P)	Ability (A)
Xiao Qiang/Jason (C1)	+1	4	3
Li Li/Kirsty (C2)	+1	3	2
Xiao Tao/Adam (C3)	-1	2	4
Ming Ming/Kim (C4)	+1	1	1

Formula of weights of relevant factors: $C1(E+4P+3A) + C2(E+3P+2A) + C3(E+2P+4A) + C4(E+P+A)$ where C = Number of actual allocation for the protagonist - Mean of allocation.

Allocation of 8 Apples

The allocation of 8 apples is shown in Table 3. As no gender differences were found in either the Chinese or the Australian data, the data for males and females were combined for the analysis. Log linear analysis showed that the result was accounted for by differences in pattern and interactions between culture and age, culture and pattern and age and pattern, as in Table 4. Neither culture alone nor age alone was significant.

Table 2. Hypothetical Patterns for Allocations of Rewards with 8 Apples.

Protagonists	C1	C2	C3	C4	Weights of Factors
PatternA	4	2	1	1	5P+2E+ 1A
Pattern B	2	2	3	1	1P + 3A
Pattern C	3	2	1	2	2P + 2E - 1A
Pattern D	3	3	1	1	4P + 2E
Pattern E	2	2	2	2	0
Pattern F	2	2	1	3	P + 2E - 3A
Pattern G	3	3	0	2	3P + 4E - 3A

Table 3. Autonomous Distribution with 8 Apples

Pattern	7yrs		9yrs		12yrs		Adults		Total	
	Ch	Aus	Ch	Aus	Ch	Aus	Ch	Aus	Ch	Aus
A(4211)	0	0	0	0	0	1	1	0	1	1
B (2231)	0	0	0	0	0	0	0	0	0	0
C (3212)	7	0	6	1	19	12	17	10	49	23
D (3311)	2	0	2	0	0	1	3	0	7	1
E (2222)	12	30	14	29	1	15	3	12	30	86
F (2213)	5	0	3	0	10	0	6	3	24	3
G (3302)	1	0	3	0	0	0	0	2	4	2
H(Others)	3	0	2	0	0	1	0	3	5	4

Table 4. Log Linear Analysis of Allocation of 8 Apples

Effect Name	df	Partial chi sq.	Prob.
Culture x age	3	23.972	.0000
Culture x pattern	7	85.918	.0000
Age x pattern	21	100.284	.0000
Culture	1	.000	1.0000
Age	3	.300	.9500
Pattern	7	361.168	.0000

Allocation of up to 12 Apples

Frequencies were obtained and the allocations categorised into the patterns according to the weights set out above. The Australian children tended to use all 12 apples, whereas the Chinese did not. The results are shown in Table 5. Log linear analysis showed that the best model had the generating class culture x pattern (Likelihood ratio chi square = 60.54432, df48, $p=.106$). The results are shown in Table 6. As in the case of 8 apples, culture alone did not account for the differences.

Table 5. Autonomous Distribution of up to 12 Apples

Group Patterns	7yrs Ch	7yrs Aus	9yrs Ch	9yrs Aus	12yrs Ch	12yrs Aus	Adults Ch	Adults Aus	Total Ch	Total Aus
A(4211)	0	0	0	1	0	0	2	0	2	1
B(2231)	0	0	0	0	0	0	0	0	0	0
C(3212)	18	1	23	4	20	10	19	8	80	23
D(3311)	3	2	1	2	2	2	7	0	13	6
E(2222)	3	23	3	12	0	11	1	11	7	57
F(2213)	4	1	2	1	8	0	1	2	15	4
G(3302)	0	1	0	2	0	1	0	3	0	7
H(Other)	2	2	1	8	0	6	0	6	3	22

Table 6. Log linear Analysis with up to 12 Apples

Effect name	df	Partiatchi square	Prob.
Culture x age	3	5.374	.1464
Culture x pattern	7	119.080	.0000
Age x pattern	21	29.461	.1034
Culture	1	.000	1.0000
Age	3	.000	1.0000
Pattern	7	273.072	.0000

It will be noted that in this allocation there are a number of entries which do not fit neatly into the seven patterns. (Some Australian children, used to mothers cutting up fruit to share out with their brothers, sisters and/or friends, and seeing that the "apples" could be undone into two halves, wanted to cut the apples in two, and even into three in one case.)

Consistency in Autonomous Allocations

An analysis of the relationship between the allocations for the two conditions showed that in the Chinese the 7- and 9-yr olds changed their allocations but the 12-yr olds and the adults were more consistent. In the Australians few of the 7-yr olds changed but over half of the 9-yr olds and 12-yr olds and the adults changed.

It is possible that this result could have been an artifact of the style of giving the instructions, but the effect is also seen in the relatively large number of responses in the H (Other) category. Further analysis is needed to resolve this question. The results are shown in Table 7.

Judgements of Fairness in Hypothetical Allocations

The results are summarised in Table 8. As no gender differences were found in either the Chinese or the Australians the data for males and females were combined for the analysis. Log linear analysis of these data showed no significant effect for culture alone and the final model with generating class culture x age, culture x pattern and age x pattern. The results are shown in Table 9.

Table 7. Consistency in Autonomous Allocations

Group	Number of consistent allocations	
	Chinese	Australian
7yrs	8	23
9yrs	10	12
12yrs	20	15
Adults	15	15

Table 8. Judgements of Fairness of Hypothetical Allocations

Group Pattern	7yrs		9yrs		12yrs		Adults		Total	
	Ch	Aus	Ch	Aus	Ch	Aus	Ch	Aus	Ch	Aus
A(4211)	12	7	6	9	0	11	6	2	24	29
B(2231)	0	5	0	0	0	1	0	0	0	6
C(3212)	19	13	17	7	25	20	23	13	94	53
D(3311)	10	9	9	10	1	14	13	6	33	39
E(2222)	16	30	12	29	2	21	7	19	37	99
F(2213)	19	11	13	9	16	10	15	6	63	36
G(3302)	16	10	13	5	1	3	0	2	30	20

Table 9. Log Linear Analysis of Judgements of Fairness of Hypothetical Allocations

Effect name	df	Partial chi square	Prob.
Culture x age	3	18.713	.0003
Culture x pattern	7	61.376	.0000
Age x pattern	21	61.731	.0000
Culture	1	.219	.6400
Age	3	16.619	.0008
Pattern	7	358.621	.0000

Ranking of Hypothetical Allocations

Chinese and Australian adults' rankings of the 7 hypothetical distributions were compared. Results showed a high level of general agreement but some differences appeared in the rankings of the Australian females. The results are shown in Table 10.

Table 10. Rank Order Correlations between Rankings for Fairness of Hypothetical Allocations

Group Patterns	Chinese Total	Rnk	Aus(M) Total	Rnk	Aus(F) Total	Rnk	Aus(M+F) Total	Rnk
A(4211)	92.5	3	72	5	70	5	142	5
B(2231)	207	7	89	7	93	7	182	7
C(3212)	49.5	1	30	1	34	2	64	1
D(3311)	91.5	2	49	2	53	3	102	3
E(2222)	144.5	6	51	3	31	1	82	2
F(2213)	111.5	4	53	4	62	4	115	4
G(3302)	143.5	5	76	6	75	6	151	6

Spearman rank order correlations: Chinese vs. Australian total samples .61; Australian male vs. Australian female .89; Australian male vs. Chinese total sample .77; Australian female vs. Chinese total sample .43.

Reasons for Judgements

The analysis of reasons is proceeding. Reasons have been classified into categories based on simple equality concepts, the major variables of effort, ability and product, with the addition of care and gender, and combinations of these variables in simple additive form and in interactive form. The data were derived from the autonomous judgements and the set of hypothetical patterns. Because of the large volume and complexity of these data, only some typical responses are treated in the Discussion.

Discussion

In both China and Australia, the task appeared to be relevant and highly engaging to the subjects in all age groups.

As predicted, the results support the notion of the development of more differentiated and integrated criteria rules of fairness, shown in the increase in the number of variables taken into account. However, there are some clear differences between the Chinese and the Australian subjects in the age at which this appears. The common trend is from favouring rewarding on the basis of equality for all to a more differentiated basis with effort and product as the major variables in interaction. As expected, the Chinese show evidence of this trend at a younger age and more strongly than the Australians, who are almost unanimously for the equality basis in the two youngest groups. So strong is this preference for the equality basis that it appears also in the adults, especially the females, whereas in the Chinese adults the effort x product basis is unanimous. These results give some support to Damon's theory, but the cultural differences limit this support.

However, the log linear analysis in each case shows that culture alone does not account for the result, although it plays a part. Nor does age alone account for the result, but it is the interaction between pattern and culture which is significant in each case.

It would appear from these results that the Chinese do have an adult model towards which children progress. This is less strong with the Australian data. Although the trend toward Pattern C is apparent, it could be that the difference is linked to the Australians' concern that the three willing workers should receive similar rewards given their relative abilities rather than that the reward should include a bias toward product. Further work on the reasons should elucidate this point. The finding is similar to that obtained in our previous work on concepts of intelligence (Keats & Fang, 1987) which showed Chinese children to be more like adults in their concepts than were the Australian children.

There is quite high correspondence between Australian and Chinese in the rank ordering of the patterns by the adults. The difference clearly comes from the high ranking given to the equality pattern by the female respondents. Otherwise there is a high priority given to the effort x product pattern and consensus against giving a high reward to the biggest boy, i.e. rewarding solely on the basis of ability.

The Australians did not approve of differentially favouring the youngest child (Pattern F, 2213). Approval of this pattern was found only with the Chinese subjects, where it occurred in all age groups but most often in the 12-yr olds. This failure to favour the youngest is also apparent in the Australians' greater preference for Pattern D (3311) which was not favoured by the Chinese in their autonomous distributions.

The analysis of reasons will explain the basis for many of these cultural differences and the developmental changes which underlie the judgements. The following excerpts illustrate some of these differences. The examples are from reasons given by some Australian subjects as to why Pattern C (3212) was not considered to be fair.

Age 7 yrs. Simplistic restatement of the situation: "Not fair." Why isn't it fair? "Because he's got 2, he's got 1, she's got 2 and he's got three." Why is that not fair? "Because he's got the most and he's got the least."(Child points to the children.) But why isn't that fair? "Because he's only got one. "

Age 9 yrs. Attribution to effort: "No, because they all tried and they should all get the same amount. "

Age 9 yrs. Attribution to ability: "But Adam's bigger and he only gets one. "

Age 12 yrs. Attribution to effort: "They both (Jason and Kirsty) did the same amount of work and they both (Adam and Kim) didn't. "

Adult Complex attribution with empathetic viewpoint: "No. There is no reason why he (Jason) should receive more than Kirsty, Kirsty two, Adam one and Kim two. The children would not see it as fair. Adam is being punished, but so are Kirsty and Kim."

Conclusion

The theoretical implications of the findings are as yet tentative, but suggest the possibility of a common developmental route in terms of development from simple to complex reasons but one that is modified by the presence of differences reflecting cultural values and child rearing practices. However, the methodology designed for this study can be readily adapted for use in other cultural settings: if apple trees are scarce a local variation can be created. The patterns used can also be varied to stress culturally relevant factors considered to be of importance.

References

Anderson, N.H. (1980). Information integration theory in developmental psychology. In F. Wilkening, J. Becker, & T. Trabasso (Eds.), *Information integration by children*. Hillsdale, N.J.: Erlbaum.

Damon, W. (1977). *The social world of the child*. San Francisco: Jossey Bass.

Damon, W. (1981). The development of justice and self-interest change in childhood. In M.J. Lerner & S. Lerner (Eds.), *The justice motives in social behavior*. New York: Plenum.

Darley, J.M., & Schultz, T.R. (1990). Moral rules: Their content and acquisition. *Annual Review of Psychology, 41,* 525-526.

Fang, F-X., & Keats, D.M. *(1992)*. The effect of modification of the cultural content of stimulus materials on social perspective taking ability in Chinese and Australian children. In S. Iwawaki, Y. Kashima, & K. Leung (Eds.), *Innovations in crosscultural psychology*. Lisse: Swets & Zeitlinger.

Keats, D.M., & Fang, F-X. (1989). "The Master and the Wolf": A study in the development of social perspective taking in Chinese and Australian children. In D.M. Keats, D. Munro, & L. Mann (Eds.), *Heterogeneity in cross cultural psychology*. Lisse: Swets & Zeitlinger.

Kohlberg, L. (1981). *Essays on moral development: Vol.1. The philosophy of moral development: Moral stages and the idea of justice*. San Francisco: Harper & Row.

Keats, D. M., & Fang, F-X. (1987). Cultural factors in concepts of intelligence. In C. Kagitcibasi (Ed.), *Growth and progress in cross cultural psychology* (pp. 236-247). Lisse: Swets & Zeitlinger.

Keats, D. M., & Jang, F-X. (1992). The effect of modification of the cultural content of stimulus materials on social perspective taking ability in Chinese and Australian children. In S. Juawaki, Y. Kashima, & K. Leung (Eds.), *Innovations in cross-cultural psychology* (pp. 319-327). Lisse: Swets & Zeitlinger.
Piaget, J. (1932). *The moral judgement of the child.* London: Routledge & Kegan Paul.
Selman, R.L.(1976). Social-cognitive understanding: A guide to educational and clinical practice. In T. Lickona (Ed.), *Moral development and behavior: Theory, research and social issues.* New York: Holt, Rinehart & Winston.

Long-Term Equity within a Group: An Application of the Seniority Norm in Japan

Masaki Yuki & Susumu Yamaguchi
The University of Tokyo, Tokyo, Japan

Adams' (1965) pioneering research on equity theory explored the concept of fairness in interpersonal relationships by outlining how resources are exchanged between two people. According to Walster, Berscheid, & Walster (1973), a relationship is considered equitable when participants' outcome/input ratio is equal and is perceived to be fair by the participants. If one's outcome/input ratio is more or less than the other's, then the relationship is considered inequitable. Individuals who experience an inequitable relationship suffer distress and are motivated to establish equity. Considerable amount of evidence has accumulated to support this model in explaining interpersonal behavior (Hatfield, Walster, & Berscheid, 1978).

Mainstream equity theory, evolved from earlier work of Adams (1965), has been reformulated by Hatfield et al., (1978) to examine outcome/input ratio of individuals in a particular timeframe and relationship (i.e., the theory focuses on short-term, one-to-one equity). Austin and Walster (1974, 1975) described such a model as the "person-specific equity" model. Austin and Walster (1974, 1975) articulate an alternative approach that extends the person-specific equity model, known as the "Equity with the World." The Equity with the World is defined as "the degree of equity present in the *totality* of a person's relationships during a given period of time" (Austin & Walster, 1974, p. 478). Austin & Walster (1974, 1975), and Moschetti & Kues (1978) found that people engage in trans-relational equity assessment, which deals with not only one particular relationship, but on two or more relationships. In the experiments, the subjects who experienced disadvantageous outcomes in one relationship showed a strong tendency to compensate and to seek greater rewards in other relationships. The results were interpreted as the subjects' attempt to maintain overall equity with the world.

The Seniority Norm

There is, however, a relationship in which the person-specific equity theory and even the extended Equity with the World model cannot adequately explain. In social groups such as the American fraternity club or sorority club, equitable relationships are not maintained in a particular timeframe and situation. In the case of fraternity clubs that maintains hierarchical relationships, not all members enjoy the same benefits and privileges. The group members are differentiated by their seniority status, and senior members receive preferential treatments and services (e.g., in some fraternity clubs, incoming freshmen are required to cook, clean, and serve the wishes of the senior members); known as the "seniority norm."

From the perspective of the person-specific equity theory, the seniority norm appears to violate the equity principle and therefore junior members should perceive the relationship to be inequitable. Senior members have access to more privileges and enjoy more benefits and junior members are in a disadvantageous position. The asymmetry of the benefits received will not usually be resolved or reciprocated since the senior members will graduate and leave the group.

Although the seniority norm appears to violate the equity principle from the person-specific, short-term perspective, the long-term equity can be maintained by considering the trans-relational aspects of a group. An incoming junior member of a group have to serve senior members, and the senior members do not reciprocate. However, as long as the group persists and the seniority norm is maintained, the junior members can expect similar privileges and benefits of current senior members when they themselves become a senior member. Although this type of short-term inequity is also found in many sports teams, recreational clubs, and interest groups, the equity is maintained in a long-term, trans-relational, and trans-situational context.

The Equity with the World model cannot adequately explain the seniority norm. Within the Equity with the World model, a person who experienced an inequitable relationship will attempt to find a non-specific other whom they could enjoy surplus benefits from. Under the seniority norm, the target person is predefined as an incoming junior member of one's ingroup. The incoming junior members will serve them when they themselves become senior members. Equity will be maintained in the seniority norm if the ingroup persists and newcomers uphold the seniority norm. This certainty of future benefits makes it possible for the junior members to willingly serve the senior members.

Long-Term Equity within a Group

The *Long-Term Equity within a Group* (L-TEG) model extends the equity theory by incorporating the long-term, and trans-relational aspects. The seniority norm is an example of the Long-Term Equity within a Group. Within this model, a new member of a group is willing to serve senior members with an expectation that the person will be served when the person becomes a senior member. In other words, one could say: "For me, the current inequity with my senior members is not a problem because I will become a senior member and enjoy similar benefits soon." By accepting the seniority norm, individual members could expect that equity will be maintained in the long run.

As discussed above, under the seniority norm, an equitable senior-junior relationship is maintained if the following two conditions are met. First, the group must remain intact. If the group is dismantled, the current junior members will not be able to enjoy the benefits since there will not be incoming junior members to serve them. Second, the seniority norm must be maintained in the group. If the seniority norm is rejected by incoming junior members, the current senior members will not receive benefits that they once provided to their former senior members. In these conditions, junior members will accept the seniority norm and consider it as fair.

Japanese Society and Culture

Japanese society is found to be relatively high in *Power Distance* and *Collectivism* compared to Western societies (Hofstede, 1980). The seniority norm, known as *Nenko-Joretsu* in Japanese, is widespread in many informal and formal groups. For example, rookies of university baseball teams are required to prepare the ground before a game, pick up stray balls, and clean up after the game for senior members. Even in informal situations, junior members are expected to be compliant and entertain the senior members. Similarly, in Japanese companies, junior workers are expected to work harder, show respect to senior workers, and receive less salary compared to senior workers.

There are several reasons why Japanese people are likely to accept the seniority norm. First, Japanese people tend to adopt a long-term perspective in forming and maintaining their interpersonal relationships. Acceptance of the seniority norm requires individuals to adopt a long-term perspective since the interval between the time of serving and served is usually several years. This is consistent with Triandis' (1990) observation that people living in collectivistic cultures tend to foster long-term interpersonal relationships. They do not expect immediate reciprocity, but emphasize the importance on long-term reciprocity.

Second, Japanese people tend to see individuals in terms of their ingroup membership. Individuals who have collectivistic tendencies are likely to see one's ingroup as being more homogeneous than outgroups (Hidaka & Yamaguchi, 1994; Triandis, 1990, 1994). In a study on Equity with the World, Austin and Walster (1975) suggested that individuals were more likely to engage in trans-relational equity with similar others. Thus, it is reasonable to assume that Japanese people, who are relatively collectivistic, are more likely to engage in trans-relational equity in an ingroup situation.

Third, anthropologists point out that Japanese social groups are stable and long-standing (Hsu, 1975; Lebra, 1976; Nakane, 1970). Thus, it may be reasonable for Japanese people to expect long-term equity maintenance. The indigenous Japanese concept of *On* (obligation) is an example of a long-term equity maintenance (Ho, 1993).

We experimentally tested the validity of the L-TEG model, and more specifically, the seniority norm, with a sample of Japanese college students. In the present study, we examined how students perceive a protester of the seniority norm in a fictitious situation as a third party. Research has demonstrated that impartial observers react to injustice in much the same way as participants (Baker, 1974; Lerner, 1965, 1970; Lerner & Matthews, 1967; Lerner & Simmons, 1966). Austin, Walster, & Utne (1976) concluded that impartial observers' reactions to injustice were similar to participants' reactions, although they are less passionate and more objective than participants. Thus, it is meaningful to examine the perception of subjects who evaluate a fictitious scenario as a third party.

The independent variable in the present study was the expectation of the persistence of a fictitious ingroup of the stimulus person. In the condition labelled as the *breakup*, the group was to be disbanded in the near future. In the *control* condition, there was no mention of the future of the group. Leventhal (1980) suggested that individuals' concern with inequity and injustice may become salient when exchanged resources change suddenly. Similarly, Hatfield, Traupmann, Sprecher, Utne, & Hay (1985) point out that concerns with equity may become salient when there are dramatic shifts in the relationship.

We also examined a contributing factor, the subjects' attitude toward the seniority norm that affects the perception of equity. Even in the collectivistic Japanese society, there are individuals who reject the seniority norm. Therefore, it was expected that those who have a positive attitude toward the seniority norm will dislike the protester of the seniority norm, whereas those who have a negative attitude toward the seniority norm will like the protester, due to similarity in attitudes (Byrne & Nelson, 1965).

Attitude toward the seniority norm may also operate as a moderating variable. It is expected that people who have a positive attitude toward the seniority norm will sympathize with the protester who cannot maintain the long-term equity within the group and accept the person's objection to the seniority norm. Therefore, it is expected that those who have a positive attitude toward the seniority norm will like the protester of the seniority norm more when his ingroup is going to be disbanded than when it is not. On the other hand, since the L-TEG model is based on the perception of fairness of people who accept the seniority norm, our experimental manipulation may not affect people who reject the seniority norm. Thus, for individuals who have negative attitudes toward the seniority norm, a direct prediction was not made. The above reasoning led us to the following two predictions.

Prediction 1: Compared to the opponents of the seniority norm, the proponents will perceive the protester as less desirable.

Prediction 2: Compared to the control condition, the proponents will perceive the protester of the seniority norm as more desirable when the protester's ingroup is to be disbanded (breakup condition).

Method

Subjects

Subjects were 90 Japanese female undergraduate students recruited from a women's college in Tokyo area. They were enrolled in introductory psychology classes; their age ranged from 18 to 25 years old; 99% of them were first year students; and their major was either English literature or Japanese literature.

Procedure

The subjects were classified into two groups according to their individual attitudes toward the seniority norm: (1) subjects who have a positive attitude toward the seniority norm (i.e., *proponents* of the seniority norm), and (2) subjects who have a negative attitude toward the seniority norm (i.e., *opponents* of the seniority norm). They were then randomly assigned to either the breakup condition or control condition.

The subjects were given a questionnaire entitled, *Research on College Club Activities*. There were two types of questionnaires corresponding to two different experimental conditions scenarios (control vs. breakup).

Scenario

The subjects were presented with the following fictitious scenario:

There is a first year college student, named Sato, who is enrolled in a college tennis club. In the club's tradition only the first-year members do the "dirty-work" such as picking up stray balls, preparing the courts, and disassembling the equipment after the game. As a first-year student, Sato has been doing the "dirty work." One day, however, Sato went to the leader of the club and complained against this norm. Sato's objection was that "this norm is not just and from now on, let us share our dirty work equally.

Experimental Conditions

Subjects were randomly assigned to one of the two conditions and read the corresponding scenario. Information about the future of the group was varied in two ways: (1) the tennis club is going to be disbanded at the end of the year (breakup condition), or (2) no mention is made about the future of the club (control condition).

Dependent Variable

After having read one of the two scenarios, the subjects were asked to rate the perceived desirability of the protester on 7-point scales that measured:

(1) likability, (2) independence, (3) rationality, and (4) reliability of the protester.

Attitude toward the Seniority Norm The subjects rated on a 5-point scale which measures their attitude toward the seniority norm (ranging from 1: very unacceptable, to 5: very acceptable). They were thoroughly debriefed at the end of the experiment.

Results

The Manipulation Check

To examine the effectiveness of the manipulation of the persistence of the protester's ingroup, the subjects were asked to rate the likelihood that the tennis club would exist a year later. The result of the one-way analysis of variance yielded a significant effect for the experimental conditions, $F(1, 89) = 41.41, p < .0001$. In other words, the perceived persistence was rated lower in the breakup condition when compared to the control condition. Thus, the effectiveness of the experimental manipulation was confirmed.

The Classification of Subjects.

Based on the subjects' attitudes toward the seniority norm, they were classified into three groups. First, those subjects who answered 1 or 2 were classified as the *opponents* of the seniority norm. Second, those subjects who answered 4 or 5 were classified as the *proponents* of the seniority norm. Third, those subjects who gave the response of 3 (i.e., they had either ambivalent or undecided feelings about the seniority norm) were not included in the analysis. The classification of the subjects into the proponents and opponents of the seniority norm and the prior random assignment of the subjects into the breakup or control condition yielded four comparison groups: (1) the proponents of the seniority norm in the control condition (*proponent-control*, a total of 17 subjects); (2) the proponents of the seniority norm in the breakup condition (*proponent-breakup*, 17 subjects); (3) the opponents of the seniority norm in the control condition (*opponent-control*, 13 subjects); and (4) the opponents of the seniority norm in the breakup condition (*opponent-breakup*, 15 subjects).

The Evaluation of the Protester

A principal component analysis was conducted on the four evaluation items, which measured the perceived desirability of the protester. It yielded one factor solution, explaining 53% of the total variance, with the item loadings varying from .61 to .80. A composite index named *Desirability* score was computed by averaging the scores based on the four items.

Table 1 shows the mean scores on the desirability of the protester by experimental conditions, differentiated by the subjects' attitude toward the seniority norm. As can be seen in Table 1, the opponents of the seniority norm perceived the protester as more desirable than the proponents of the seniority norm. A 2 x 2 analysis of variance was conducted on the Desirability score. The main effect for attitude toward the seniority norm was statistically significant, $F (1, 58) = 9.98, p < .005$. This result supports Prediction 1.

Table 1. The Effect of Subject's Attitude Toward the Seniority Norm and Future Persistence of the Stimulus Group on Perceived Desirability of the Protester of the Seniority Norm

Attitude toward the seniority norm	Group's fate	
	Control	Breakup
Pro	3.16	3.92
Con	4.55	3.83

The higher the number, the more desirable the protester was perceived.
Minimum value = 1, Maximum value = 7.

The interaction effect for experimental conditions x attitude toward the seniority norm was also statistically significant, $F (1, 58) = 6.98, p < .05$. The planned comparison between the proponent-control and proponent-breakup condition yielded a statistically significant result, $F (1, 32) = 5.08, p < 05$. This result supports Prediction 2, since the proponents of the seniority norm perceived the protester as more desirable in the breakup condition than in the control condition.

Although it was not predicted, a follow-up comparison between the opponent-breakup and opponent-control condition yielded another significant result, $F(1, 26) = 5.28, p < .05$. The opponents of the seniority norm perceived the protester as less desirable in the breakup condition than in the control condition.

Discussion

The present results clearly support our predictions. The proponents of the seniority norm liked the protester of the seniority norm to a lesser extent than the opponents of the seniority norm (Prediction 1). When the ingroup of the protester was to be disbanded, the protester was more liked by the proponents of the seniority norm (Prediction 2). Thus, the predictions derived from the Long-Term Equity within a Group model were supported. It appears that the proponents sympathized with the protester who could not maintain the L-TEG.

We found an unexpected, but interesting result. The opponents of the seniority norm liked the protester less when the person's ingroup was expected to be disbanded than when the group was not going to be disbanded. It appears that the opponents of the seniority norm are objecting to the long-term, trans-relational equity principle, in favor of the short-term, person-specific equity principle. This result suggests that the opponents liked the protester's *heroic* efforts, in the control condition, to reject the L-TEG and establish the person-specific equity in the ingroup. When the ingroup is to be disbanded, on the other hand, the protester's behavior is considered as reasonable and not at all *heroic*, since the protester could experience both long-term and short-term inequity if the person accepts the seniority norm.

The results of the present experiment demonstrate the usefulness of the L-TEG model in explaining the seniority norm. In a situation in which one's ingroup is expected to persist, a protester of the seniority norm would be disliked, since the person should tolerate the seniority norm as a junior member and expects benefits as a future senior member. In a situation in which one's ingroup will be disbanded, protesting against the seniority norm is considered legitimate since the person will not receive any benefits and thus experience inequity. For a stable group, the seniority norm could be considered equitable from the long-term, trans-relational perspective. Even if the short-term person-specific inequity exists in the group, the persistence of the group could ensure equitable relationships in the long run.

It should be noted, however, that there is a limitation in generalizability of the results obtained in this study. The subjects employed in this study were Japanese female undergraduate students. Thus it can not be exclusively concluded that these results show that any people who have positive attitude toward the seniority norm employ the L-TEG framework. Further research should be done using diversified subjects who are different in sex, age, social class, and culture.

The L-TEG model has important implications for people's decision making processes and group participation. Individual beliefs about the nature of group stability can affect their decision to join the group which maintain the seniority norm. If they consider a social group as a permanent entity which will exists over a long period, then they are more likely to participate in the group and sacrifice for the group, expecting long-term equity. If, on the other hand, they believe a group to be a transient aggregate of individuals, then they may not be willing to participate in and sacrifice for the group. In Japan, many social groups and companies foster the ideas inherent in the L-TEG model. The concept of lifetime employment, paternalism, filial piety, and *On* (obligations) are examples of the L-TEG model.

References

Adams, J.S. (1965). Inequity in social exchange. In L. Berkowitz (Ed.), *Advances in experimental social psychology* (vol. 2). New York: Academic Press.

Austin, W., & Walster, E. (1974). Participants' reactions to "Equity with the World." *Journal of Experimental Social Psychology, 10*, 528-548.

Austin, W., & Walster, E. (1975). Equity with the World: The trans-relational effects of equity and inequity. *Sociometry, 38*, 474-496.

Austin, W., Walster, E., & Utne, M.K. (1976). Equity and the law: The effect of a harmdoer's "suffering in the act" on liking and assigned punishment. In L. Berkowitz (Ed.), *Advances in experimental social psychology* (vol. 9). New York: Academic Press.

Baker, K. (1974). Experimental analysis of third-party justice behavior. *Journal of Personality and Social Psychology, 30*, 307-316.

Byrne, D., & Nelson, D. (1965). Attraction as a linear function of proportion of positive reinforcements. *Journal of Personality and Social Psychology, 1*, 659-663.

Hatfield, E., Walster, G.W., & Berscheid, E. (1978). *Equity: Theory and research.* Boston: Allyn and Bacon.

Hatfield, E., Traupmann, J., Sprecher, S., Utne, M., & Hay, J. (1985). Equity and intimate relations: Recent research. In W. Ickes (Ed.), *Compatible and incompatible relationships* (pp. 91-117). New York: Springer-Verlag.

Hidaka Y., & Yamaguchi, S. (1994). *Influence of the perception of difference between groups upon consensus estimation in Japan.* Paper presented at XII International Congress of Cross-Cultural Psychology, Pamplona, Spain.

Ho, D. Y. (1993). Relational orientation in Asian social psychology. In U. Kim & J.W. Berry (Eds.), *Indigenous psychologies: Research and experience in cultural context* (pp. 240-259). Newbury Park, CA: Sage.

Hofstede, G. (1980). *Culture's consequences: International differences in work-related values.* Beverly Hills, CA: Sage.

Hsu, F. L.K. (1975). *Iemoto: The heart of Japan.* New York: Schenkman.

Lebra, T.S. (1976). *Japanese patterns of behavior.* Honolulu, HI: University of Hawaii Press.

Lerner, M.J. (1965). Evaluation of performance as a function of performer's reward and attractiveness. *Journal of Personality and Social Psychology, 1*, 355-360.

Lerner, M.J. (1970). The desire for justice and reactions to victims. In J. Macaulay and L. Berkowitz (Eds.), *Altruism and helping behavior* (pp. 205-229). New York: Academic Press.

Lerner, M.J., & Matthews, G. (1967). Reactions to the suffering of others under conditions of indirect responsibility. *Journal of Personality and Social Psychology, 5*, 319-325.

Lerner, M.J., & Simmons, C.H. (1966). Observer's reaction to the "Innocent victim": Compassion or rejection? *Journal of Personality and Social Psychology, 4*, 203-210.

Leventhal, G.S. (1980). What should be done with equity theory?: New approaches to the study of fairness in social relationships. In K.J. Gergen, M.S. Greenberg, & R.H. Willis (Eds.), *Social exchange: Advances in theory and research* (pp. 27-55). New York: Plenum.

Moschetti, G.J., & Kues, J.R. (1978). Transrelational equity comparisons: Extensions to the third partner relationship and a decision-making model. *Journal of Personality and Social Psychology, 36,* 1107-1117.

Nakane, C. (1970). *Japanese society.* Rutland, VT: Charles E. Tuttle.

Triandis, H.C. (1990). Cross-cultural studies of individualism and collectivism. In J.J. Berman (Ed.), *Nebraska symposium on motivation, 1989: Cross-cultural perspectives* (pp. 41-133). Lincoln, NB: University of Nebraska.

Triandis, H.C. (1994). Theoretical and methodological approaches to the study of collectivism and individualism. In U. Kim, H.C. Triandis, C. Kagitcibasi, S.C. Choi, & G. Yoon, (Eds.), *Individualism and collectivism: Theory, method, and applications* (pp. 41-51). Thousand Oaks, CA: Sage.

Walster, E., Berscheid, E., & Walster, G. (1973). New directions in equity research. *Journal of Personality and Social Psychology, 25,* 151-176.

Note

This paper is originally based on the first author's Master's thesis at The University of Tokyo. We would like to thank Uichol Kim for providing helpful comments. We also thank Taeko Furuya and Minoru Yabuuchi for their assistance.

Distinctiveness Effects in Intergroup Perceptions: An International Study

Walter G. Stephan & Cookie White Stephan, New Mexico State University, Las Cruces, New Mexico, U.S.A
Marina Abalakina & Vladimir Ageyev, Moscow State University, Moscow, Russia
Amalio Blanco, Universidad Autonoma de Madrid, Madrid, Spain
Michael Bond, The Chinese University of Hong Kong, Hong Kong
Isamu Saito, Rissho University, Tokyo, Japan
Petar Turcinovic, University of Rijeka, Rijecka, Croatia
Brenda Wenzel, New Mexico State University, Las Cruces, New Mexico, U.S.A

For decades psychologists have been interested in the factors that influence intergroup perceptions (Allport, 1954). Much of this literature is concerned with stereotypes of social groups within nations, but a considerable international literature now exists as well (Bond, 1988; Herrman, 1985; Holt & Silverstein, 1989; LeVine & Campbell, 1972; Peabody, 1985). One set of biases that has received considerable attention concerns factors that influence perceptions of group differences. Two theories positing different mechanisms, social identity theory and self categorization theory, make similar predictions regarding intergroup comparisons. Both theories predict that people will make predominantly favorable distinctions between the ingroup and outgroups on dimensions relevant to the intergroup comparisons.

Social identity theory offers a predominantly motivational explanation for such perceived intergroup differences (Tajfel, 1978; 1982; Tajfel & Turner, 1986). A major premise of the theory is that the process of categorizing oneself as an ingroup member, and others as outgroup members, creates and maintains attitudinal and behavioral distinctions favoring the ingroup (Tajfel, 1978; 1982; Tajfel & Turner, 1986). The theory posits a linkage between individuals' attempts to maintain or enhance their self-esteem and ingroup favoritism. Tajfel assumed that important aspects of an individual's self-image are derived from his or her memberships in social groups (Tajfel, 1978; 1982; Tajfel & Turner, 1986). He argued that making positive comparisons between the groups with which one identifies and relevant outgroups is a primary means of maintaining or enhancing self-esteem. Succinctly put, the argument is that "individuals are motivated to establish positively valued differences (positively discrepant comparisons) between the ingroup and a relevant outgroup to achieve positive social identity" (Turner, Brown & Tajfel, 1979, p. 190). Social identity theorists have also argued that in order for these biased intergroup judgments to have the desired effect on self-esteem, the comparison must be relevant to the distinction between groups. The social situation must also allow for comparisons on variables which can be perceived to favor the ingroup over relevant outgroups.

Self categorization theory, proposed by Turner and his colleagues, differs from social identity theory in at least two major respects (Turner, Hogg, Oakes, Reicher, & Wetherell, 1987). First, it is a much broader theory, encompassing more aspects of human behavior. Second, it is strictly a cognitive theory: It does not make reference to motivational factors such as self-esteem maintenance (Abrams & Hogg, 1990; Hogg & McGarty, 1990; Turner, 1985; Turner et al., 1987). According to Turner, the self is composed of three levels: the self as a human being, the self as a member of a variety of social groups, and the self as a unique individual. Which of these three components of the self is activated depends upon the specific situation. Turner argues that factors enhancing the salience of ingroup-outgroup categorizations or increasing identification with the ingroup evoke the group self.

According to this theory, the activation of the group self elicits perceptions of similarity to the ingroup and dissimilarity from the outgroup (Haslam & Turner, 1992; Haslam, Turner, Oakes, McGarty & Hayes, 1992). Perceived ingroup similarity, as well as a variety of cognitive information processing biases, ensures that the more the individual identifies with the ingroup, the more favorably the ingroup will be evaluated. Positive ingroup differentiation from the outgroup is most likely to occur when the characteristic being evaluated is relevant to the groups being categorized (Oakes, Turner, & Haslam, 1991). The more important the characteristic to defining group differences, the more extreme is the tendency to accentuate a favorable comparison. However, if an outgroup is clearly superior to the ingroup on a characteristic, this superiority is likely to be acknowledged, although its importance may be de-emphasized.

A number of tests of social identity and self categorization predictions concerning intergroup perceptions have been conducted but many of them suffer from a number of limitations. These tests have tended to use overall group evaluations or they have contrasted groups on only one dimension (Oakes, Turner, & Haslam, 1991), rather than employing multiple dimensions. Typically, only two groups are compared (Tajfel, Sheikh, & Gardner, 1964, Van Knippenberg, Pruyn, & Wilke, 1982) rather than employing multiple groups. And, only a limited number of types of groups have been investigated. For instance, to our knowledge there are no studies applying this idea to perceptions of national differences. The present study addresses these shortcomings by using a variety of different characteristics to make multiple international comparisons.

Both social identity and self-categorization theory would predict that favorable distinctions should be created on relevant traits. Favorable distinctions can be created either by perceiving one's group to possess higher levels of positive traits or lower levels of negative traits (Diehl & Jonas, 1991). In accordance with social identity and self categorization theory, relevant and nonrelevant traits were examined in the present study, as was trait valence. The previous literature on this issue suggests that traits are relevant to intergroup comparisons when they are part of "the culturally accepted stereotype of their group," (Tajfel, 1981, p. 115), or consist of "dimensions which are considered to be associated (or correlated) with the categorization," (Hogg & Abrams, 1988). We defined relevant traits as those that were attributed to the majority of members of the ingroup, since this attribution indicates that the trait is believed to characterize the group and is associated with the group category.

The predictions regarding favorable intergroup distinctions were tested by examining perceptions of traits possessed by people in Japan, Russia, and the United States from the perspective of students from Hong Kong, Japan, Russia, Spain, the United States, and the former Yugoslavia. The target countries were chosen because they are so well known that it seemed likely that students in all of these countries could easily use them in making comparisons with their own country.

Method

The subjects consisted of university students from 6 countries, the Chinese University in Hong Kong (78); Risshou University in Japan (118); Universidad Autonoma de Madrid in Spain (102); Moscow State University in Russia (58); New Mexico State University in the United States (107), and the University of Rijeka in Croatia (112). The study was conducted in 1991, when there was considerable political, social, and economic turmoil in the USSR and Yugoslavia, but before these countries were formally divided. All the subjects were undergraduates. Those in Hong Kong, Japan, Spain, the United States, and Yugoslavia were students in psychology classes, while those from Russia were students in language classes.

Each subject completed a questionnaire form containing questions concerning the traits of people from Japan, Russia, and the U.S. This scale contained 10 traits: competitive, materialistic, rigid, group-oriented, emotional, traditional, passive, concerned with dignity, hard-working, and respectful. These traits were derived from previous studies of the stereotypes of the 3 target countries. The traits competitive, materialistic, and emotional were taken from an earlier study of American and Russian stereotypes (Stephan et al., 1993). The traits group-oriented, traditional, passive, concerned with dignity, and respectful were adapted from discussions of values that predominate in Eastern cultures (The Chinese Cultural Connection, 1987; Reischauer, 1981). Each trait was rated on a 10-point scale, with percentile increments of 10 percentage points (e.g., 0-10%) indicating the percentage of people from the country perceived to have the characteristic in question (Brigham, 1971). The subjects were asked to use these traits to rate the Japanese, the Russians, and the Americans. If the subjects were not from one of these 3 countries, they were also asked to rate people from their own country.

To determine the valence of these traits in each country, positive-negative ratings of the traits were collected using a second sample of subjects taken from the same populations of university students (n's ranged from 24 to 46). The response format consisted of a 10-point scale running from "extremely negative" to "extremely positive."

Results

A descriptive analysis examining the overall ratings of the 3 target countries (Japan, Russia, and the U.S.) by all 6 countries was conducted first. The Target main effects for the 10 traits were analyzed using 3 (Target Country) x 6 (Country of Evaluator) analyses of variance (Table 1). Each of these main effects was highly significant (p's < .001). The Japanese were rated as the most traditional, group-oriented, respectful, concerned with dignity, and hard-working. The Russians were rated as the most passive and were approximately tied with the Japanese for most rigid. The Americans were rated as the most competitive, materialistic, and emotional. These ratings are generally consistent with the stereotypes of these groups obtained in previous research (Eagly & Kite, 1987; Karlins, Coffman, & Walters, 1969; Peabody, 1985; Stephan, et al., 1991).

Table 1. Characterizations of Japan, Russia, and USA across all Six Samples

	Japan	Russia	U.S.
Traditional	6.58**	4.83	3.41
Group-oriented	6.12**	4.98	3.41
Respectful	6.45**	4.87	4.06
Concerned with dignity	6.16**	4.84	4.92
Competitive	6.08	4.23	6.54**
Materialistic	5.05	4.04	6.99**
Emotional	4.24	4.30	5.56**
Hard-working	7.72**	4.82	4.98
Rigid	5.43**	5.36**	3.73
Passive	3.60	4.31**	2.80

** Significantly greater than other countries or country p < .01.
Note: Scores greater than 5.0 indicate that over 50% of the group is viewed as having this characteristic.

The hypothesis tested was that subjects would make predominantly favorable distinctions between their own country and the comparison countries on relevant traits. Two assumptions regarding relevance were made. First, asking subjects to compare the people from their country with the people from these target countries makes these comparisons salient. Second, those traits considered to be characteristic of the ingroup constitute relevant dimensions for comparison. Using these assumptions, we defined traits as relevant if the ingroup perceived the trait to be possessed by over 50% of the citizens of their country. The hypothesis was tested by creating a Chi square matrix based on the results of a set of one-way, within-subjects analyses of variance run on each trait from each country separately. These analyses were run for both the relevant and the irrelevant traits. In these analyses we determined if the ingroup made a distinction between itself and the comparison countries for each trait. A comparison for a given trait was considered to be distinctive if the ingroup differed significantly from all of the outgroups combined. A distinction was considered to be favorable if the ingroup rated itself significantly higher on a positive trait or significantly lower on a negative trait than the comparison countries. A distinction was considered to be unfavorable if the ingroup rated itself significantly lower on a positive trait or significantly higher on a negative trait than the comparison countries. For instance, the Japanese viewed themselves as significantly more *respectful* than the other countries, a positive distinction (since they rated *respectful* positively), but they also viewed themselves as significantly more *traditional*, a negative distinction (since they rated *traditional* negatively).

Table 2. Favorability of Distinctions

Ingroup-Outgroup distinction	Relevant traits	Nonrelevant traits
Favorable	7	3
Unfavorable	8	12
No distinction made	17	13

There were 60 possible trait comparisons (6 countries rating 10 traits). Separating the traits by relevance yielded a total of 32 traits that were relevant and 28 that were not. Forty-seven percent of the distinctiveness comparisons for the relevant traits were significant (Table 2). Of these significant distinctions slightly less than half (47%) favored the ingroup. For the nonrelevant traits, slightly more than half (54%) of the comparisons were significant, and in this case only 20% of the significant distinctions favored the ingroup. The Chi square for the contingency table created by these distinctiveness comparisons was non-significant $x^2 = 2.68$, *n.s.* (Table 2). The results do not support the social identity and self-categorization theory predictions. First, there was a nearly even split between favorable and unfavorable comparisons for the relevant traits. Second, although there was a substantial number of traits on which the students in these countries perceived their countries to be distinctive, these distinctions were not predominantly favorable toward the ingroup. Surprisingly, the overall frequency of significant unfavorable comparisons was almost double (33% vs. 17%) the frequency of significant favorable comparisons. In none of the 6 countries did favorable comparisons outnumber unfavorable comparisons.

The results from Russia and the U.S. can be used to illustrate the general pattern of results. Table 3 shows that the Russian students did not consider Russians to be positively distinctive on the relevant traits (those with means higher than 5.00). The only distinction they made on the relevant traits was a negative one for the trait "passive." They did perceive Russians to be distinctive on a number of the non-relevant traits, but nearly all of these distinctions, 4 out of 5, were negative ones. Thus, while the Russian students often regard Russians as distinct from people from the comparison countries, they did not make predominantly favorable distinctions on either the relevant or the irrelevant traits.

Table 3. Distinctiveness Effects of Russian Students

	Japan	Russian Sample Russia	U.S.
(Traditional) -	7.33	3.90**F	4.07
Group-Oriented -	6.39	6.17	3.12
(Respectful) +	7.26	3.86**U	5.29
(Concerned With Dignity) +	6.78	3.29**U	6.98
(Competitive) +	6.23	2.64**U	5.78
Materialistic -	5.39	5.34	6.12
Emotional +	3.95	5.72	6.00
(Hard-Working) +	7.83	3.26**U	6.05
Rigid -	5.32	5.07	4.16
Passive -	2.50	6.19**U	2.10

Scores 5.0 Or Greater Indicate Over 50% Of The Group Is Viewed As Having This Characteristic.
Traits In Parentheses Are Non-Relevant Traits.
** = Significant At $P < .01$
+ = Russian Subjects Rated This Trait As Positive
- = Russian Subjects Rated This Trait As Negative
F = Comparison Favorable To The Ingroup
U = Comparison Unfavorable To The Ingroup

The results for the Americans were similar (Table 4). Here the results show that for 2 of the relevant traits (*emotional, rigid*) the American students made favorable distinctions between people in their own country and people from the comparison countries, but for 4 of relevant traits they made negative distinctions. In addition, for 1 of the non-relevant traits they made a favorable distinction, and for 1 of the non-relevant traits they made an unfavorable distinction. Thus, the American students did regard Americans as distinct from people from the comparison countries, but they did not make predominantly favorable distinctions on either the relevant or the non-relevant traits. In fact, like the Russians, they made more unfavorable distinctions (5) than favorable distinctions (3).

Discussion

The social identity and self-categorization hypothesis - that subjects would make favorable distinctions between people from their country and people from other countries on relevant traits - was not supported by the data. A nearly equivalent percentage of the statistically significant comparisons for the relevant traits were favorable (47%) and unfavorable (53%). For traits that were not perceived as characterizing the ingroup, there was a greater percentage of unfavorable (80%) than favorable (20%) statistically significant comparisons. Subjects from countries with great political power and economic wealth were no more likely to make favorable comparisons than those from countries with less power and wealth. This pattern of results suggests that for national groups there may not be a tendency for people to make predominantly favorable distinctions between the ingroup and the outgroup.

A substantial number of distinctiveness effects were found. The traits that were considered to be distinctive differed considerably from country to country. How can these disparate patterns of distinctiveness effects be explained? It appears that in some instances the subjects were responding to centuries-old cultural influences, while in other instances they were attending to traits made salient by recent social, political, and economic events in their countries.

Table 4. Distinctiveness Effects of American Students

	American Sample		
	Japan	Russia	U.S.
(Traditional) +	6.86	5.32	3.98**U
(Group-Oriented) +	5.74	4.65	4.98
(Respectful) +	7.17	5.61	4.72**U
Concerned With Dignity +	7.11	5.93	5.33**U
Competitive +	6.83	5.50	6.93**F
Materialistic -	4.38	3.73	7.60**U
Emotional +	4.46	3.84	6.18**F
Hard-Working +	7.74	6.44	5.50**U
(Rigid) -	4.72	5.74	4.62*F
(Passive) -	4.33	3.83	4.09

Scores Of 5.0 Or Greater Indicate Over 50% Of The Group Is Viewed As Having This Characteristic.
Traits in parentheses are non-relevant traits.
** = significant at $p < .01$
* = significant at $p < .05$
+ = American subjects rated this trait as positive
- = American subjects rated this trait as negative
F = comparison favorable to the ingroup
U = comparison unfavorable to the ingroup

For instance, consider the results for Russia. This study was conducted before the dissolution of the Soviet Union, during a time of political and economic turmoil. The decline of the Soviet economic and political systems and the failure of the Soviet government to find adequate solutions to these problems may have caused Russian subjects to emphasize their lack of competitiveness, reluctance to work hard (at least in state assigned jobs), loss of tradition and respect for authority, and lowered concern with dignity. In additoin, the lack of organized public outcry against the decline of the Soviet state (the study was conducted before 1991 before the events of Aug. 1992) may have highlighted the passivity of the Russian citizenry.

The declining economic fortunes of the United States have made global economic issues salient for Americans students. Although American students believe that Americans are highly competitive, they appear to attribute the loss in economic power to the fact that Americans are less hard-working than people in other countries. Large-scale changes in the traditional structures of authority in the U.S. (the family, the state, the social class structure, the ethnic dominance hierarchy) may have led the American students to perceive Americans as less traditional, respectful, rigid, and concerned with dignity than people in other countries.

Explaining these results in terms of recent social, political, and economic changes suggests that salience may be an important mediator of distinctiveness effects (see also Haslam et al., 1992). Ingroups may perceive themselves to be distinctive from other groups primarily on attributes that are currently salient. Thus, it appears that salience plays a greater role than motivational factors in creating the distinctiveness effects observed in the comparisons these subjects made between people from their own country and people from other countries.

Recent research in social identity theory also suggests that the role of motivational factors in intergroup perception is more limited than previously supposed. The self-esteem hypothesis for the perception of group differences has not been supported or has been only partially supported in many studies (Hogg & Abrams, 1990; Hogg et al., 1986; Lemyre & Smith, 1985; Oakes & Turner, 1980). Also, the self-esteem hypothesis has been supported more often in minimal group situations than in other, more ecologically valid, situations (Hogg & Abrams, 1990). Some studies have found an interaction of category salience, discrimination, and self-esteem, in which neither category salience nor discrimination alone was sufficient to elicit increased self-esteem (Hogg & Abrams, 1990; Lemyre & Smith, 1985; Hogg et al, 1986). For this reason, it has been argued that self-esteem maintenance is only one of several motives that direct intergroup perceptions. The others include self-knowledge, meaningful interpretations of the world, power and control, self-efficacy, and self-evaluation (Abrams & Hogg, 1988; Hogg & Abrams, 1990). To this list we would add the salience of recent events in the history of the group.

Previous tests of social identity and self-categorization theories have not mirrored the complex stereotypes of multiple groups that exist in the real world. The importance of the current study lies in its attempt to test these ideas in the context of our global culture. One interpretation of the lack of support for the favorable distinctiveness hypothesis is that in times of peace, national groups may not fulfill distinctiveness needs. Also, if people do not identify with their nationality groups, they may not be motivated to make favorable distinctions between their own and other countries. However, recent ethno-nationalist movements in the former USSR, the former Yugoslavia, and other parts of the world provide evidence that nationality can rapidly become a powerful source of identity during times of social instability and intergroup conflict. It may be that cultural and national identities are latent and implicit unless those group identities are threatened, at which time the processes posited by social identity and self-categorization theories come into play.

References

Abrams, D., & Hogg, M. A. (1988). Comments on the motivational status of self-esteem in social identity and intergroup discrimination. *European Journal of Social Psychology, 18*, 317-334.

Abrams, D., & Hogg, M. A. (1990). *Social identity theory: Constructive and critical advances.* New York: Springer-Verlag.

Allport, G. W. (1954). *The nature of prejudice.* Cambridge, MA: Addison-Wesley.

Bond, M. H. (1988). *The cross-cultural challenge to social psychology.* Newbury Park, CA: Sage.

Brigham, J. C. (1971). Ethnic stereotypes. *Psychological Bulletin, 76*, 15-38.

Chinese Cultural Connection. (1987). Chinese values and the search for culture-free dimensions of culture. *Journal of Cross-Cultural Psychology, 18*, 143-163.

Diehl, M., & Jonas, K. (1991). Measures of national stereotypes as predictors of the latencies of inductive versus deductive stereotypic judgments. *European Journal of Social Psychology, 21*, 317-330.

Eagly, A. H., & Kite, M. E. (1987). Are stereotypes of nationalities applied to both men and women? *Journal of Personality and Social Psychology, 53*, 451-462.

Haslam, S. A., & Turner, J. C. (1992). Context-dependent variation in social stereotyping: The relationship between frame of reference, self-categorization, and accentuation. *European Journal of Social Psychology, 22*, 251-277.

Haslam, S. A., Turner, J. C., Oakes, P. J. McGarty, C., & Hayes, B. K. (1992). Context-dependent variation in social stereotyping: The effects of intergroup relations as mediated by social change and frame of reference. *European Journal of Social Psychology, 12*, 3-20.

Herrman, R.K. (1985). Analyzing Soviet images of the United States. *Journal of Conflict Resolution, 29*, 665-697.

Hogg, M.A., & Abrams, D. (1988) *Social identification: A social psychology of intergroup relations and group processes.* New York: Routledge.

Hogg, M. A., & Abrams, D. (1990). Social motivation, self-esteem, and social identity. In D. Abrams & M.A. Hogg (Eds.), *Social identity theory: Constructive and critical advances.* New York: Springer-Verlag.

Hogg, M.A., & McGarty, C. (1990). Self-categorization and social identity. In D. Abrams & M.A. Hogg (Eds.), *Social identity theory: Constructive and critical advances*. New York: Springer-Verlag.

Hogg, M.A., Turner, J.C., Nascimento-Schulze, C., & Spriggs, D. (1986). Social categorization, intergroup behavior, and self-esteem: Two experiments. *Revista de Psicologia Social, 1,* 23-37.

Holt, R.R., & Silverstein, B. (1989). The image of the enemy: U.S. views of the Soviet Union. *Journal of Social Issues, 45.*

Karlins, M., Coffman, T. L., & Walters, G. (1969). On the fading of stereotypes: Studies of three generations of students. *Journal of Personality and Social Psychology, 13,* 1-16.

Lemyre, L., & Smith, P. M. (1985). Intergroup discrimination and self-esteem in the minimal group paradigm. *Journal of Personality and Social Psychology, 49,* 660-670.

LeVine, R.A., & Campbell, D.T. (1972). *Ethnocentrism*. New York: John Wiley & Sons.

Oakes, P.J., & Turner, J.C. (1980). Social categorization and intergroup behavior: Does minimal intergroup discrimination make social identity more positive? *European Journal of Social Psychology, 10,* 295-301.

Oakes, P.J., Turner, J.C., & Haslam, S.A. (1991). Perceiving people as group members: The role of fit in the salience of social categorizations. *British Journal of Social Psychology, 30,* 125-144.

Peabody, D. (1985). *National characteristics*. Cambridge: Cambridge University Press.

Reischauer, E.O. (1981). *Japan: The story of a people*. New York: Knopf.

Stephan, W.G., Ageyev, V.S., Stephan, C.W., Abalakina, M., Stefanenko, T., & Coates-Shrider, L. (1993). Measuring stereotypes: A comparison of methods using Russian and American samples. *Social Psychology Quarterly, 56,* 54-64.

Stephan, W.G., Walter, G., Stephan, C.W., Wenzel, B., & Corneius, J. (1991). Intergroup interaction and self-disclosure. *Journal of Applied Social Psychology, 21,* 1370-1378.

Tajfel, H. (1978). *Differentiation between social groups: studies in the social psychology of intergroup relations*. London: Academic Press.

Tajfel, H. (1982). *Social identity and intergroup relations*. Cambridge: Cambridge University Press.

Tajfel, H., Sheikh, A.A., & Gardner, R.C. (1964). Content of stereotypes and the inference of similarity between members of stereotyped groups. *Acta Psychologica, 22,* 191-201.

Tajfel, H., & Turner, J.C. (1986). The social identity theory of intergroup relations. In W. Austin, & S. Worchel (Eds.), *The social psychology of intergroup relations* (pp. 7-24). Monterey, CA: Brooks/Cole.

Turner, J.C. (1985). Social categorization and the self-concept: A social cognitive theory of group behavior. In E.J. Lawler (Ed.), *Advances in group processes: Theory and research* (Vol. 2). Greenwich, CT: JAI Press.

Turner, J.C., Brown, R.J., & Tajfel, H. (1979). Social comparison and group interest in ingroup favoritism. *European Journal of Social Psychology, 9,* 187-204.

Turner, J.C., Hogg, M., Oakes, P., Reicher, S., & Wetherell, M. (1987). *Rediscovering the social group: A self-categorization theory*. Oxford: Basil Blackwell.

Van Knippenberg, A., Pruyn, A., & Wilke, H. (1982). Intergroup perception in individual and collective encounters. *European Journal of Social Psychology, 12,* 187-193.

Concept of "Mien Tzu" (Face) in East Asian Societies: The Case of Taiwanese and Japanese

Hsiao Ying Tsai
Osaka University, Osaka, Japan

"Mien Tzu" as a Significant Term in East Asian Culture

The term "Mien Tzu" as a significant term has its roots in East Asian culture. Originally Chinese, this term was borrowed by Japanese, retaining a similar meaning and pronunciation (*mentsu*; Iwanami Japanese Dictionary; 4th edition). The Chinese characters of *Mien Tzu* are also used in Japanese. This concept has been translated as *face* by English residents in China (Kenkyusha's New English-Japanese Dictionary; 5th edition).

Even though it is a human universal, the concept of *face* is particularly meaningful for the Chinese and is argued to be a key in explaining much of their behavior (Stover, 1974; Lin, 1977). The important point to note, in considering the way in which face-related values occur in the Asia, is that in almost every case the values are described as of central importance, and often as the most important of all (Redding, 1982).

To illustrate the above point of view, an event which happened in 1994 could be mentioned. This is the news about the North Korean government's hostile reaction to the International Atomic Energy Association's report concerning their nuclear development (Mainichi Daily News, 1994 June 15). In this case, the former American President,. Jimmy Carter was regarded as a successful mediator due to his softening hostile reactions. The whole event could be analyzed by using the key concept of *Mien Tzu*. The negotiations succeeded because Carter was the right person to allow the North Korean government to regain or save its *Mien Tzu* when confronted by officials from the I.A.E.A. .

On the other hand, according to the author (Tsai, in press), Asian foreigners were found to be more nativistic to their host culture (Japan) than Non-Asian foreigners in their adjustment to Japan. In this study, *nativism*, defined as one kind of extreme ethnocentrism, included both attitudes to and repercussions from the changing image of the home culture and its effect upon the sojourners themselves. It also included the foreigners' tendency to define themselves vis-a-vis the host nations by considering themselves to be superior. At the same time, progress in socio-cultural adjustment was found in social skills, which did not necessarily mean the development of psychological adjustment in terms of positive attitudes. In addition, according to Oberg's (1960) Culture Shock theory, sojourners tend to show hostility to their host nation. Thus, the above finding could be considered incorporating

Asian *mien tzu* action as well. Faced with culture shock in Japan, Asian foreigners, in comparison with non-Asians, tend to define themselves more aggressively vis-a-vis Japanese. This study aimed to analyze *mien tzu* in relation to self-respect for East Asians, in the case of Taiwanese and Japanese by looking at intercultural communication problems involving *mien tzu* of Taiwanese in Japan.

Mien Tzu in the East Asian Cultural Context

As mentioned above, *face* is the accepted English translation for *mien tzu*. A review of studies concerning *face* shows that the earliest study of *mien tzu* compared it to the related concept *lien* (also spelled *lian*) in Chinese (Hui,1944). *Mien tzu* is a reputation achieved through getting on in life while *lien*, on the other hand, represents the confidence of society in the integrity of a person's moral character. Later, *face* was defined by Goffman (1955) as "the positive social value a person effectively claims for himself by the line others assume he has taken during a particular contact. Face is an image of self, delineated in terms of approved social attributes.". Furthermore, other researchers are trying to extend our understanding of *mien tzu*. Stover (1962) considered *face* as being one kind of "other-directed self-esteem". Still, something is lacking in their explanation of *face* in Chinese or other Asian societies. Finally, Ho (1976) wrote, " A man who has *mien tzu* is in a position to exercise considerable influence, even control, over others in both direct and indirect ways. At the same time, however, he is under strong constraint to act in a manner consistent with the requirements for the *mien tzu* of others. Thus the concern for *face* exerts a mutually restrictive, even coercive, power upon each member of the social network". Thus, "mien tzu" is a kind of reciprocal social control, but not a standard of behavior, personality variable, status, dignity, or honor. In addition, concerning Chinese *mien tzu*, Redding (1982) found that "Justification of face-related behavior is normally in terms of group harmony, and hierarchical perceptions of the social order influence face transactions." In Chinese society, Hwang (1985) interpreted *mien tzu* management as a necessary process of the Chinese power game. Chu (1991) found that ability or social status was perceived as the greatest threat to *mien tzu* of Taiwanese college students. Consequently, the concept of *mien tzu* could be considered as being related to the self-evaluation or self-respect of East Asians. It is not exactly equal to the Western concept of self-esteem or assessment of an individual by those close to him. Redding (1982) tried to distinguish between face and self-esteem, defining *self-esteem* as the individual's view of himself and *face* as the individual's assessment of how others close to him to see him. Such a distinction is however based on Western individualism. For example, in actions such as those referred to by the Chinese expression *tiu mien tzu* (also spelled diu-mianzi: losing mien tzu) or the Japanese expression *mentzu o tsubushita* (crushing or losing mien tzu), or the action of regaining mien tzu, *mien tzu* itself not only depends on others' assessment, but there is a subjective element at the same time. The concepts of psychosocial homeostasis and *jen* (also spelled *ren*) proposed by Hsu (1971) may be applied when analyzing Mien Tzu in its East Asian cultural context concerning self-evaluation. The Chinese character of *jen* is similar to but not equal to the English word of *person* in the conceptual content. Instead of the Western concept of personality, *psychosocial homeostasis* and *jen* are designed to illustrate the concept of *self*. The first describes the process whereby every human individual tends to seek certain kinds of affective involvement with some of his fellow humans. The second refers to the internal and external limits of the individual's affective involvement (Hsu, 1971). These have developed into the concept of *jen-ism* (or contextualism), as opposed to collectivism as employed by Hamaguchi (1982) for describing East Asian societies, especially Japanese. This concept emphasizes the interdependent relationship of individuals in East Asian societies, which is based on ties between people. Thus, the following assumption can be introduced: in East Asian, non-individualistic societies, the interaction and integration of one's own *public-self* and *private-self* (Buss, 1980) is considered to be more necessary than in Western individualistic societies. Also, *social-esteem* affects one's daily life more than *self-esteem*. Just as *mien tzu* is itself a meaningful concept in East Asian culture, *mien tzu-action* could be considered as the very process of winning respect in life.

The Effect of Culture Shock on the "Mien Tzu" of Taiwanese in Japan

Even though the concept of mien tzu is written with the same characters and has a similar connotation in Taiwan and Japan, the contents are quite different. In other words, even though *mien tzu* is not used as much in the daily life of Japanese when compared to that of most Taiwanese, it has been somehow Japanized. For example, in comparison with Taiwanese usage, various Japanese concepts, such as *sunao* (meek/modest), *kanshya* (gratitude), *nakama ishiki* (belongingness/congruence) and *shinraikan* (mutual reliance), are more important ingredients of the Japanese concept of Mien Tzu. As a result, some Taiwanese in Japan suffer culture shock when faced with Japanese Mien Tzu. The same situation could be expected for Japanese going to Taiwan or to Chinese societies.

According to a field study the concerning the culture shock which Taiwanese tend to experience in Japan (Tsai, 1993), the following situations demonstrate the differences between these two cultures.

Situation 1: " I would like to do it, but I can't."

Japanese tend to say "I would like to do it, but I can't", (*sitaikedo, dekinai.*) directly as an acceptable way of saving Mien Tzu. In contrast, Taiwanese tend to hide their limitation; they pretend that they can do something but they just don't want to. This may come partly from the influence of Chinese culture relating to mien tzu. *Mien tzu* is closely tied to one's ability or *achievement* in Chinese culture (Hu, 1944; Ho, 1976; Chu, 1991). In this case, many Taiwanese perceive the Japanese as not caring much about their own "Mien Tzu", which demonstrates the existence of very different values between these two cultures. Versatility or omnipotence is not as expected in the Japanese as in the Chinese Culture. The popular phenomenon of so-called education of gifted children found in the elementary schools of Taiwan is not found in Japan as well. Taiwanese are encouraged to show their talent, whereas Japanese society may not appreciate individual qualities that much (White, 1988). On the other hand, Taiwanese asking for help from Japanese friends may be hurt or feel their *mien tzu* threatened by responses such as, "I would like to help you, but I can't." This is because really good persons should try hard on behalf of their friends instead of showing their incompetence immediately, no matter how difficult a task may be. This is the value of *i chi* (also spelled *yi-qi*: being loyal to one's friend) in Chinese society. If the task is not achieved as expected, they would lose *mien tzu* in front of their friends, whereas Japanese would have to face a mien tzu threat if they didn't admit their limitation from the beginning. They would lose their friends' trust by not carrying out a promise (Hamaguchi, 1982).

Situation 2: Sumimasen

Sumimasen is not simply equal to "Excuse me" or "I'm sorry" for the Japanese. Depending on the situation, it may also include an expression of guilt and/or gratitude (Lebra, 1976). The very custom of apologizing by saying "sumimasen" is certainly the reciprocity of mien tzu "saving" or "regaining" in Japan, while Taiwanese may feel a loss of, threat to their mien tzu if they were to apologize as much as Japanese do. For example, Taiwanese are generally not willing to say "sumimasen" when declining to follow someone else's advice. They tend not to take the action of apologizing easily without knowing how to save Japanese mien tzu, which is strongly related to mutual trust in Japanese culture. For Japanese, when a friend accepts their advice, they feel trusted and receive mien tzu at the same time.

Situation 3: "You look like a Japanese."

In a relatively homogeneous society like Japan, being the same as others is regarded as a virtue (Kidder, 1992). Some Japanese like to show their acceptance of Taiwanese friends by making them feel that they are like Japanese. This action gives their Taiwanese friends mien tzu from the Japanese viewpoint. Taiwanese generally tend to feel foreign and not respected in Japanese society. Sojourners particularly may suffer culture shock - may even face a "mien tzu" threat. In attempting to protect their mien tzu, Taiwanese tend to act in an exaggerated Taiwanese way by aggressively showing their differences from Japanese (Tsai, in press).

"Mien Tzu" as a Subject of East Asian Self-Respect

The concept of *tatemae* is a key point of Japanese social life. It is very close to the "front stage" in Goffman's dramaturgical theory concerning "face" (1955) regarding individuals appearing in front of others. Individuals are concerned with maintaining the impression that they are living up to the standards by which they are judged by others; thus they are able to separate themselves into two parts which relate to the presentation of themselves to others. This is also exactly one kind of *mien tzu*. For example, when considering the use of *tatemae* in Situation 1, Japanese tend to say, "I would like to do it, but I can't" directly as an acceptable way of saving mien tzu. Actually they might be saying "no" in their hearts instead of "yes". In Situation 2, in many cases when Japanese say "sumimasen", a real apology or gratitude is not necessary. This phrase is a process of saving or regaining mien tzu. In Situation 3, by saying, "You look like a Japanese" to a Taiwanese or, "You can use chopsticks well", to Westerners, Japanese want to indicate their feeling that their foreign companions are not incongruous. This is may not reflect their real feelings, but is often said just for the sake of "tatemae". That is why the action of ascribing the characteristic of *mien tzu* to their foreign friends may not necessarily result in positive feedback. Those foreigners who are not able to follow or accept the Japanese Mien Tzu concerning "tatemae" may be especially offended.

Even though *tatemae* may also be considered as an obstacle to mutual understanding, even by the Japanese, it has an important function in the Japanese cultural context (Doi, 1985). According to the Japanese concept *wa* (group harmony), particularly among people who are only casually related, *tatemae* is considered a necessity for mutual respect at the beginning of the relationship. It also has the meaning of self-control and is strongly linked to social trustworthiness. These traits are required not only from others but also from oneself. Thus *tatemae* constitutes an important subdiscipline of self-respect in the Japanese culture.

Mien Tzu can be seen as one of the common dominators of life in Chinese society (Lin, 1977), and has been found to cause stress in daily life (Chu, 1991). On the other hand, most overseas Chinese take Mien Tzu particularly seriously. That is because they long to "i jin huan shiang" (return home after making good) to prove their achievement to their fellow countrymen. Because of this, some tend to become more creative, which is one consequence of mien tzu. Thus, mien tzu could be regarded as the main requirement for self-respect for Chinese.

Mien tzu exerts considerable influence in Chinese culture. In Taiwan, mien tzu seems to be somehow linked to the purpose of being heroic for others, whereas in Japan, mien tzu is linked more to the value of being trusted by others. In addition, from the individualistic Westerners' viewpoint, "mien tzu-behavior" may be considered as being closed to an over managed "public self" (Buss, 1980), which might unexpectedly result in an increase of one's social anxiety. "Mien Tzu-behavior" could be considered as resulting from the requirements of the social roles emphasized by the Confucian ethics which have a long history in East Asia. In other words, these social roles which an individual should play somehow lead to the management of the East Asians' lives (Yu, 1994). However, in comparison with the Taiwanese, the Japanese "public self" could be assumed as being more split off from the "private self". Furthermore, this splitting seems to be regarded as one kind of "anmoku no ryokai" (tacit understanding) which does function as a key point in the social interaction of Japanese society. For example, the existence of "tatemae" (the front) and "honne" (the back) (Doi, 1985), seems to be more acceptable and rational to the Japanese, whereas it may lead to more resistance in Taiwanese, even if they have no alternative. Some Taiwanese or other foreigners in Japan might be embarrassed if they have never learned to decline an invitation which was made just for the sake of "tatemae" by their Japanese friends. In this sense, a self consciousness which is able to clarify the "public self" and "private self" is required in social interaction.

From an East Asian indigenous psychological point of view, especially Confucianism, further study of East Asians *mien tzu* concerning their self-respect in life seems merited.

References

Buss, A.H. (1980). *Self-conciousness and social anxiety*. San Francisco: Freeman,
Chu, R.L. (1989). Face and achievement: The examination of social oriented motives in Chinese society. *Chinese Journal of Psychology, 31,* 79-90.
Doi, T. (1985). *Omote and Ura*. Tokoy: KOBUNDO (in Japanese).
Furnham, A., & Bockner, S. (1986). *Culture shock: Psychological reaction to unfamiliar environments*. London: Methuem&Co.
Goffman, E. (1955). On face work. *Psychiatry, 18,* 212-23.
Hamaguchi, E. (1982). *Japan, A society of Jen-ism*. Tokyo: TOYOKEIZAI(in Japanese).

Ho, D.Y.F. (1976). On the concept of face. *American Journal of Sociology*, *81*, 867-884.
Hsu, F.L.K. (1976). Psychosocial homeostasis and Jen: Conceptual tool for advancing psychological anthropology. *American Athropologist, 73*, 23-44.
Hu, H.C. (1944). The Chinese concepts of face. *American Anthropologist 46*, 45-64.
Hwang, K.K. (1985). "Favour" and "Mien Tzu": The Chinese power game. In K. S. Yang, I. I. Lee, & T. I. Uen, (Eds.), *Modernization and indigenous* (pp. 125-154). Taipei: GUEI GUAN, (in Chinese).
Iwanami Japanese Dictionary (1987). (4th ed.). Tokyo.
Kenkyusha's New English-Japanese Dictionary (1988). (5th edition). Tokyo.
Kidder, L.H. (1992). Requirements for being "Japanese": Stories of returness. *International Journal of Intercultural Relations, 16*, 383-393.
Lebral, S.T. (1976). *Japanese patterns of behavior*. Honolulu: The University press of Hawaii,.
Lin, Y.T. (1977). *My country and my people*. Hong Kong: Heinemann.
Mainichi Daily News, June, 15, 1994. Japan.
Oberg, K. (1960). Culture shock: Adjustment to new culture enviroment. *Practical Anthropology, 7*, 177-182.
Redding, S.G., (1982). The role of 'face' in the organizational perceptions of Chinese managers. *Organization Studies, 3*, 201-219.
Stover, L.E., (1974). *The cultural ecology of Chinese civilization*. Mentor, New York.
Tsai, H.Y. (1993). *Research concerning Sojourner Adjustment*. Master thesis, Educational Department, Osaka University (In Japanese).
Tsai, H.Y. (1994). Perspcetive research for the concept of "Mien Tzu"(face) in collectivistic societies: An introduction in relation to self-esteem. *Annual of Educatioanl Psychology in Osaka University, 3*, 71-78 (in Japanese).
Tsai, H.Y. (In press). Sojourner adjustment: the case of foreigners in Japan. *Journal of Cross-Cultural Psychology*.
White, M. (1988). *The Japanese overseas: Can they go home again?* Free Press, New York.
Yu, A.B. (1994). The self and life goals of traditional Chinese: A philosophical and psychological analysis. In A. M. Bouvy, F. J. R. Van de Vijver, P. Boski, & P. Schmitz, (Eds.), *Journeys Into Cross-cultural Psychology* (pp. 50-67), Lisse: Swets&Zeitlinger.

Notes

1. According to Ho's review (1976).
2. See Tsai, 1994, Perspective research for the Concept of "Mien Tzu" (face) in collectivistic societies.

Part VI

Personality

Developmental Psychology

Health Psychology

Subjective Well-Being in Cross-Cultural Perspective

Ed Diener
University of Illinois, Champaign-Urbana, Illinois, U.S.A.

Introduction to Subjective Well-Being

Subjective well-being (SWB) is a growing area of psychology that focuses on how people evaluate their lives. Subjective well-being is often divided into three major components: life satisfaction, the presence of frequent positive affect, and the relative absence of negative affect. The relative preponderance of positive over negative affect is referred to as hedonic balance.

Subjective well-being is, as the name implies, subjective - it is the person's own emotional and cognitive evaluation of how his/her life is going. The field of SWB covers the full-range of affect from depression and dissatisfaction to euphoria and high satisfaction. Although many studies are conducted on the correlates and causes of SWB others have focused on the consequences of well-being. For example, people with high SWB are found to be more helpful, forgiving, energetic, creative, sociable, and less vulnerable to disease (Myers, 1992; Veenhoven, 1988). For general reviews of the field, the reader is referred to Argyle (1987), Diener (1984), Diener and Larsen (1993), Myers (1993), Myers and Diener (1995), and Veenhoven (1984).

The study of subjective well-being is of substantial applied importance. When policy makers seek to understand how to improve the quality of life, measures of SWB are necessary to complement more objective measures such as economic indices. Indeed, measures of happiness and life satisfaction are now collected in highly industrialized nations to monitor the well-being of these societies. If only objective social indicators are considered (e.g., per capita income and crime rates), valuable information is lost about how people weigh and react to their life circumstances. Furthermore, objective indicators do not capture important aspects of life such as interpersonal relationships.

Some Replicable Subjective Well-Being Findings

Levels of Well-Being

Most respondents report a mix of both positive and negative affect, but more positive affect (Diener & Diener, 1995). Similarly, most respondents report positive satisfaction with their lives and with most domains of their lives (Diener, Fujita, & Sandvik, 1994). Although there are differences in reported happiness between individuals and between nations, most people appear to be above neutral in terms of happiness and satisfaction. Veenhoven's summary of surveys in 56 nations from 1946 to 1992 reveals that the majority of respondents in virtually all nations are above the neutral point in terms of SWB. Data from the U.S.A., Japan, and France reveal that SWB in these nations has remained in the slightly or moderately positive range for many years. The data presented in Table 1 for SWB in the last national survey conducted in 43 nations indicates that only four of the countries fell below the neutral point of the scale. Perhaps people in very poor societies as well as those living under extremely adverse circumstances are likely to be unhappy, although evidence on this issue is scanty.

Although people report a positive level of well-being, they also always report some negative affect. Furthermore, most respondents say they are mildly or somewhat happy, not ecstatic. Diener, Colvin, Allman, & Pavot (1991) review why it is not possible to remain elated over long periods of time. For example, people who experience many positive events also tend to experience many negative events. Thus, the picture of SWB is one in which people are on the whole positive, but say there is room for improvement. It is possible that there is a biological set-point for mild positive affect. Good and bad events can move the person temporarily above or below their set-point, but over time they adapt and return to their baseline. The evolutionary advantage to a mildly positive hedonic set-point are manifold: it allows for approach tendencies to dominate the person's behavior, it allows room for positive events to be rewarding because there is still room to become more positive, and it allows negative events to seize the person's attention and have maximal information value because they are perceived as figure against ground. Much more data on people in extremely difficult circumstances (e.g., prison) and diverse cultures is needed before we can know the viability of the positive set-point hypothesis.

Several temperament variables are strong correlates of subjective well-being. For example, extraversion covaries with positive affect (Costa & McCrae, 1980; Emmons & Diener, 1985; Diener, Sandvik, Pavot & Fujita, 1992). Extraverted people experience more positive affect both when alone and with others (Pavot, Diener, & Fujita, 1990), whether living alone or with others, whether working in a social or solitary job, and whether living in a small town or urban area (Diener, Sandvik, Pavot, & Fujita, 1992). The relation between positive affect and extraversion is particularly strong when measurement error is controlled (Fujita, Diener, & Pavot, 1994). It is not yet known why this relation occurs.

Personality

It may be that relatively high amounts of dopamine in the brain lead to both approach behaviors such as extraversion and also to feelings of positive affect. It could also be possible, however, that extraverts receive more rewards in active, industrialized nations and therefore are more likely to be in a good mood in those environments. Cross-cultural data on the extraversion and positive affect relation is essential in determining the causes of this correlation. Neuroticism has been repeatedly shown to be related to the experience of negative affect (e.g., Costa & McCrae, 1980; Emmons & Diener, 1985). Indeed, neuroticism is virtually synonymous with negative affectivity (Watson & Clark, 1984). Optimism and self-efficacy covary with subjective well-being as well (Myers & Diener, 1995). Although self-esteem is a strong predictor of SWAB in western nations, Diener and Diener (1995) argued that this connection is much weaker in collectivist societies. Lucas, Diener, and Suh (1995) found that subjective well-being variables show discriminant validity from personality variables such as optimism and self-esteem.

Personality is a relatively strong predictor of long-term average levels of affect, but does not predict momentary moods well. Long-term *average* affect is relatively consistent across situations (Diener & Larsen, 1984) and stable across time (Diener, 1994; Magnus & Diener, 1991) and therefore can be predicted by stable personality traits.

People Adapt to their Life Circumstances

A surprising finding is that life conditions often do not strongly predict people's happiness or life satisfaction. For example, income in the U.S.A. correlates only .12 with SWB (Diener, Sandvik, Seidlitz, & Diener, 1992).

Okun and George (1984) found in an elderly sample that objective health as rated by physicians correlated about .10 with reported SWB. And Diener, Wolsic, and Fujita (1995) found that physical attractiveness, a powerful resource in western society, correlated only about .13 with SWB. Variables such as education, ethnic group, and gender barely correlate with SWB (e.g., Campbell, Converse, & Rodgers, 1976). Costa, McCrae, and Zonderman (1987) discovered that people going through life transitions (e.g., divorce or retirement) during a 10 year period did not vary in their level of SWB more than people who did not go through important life transitions.

Recent life events sometimes have a large impact on one's immediate SWB. For example, Silver (1980) found that people who acquired spinal cord injuries were extremely unhappy after their accidents. Surprisingly, however, the moods of these people improved in the weeks following their accidents so that after eight weeks the positive mood of happiness was stronger than their negative feelings. In our own research we found that life events occurring during the past three months can influence one's SWB. Life events occurring several years before, however, did not influence a person's SWB (Suh, Diener, & Fujita, 1995). Similarly, Diener, Sandvik, Seidlitz, & Diener (1992) found that people whose income over a 10-year period either markedly increased or decreased relative to other people's income did not differ in SWB. Apparently, people react strongly to events, but then adapt and return to a baseline that is set by their temperaments (Headey & Wearing, 1992) and socialization. This may explain why even people in bad circumstances such as abused women and individuals who are chronically mentally ill often report positive levels of well-being (Diener & Diener, 1995).

Although people's ability to adapt may be immense, we cannot conclude that life circumstances never make any difference to long-term SWB. For example, we know that people in wealthier nations report more SWB than those in poorer countries (Diener, Diener, & Diener, 1995). Especially at the level of poverty at which many people cannot meet their basic physical needs, lower levels of SWB are reported. Similarly, people in prison report lower levels of life satisfaction (Joy, 1990). Finally, family members who care for patients with Alzheimer's disease do not appear to adapt to their burden (Vitaliano, Russo, Young, Becker, & Maiuro, 1991). Thus, some conditions may enhance or detract from SWB, but we as yet do not know under what circumstances people may adapt to the conditions of their lives.

Validity of Measures

A large number of survey instruments are available for measuring SWB (see Andrews & Robinson, 1992). Some of the measures emphasize judgments of life satisfaction, such as our Satisfaction with Life Scale (Pavot & Diener, 1993), whereas others emphasize the emotional component of well-being, such as the Affect Balance Scale (Bradburn, 1969) and our Frequency of Affect Survey. Some of the scales are single items, such as Andrews and Withey's (1976) D-T scale, while others are multi-item such as the MUNSCH (Kozma & Stones, 1980). Watson, Clark, and Tellegen's (1988) PANAS separately measures activated forms of pleasant and unpleasant affect. The single item scales are useful in large national surveys, whereas the multi-item scales are appropriate to more intensive studies.

Work on the psychometric properties of SWB scales reveals that they have substantial validity, reliability, sensitivity to change, and factorial invariance across groups (e.g., Andrews & Withey, 1976; Larsen, Diener, & Emmons, 1985; Sandvik, Diener, & Seidlitz, 1993). For example, Sandvik et al. found that self-reported SWB measures converge substantially with interviewer ratings, informant reports, experience sampling of affect, and the recall of positive versus negative events. Similarly, Diener, Diener, and Diener (1995) report substantial convergence across surveys between the mean levels of happiness and life satisfaction of nations. They found an average correlation of .71 between mean levels of SWB in three national surveys of approximately 30 countries.

Reservations are often expressed about using the SWB self-report scales in a cross-cultural context because differences in responses might be due, not to substantive differences in SWB, but rather to differences in humility, in use of the scale, or language translation. For example, although Balatsky and Diener (1993) reported that Russian students were much less satisfied than students in the U.S.A., was this due to actual satisfaction or to measurement artifacts? Veenhoven (1993) and Shao (1993) present evidence that demonstrates that language translation is unlikely to produce substantial differences on SWB measures. Veenhoven also reviews data that suggest that differences in response styles, familiarity with the concept, and differences in the desirability of SWB are not the cause of variations across countries in reported SWB. It must be noted, however, that his analysis is based solely on highly westernized European nations. Diener, Suh, Smith, and Shao (1994) report evidence based on the People's Republic of China, South Korea, Japan and the U.S.A. that suggests that artifacts such as humility and using the middle of the scales are not responsible for the differences in SWB reported between these societies. They found that norms for the experience and expression of affect did mirror the mean levels of positive and negative affect reported in these three nations. Initial work on measurement artifacts indicates that they are not necessarily the causes of variations in well-being reported between one nation and another. Nevertheless, additional work is certainly needed on this question. Situational factors can influence life satisfaction responses (Schwarz & Strack, 1991) and there are a number of conceptual issues and interpretive problems that may apply to the measurement of SWB (Diener, 1994; Diener & Fujita, 1994). Thus, although initial findings are promising in terms of making cross-cultural comparisons of SWB, all the issues are not yet resolved.

Nations Differ in SWB

Table 1 shows the SWB values from the last national survey conducted in each of 43 nations. Different response formats were used in the surveys, and therefore Veenhoven (1993b) computed a Thurstone value for the responses in each survey, and then calculated a SWB value that was comparable across surveys. In order to maximize the number of nations, I included surveys of both happiness and life satisfaction. Although most nations fall in the slightly to very positive range, it can be seen that there are substantial differences in SWB between societies. For example, several Scandinavian nations have very high means, whereas several Latin American countries have low SWB.

Predictors of SWB

What are the correlates of SWB across nations? Table 2 shows the correlation between a number of predictor variables and the SWB scores shown in Table 1. In addition, two other SWB surveys were included when they were available: the most recent surveys (not counting the very most recent one) on happiness and on life satisfaction. Homicide is the rate of murders in the nation, and homogeneity includes the sum of three variables: the percent of people speaking the dominant language, the percent of people sharing the dominant religion, and the percent of people sharing the dominant ethnic/racial group. It can be seen that homicide rate and cultural homogeneity do not correlate with SWB.

Table 1. *Results of Last National Subjective Well-being Survey (0 to 10 Thurstone scaling value by Veenhoven, 1994b)*

Country	Year	SWB	Country	Year	SWB
Denmark	1982	8.1	Philippines	1979	6.4
Australia	1984	8.0	India	1979	6.2
Iceland	1990	8.0	S. Korea	1980	6.2
Switzerland	1976	8.0	Puerto Rico	1963/4	6.2
Sweden	1990	7.9	Spain	1992	6.1
Finland	1981	7.7	France	1992	6.0
Netherlands	1992	7.7	Israel	1984	6.0
Canada	1982	7.6	Italy	1992	6.0
N. Ireland	1990	7.6	E. Germany	1992	5.9
Luxembourg	1992	7.5	Hungary	1982	5.9
United States	1989	7.3	Japan	1987	5.9
Austria	1974	7.1	Portugal	1992	5.8
Singapore	1979	7.1	S. Africa	1983	5.8
Belgium	1992	7.0	Egypt	1960	5.5
Brazil	1979	7.0	Greece	1992	5.3
W. Germany	1992	7.0	Poland	1984	5.3
Ireland	1992	7.0	Yugoslavia	1962	5.0
Britain	1992	6.8	Nigeria	1963	4.8
Norway	1991	6.7	Panama	1962/3	4.8
Malaysia	1965	6.5	Mexico	1981	4.3
Thailand	1965	6.5	Dom. Republic	1962	1.6
Cuba	1960	6.4			

5.0 is neutral on this 11-point scale.

Several variables correlate significantly and across surveys with SWB. First, the income (GDP per person) of nations covaries strongly with the mean SWB of nations in all three surveys. The basic needs variable is composed of the mean of five standardized variables: longevity, infant mortality (reverse scored), available food calories, percent with sanitary facilities, and percent with clean drinking water. The fulfillment of basic needs also correlated highly with SWB, which should not be surprising because basic needs were strongly correlated with income (r (53) = .76, p < .001). Two other variables that are related to income were also correlated with SWB - Individualism and Human Rights. Individualism was a score assigned to the nations by Harry Triandis, reflecting the degree of individualism versus collectivism of the society. Human rights was a weighted score for nations reflecting the degree to which their citizens enjoy a weighted sum of 40 different political and civil rights (Gupta, Jongman, & Schmid, 1994). It appears that income, rights, individualism, and basic need fulfillment are so highly intertwined that they all predict SWB. Further, their intercorrelations are so high (usually about .75), that it is difficult to separate their influence with samples of this size. Although citizen rights may lead to higher well-being, it is possible that the causal direction is reversed - that happy citizens are more likely to create democratic institutions.

It is interesting to note that suicide rates also correlate with individualism! In fact, reported well-being and suicide rates covary across societies, although this correlation disappears when individualism is held constant. Thus, individualistic cultures may offer more opportunities for rewards, but also less support in difficult times. Furthermore, attributions for both success and failure to the individual may heighten positive affect in good times and negative affect in bad times in individualistic nations. The relation between the characteristics of a society and various measures of well-being deserves intensive study.

Finally, the number of hours of sun a country receives during an average year tended to correlate inversely with SWB. This surprising result probably occurs because of nations further from the equator are also wealthier. Thus, the correlation between hours of sun and income is -.40 (41), $p < .01$. When income is partialled out of the relation between hours of sun and the three SWB measures, all drop to nonsignificance, r (30) = -.18, r (23) = -.20, and r (19) = -.18. This finding suggests that less sunny climates score higher on SWB because nations in higher latitudes are wealthier.

Income and SWB

It appears that income and its correlates are replicable predictors of SWB. Easterlin (1974) maintained that the relation between income and happiness is relative - it depends on how much income other people have, and on other changeable standards. In response, Veenhoven (1991) argued that the relation between income and happiness is absolute - it depends on income's ability to satisfy universal needs. Veenhoven maintained that income is connected to SWB only insofar as it helps people meet their innate needs.

In support of the absolute position, Veenhoven pointed out that the relation between income and happiness across nations is curvilinear - income makes little difference to SWB once subsistence needs are met. In further support of his position, Veenhoven showed that income is most highly correlated with income in poor nations. In wealthier societies where meetings one's physical needs is an issue for only a relatively small percentage of the population, income shows only a small correlation with life satisfaction. Veenhoven argued that the reason that SWB has not increased in the U.S.A., France, Japan and similar nations over the period in which surveys have been conducted is that the citizens of these nations could already meet their physical needs even at the time of the first surveys after World War II. Another piece of evidence in support of the absolute position is the finding of Diener, Sandvik,

Seidlitz and Diener (1992) that income change in the United States did not influence people's SWB.

There is also evidence to support the relative position. For example, Diener and Fujita (1995) found that income influenced the life satisfaction of people with different goal structures in different ways. For individuals who had goals related to money, income was more highly correlated with SWB than it was for those for whom money was less relevant to their goals. In addition, the income of nations still correlates significantly with SWB even after the level of basic needs is partialled out, suggesting that there are effects of wealth that transcend the meeting of physical needs. Furthermore, the data show that even for college students the relation between income and SWB is stronger in poorer nations. Because college students are an elite in most countries whose basic physical needs are usually met, this finding indicates that the fulfillment of physiological needs may not be the sole reason that income and SWB correlate more highly in poor nations. It could be that people are more concerned with material wealth in nations that are developing than they are in post-industrial societies where many individuals already possess many of the material goods that they desire. Finally, the relative position is bolstered by the fact that culture seems to mediate the effects of virtually all objective circumstances.

Table 2. Correlates of SWB Across Nations

Predictor Variables	Last SWB Survey	Prior Happiness Survey	Prior Satisfaction Survey
Homicide rate	-.18	-.08	-.10
Homogeneity	.13	-.09	.21
Income	.65***	.43*	.57**
Basic needs fulfilled	.55***	.37*	.60***
Individualism	.58***	.40*	.59***
Human Rights	.40*	.52**	.65***
Yearly hours of sun	-.44*	-.40	-.41*

Note: * $p < .05$; ** $p < .01$; *** $p < .001$

Although there is evidence to support both the relative and absolute positions, these sides are not necessarily contradictory. It could be that income most influences SWB when basic physical needs are at issue, but that there are effects of income beyond these basic needs. Further, it could be that income is likely to influence most people's happiness if they don't have enough to eat, but that relative effects such as the influence of people's goals can also influence the amount of money that makes a person happy. Furthermore, a resolution of the debate does not seem possible unless the approaches are very clearly defined. What are the basic needs? For example, do interest and self-esteem count? And what are the relative factors that influence the impact of wealth?

My hypothesis is that the effects of wealth depend on people's goals. In a poor society, increasing one's income may be a very compelling goal for many people because physical drives as well as cultural ideals may push one to adopt such a goal. In contrast, in wealthier societies, the goal of making a lot of money may differ dramatically across individuals, and other goals related to social relations and so forth may be more important for many people. In the post-materialist society young people may adopt goals such as self-development, and therefore income may become less important to their subjective well-being.

National Differences in the Causes of SWB

SWB researchers have been searching for *the* causes of happiness. In part, the belief that there are universal causes of SWB rests on the assumption that there are universal human needs and that meeting these needs leads to SWB (Veenhoven, 1993b). An approach, however, that recognizes the key importance of culture is that progress towards one's goals makes one happy, but that people's goals can vary dramatically. Support for this hypothesis was found by Diener and Fujita (1995) who showed that different resources predict happiness for different people, depending on the relevance of these resources for achieving their particular goals. In a similar manner, the causes of SWB may vary from culture to culture, depending on what goals are most relevant.

There is now compelling evidence that the correlates of SWB are different in different cultures. Veenhoven (1991) found that income was much more strongly correlated with SWB in poorer than in wealthier nations. A related finding is that financial satisfaction is more closely related to life satisfaction in poor than in rich societies (Diener & Diener, 1995). Even a cause of happiness as seemingly fundamental as self-esteem does not seem to be a universal correlate of happiness. Diener and Diener found that in some nations self-esteem was not a significant predictor of life satisfaction. Furthermore, they showed that the correlation between self-esteem and life satisfaction was much stronger in individualistic than in collectivistic cultures. Thus, the causes of SWB are likely to differ depending on people's values, goals, and cultural norms.

Future Research Directions

The intersection of SWB and cross-cultural psychology is exciting because there are so many open questions. How do we measure SWB in comparable ways across cultures? What effects do norms, values, and worldview have on SWB? Do childrearing practices lead to different levels of SWB in different societies? Is the average work in more developed nations more interesting than the work in agricultural societies? How do aspirations and expectations influence the SWB of a country? Why do people in wealthier and more individualistic nations report higher SWB? These questions and many more like them are a rich vein for scientists to mine.

Much of our knowledge about SWB comes from highly westernized nations. Thus, a most important task is to replicate and extend these findings to other nations. For example, the finding that extraverts experience more positive affect must be examined in other cultures. Only by examining SWB in a cross-cultural context can we hope to truly understand it.

References

Andrews, F.M., & Robinson, J.P. (1992). Measures of subjective well-being. In J.P. Robinson, P.R. Shaver, & L. S. Wrightsman (Eds.), *Measures of personality and social psychological attitudes* (pp. 61- 114). San Diego, Academic Press.

Andrew, F.M., & Withey, S.B. (1976). *Social indicators of well-being: America's perception of life quality.* New York: Plenum Press.

Argyle, M. (1987). *The psychology of happiness.* London: Methuen.

Balatsky, G., & Diener, E. (1993). Subjective well-being among Russian students. *Social Indicators Research, 28,* 225-243.

Campbell, A., Converse, P.E., & Rodgers, W. L. (1976). *The quality of American life.* New York: Russell Sage Foundation.

Costa, P., & McCrae, R.R. (1980). Influence of extraversion and neuroticism on subjective well-being: Happy and unhappy people. *Journal of Personality and Social Psychology, 38,* 668-678.

Costa, P., McCrae, R., & Zonderman, A. (1987). Environmental and dispositional influences on well-being: Longitudinal follow-up of an American national sample. *British Journal of Psychology, 78,* 299-306.

Diener, E. (1984). Subjective well-being. *Psychological Bulletin, 95,* 542-575.

Diener, E. (1994). Assessing subjective well-being: Progress and opportunities. *Social Indicators Research, 31,* 103-157.

Diener, E., Colvin, C. R., Allman, A., & Pavot, W. (1991). The psychic costs of intense positive emotions. *Journal of Personality and Social Psychology, 61,* 492-503.

Diener, E., & Diener, C. (1995). *Most people are happy.* Manuscript submitted for publication, University of Illinois.

Diener, E., Diener, C., & Diener, M. (in press). Factors predicting the subjective well-being of nations. *Journal of Personality and Social Psychology.*

Diener, E., & Diener, M. (1995). Cross-cultural correlates of life satisfaction and self-esteem. *Journal of Personality and Social Psychology, 68,* 653-663.

Diener, E., & Fujita, F. (1995). Resources, personal strivings, and subjective well-being: A nomothetic and ideographic approach. *Journal of Personality and Social Psychology, 68,* 926-935.

Diener, E., Fujita, F., & Sandvik, E. (1994). What subjective well-being researchers can tell emotion researchers about affect. In N. H. Frijda (Ed.), *Proceedings of the eighth conference of the International Society for Research on Emotions* (pp. 30-36). Storrs, CT: ISRE Publications.

Diener, E., & Larsen, R. J. (1984). Temporal stability and cross-situational consistency of affective, behavioral, and cognitive responses. *Journal of Personality and Social Psychology, 47,* 871-883.

Diener, E., & Larsen, R. J. (1993). The experience of emotional well-being. In M. Lewis & J. M. Haviland (Eds.), *Handbook of emotions* (pp. 405-415). New York: Guilford Press.

Diener, E., Sandvik, E., Pavot, W., & Fujita, F. (1992). Extraversion and subjective well-being in a U.S. national probability sample. *Journal of Research in Personality, 26*, 205-215.

Diener, E., Suh, E., Smith, H., & Shao, L. (1994). National and cultural differences in reported subjective well-being: Why do they occur? *Social Indicators Research, 34*, 7-32.

Diener, E., Sandvik, E., Seidlitz, L., & Diener, M. (1992). The relationship between income and subjective well-being: Relative or absolute? *Social Indicators Research, 28*, 195-223.

Diener, E., Wolsic, B., & Fujita, F. (in press). Physical attractiveness and subjective well-being. *Journal of Personality and Social Psychology*.

Easterlin, R.A. (1974). Does economic growth improve the human lot? Some empirical evidence. In P.A. David & W.R. Melvin (Eds.), *Nations and households in economic growth* (pp. 98-125). Palo Alto, CA: Stanford University Press.

Emmons, R.A., & Diener, E. (1985). Personality correlates of subjective well-being. *Personality and Social Psychology Bulletin, 11*, 89-97.

Fujita, F., Diener, E., & Pavot, W. (1993). *An investigation of the relation between extraversion, neuroticism, positive affect, and negative affect*. Unpublished manuscript, University of Illinois.

Gupta, D.K., Jongman, A.J., & Schmid, A.P. (1994). *Assessing country performance in the field of human rights: Proposal for a new methodology to create a composite index*. Paper presented at the 13th International Sociological Association Congress, Bielefeld, Germany.

Headey, B., & Wearing, A. (1992). *Understanding happiness: A theory of subjective well-being*. Melbourne: Longman Cheshire.

Joy, R.H. (1990). *Path analytic investigation of stress-symptom relationships: Physical and psychological symptoms models*. Unpublished Doctoral dissertation, University of Illinois.

Kozma, A., & Stones, M.J. (1980). The measurement of happiness: Development of the Memorial University of Newfoundland Scale of Happiness (MUNSCH), *Journal of Gerontology, 35*, 906-912.

Larsen, R.J., Diener, E., & Emmons, R.A. (1985). An evaluation of subjective well-being measures. *Social Indicators Research, 17*, 1-18.

Lucas, R., Diener, E., & Suh, E. (1995). *Discriminant validity of subjective well-being, self-esteem, and optimism*. Manuscript submitted for publication.

Magnus, K., & Diener, E. (1991). *A longitudinal analysis of personality, life events, and subjective well-being*. Paper presented at the 63rd Annual Meeting of the Midwestern Psychological Association, Chicago.

Myers, D.G. (1992). *The pursuit of happiness: Who is happy - and why*. New York: William Morrow.

Myers, D.G., & Diener, E. (1995). Who is happy? *Psychological Sciences, 6*, 10-19.

Okun, M.A. & George, L.K. (1984). Physician- and self-ratings of health, neuroticism, and subjective well-being among men and women. *Personality and Individual Differences, 5*, 533-539.

Pavot, W., & Diener, E. (1993). Review of the Satisfaction with Life Scale. *Psychological Assessment, 5*, 164-172.

Pavot, W., Diener, E., & Fujita, F. (1990). Extraversion and happiness. *Personality and Individual Differences, 11,* 1299-1306.

Sandvik, E., Diener, E., & Seidlitz, L. (1993). Subjective well-being: The convergence and stability of self-report and nonself-report measures. *Journal of Personality, 61,* 317-342.

Schwarz, N., & Strack, F. (1991). Evaluating one's life: A judgment model of subjective well-being. In F. Strack, M. Argyle, & N. Schwarz (Eds.), *Subjective well-being: An interdisciplinary perspective* (pp. 27-48). Oxford, Britain: Pergamon Press.

Shao, L. (1993). *Multilanguage comparability of life satisfaction and happiness measures in mainland Chinese and American students.* Unpublished master's thesis, University of Illinois.

Silver, R.L. (1980). *Coping with an undesirable life event: A study of early reactions to physical disability.* Doctoral dissertation, Northwestern University, Evanston, Illinois.

Suh, E., Diener, E., & Fujita, F. (1995) Events and subjective well-being: Only recent events matter. *Journal of Personality and Social Psychology,* (submitted).

Veenhoven, R. (1988). The utility of happiness. *Social Indicators Research, 20,* 333-354.

Veenhoven, R. (1991). Is happiness relative? *Social Indicators Research, 24,* 1-34.

Veenhoven, R. (1993a). *Bibliography of happiness: 2472 contemporary studies on subjective appreciation of life.* Rotterdam: RISBO.

Veenhoven, R. (1993b). *Happiness in nations: Subjective appreciation of life in 56 nations 1946-1992.* Rotterdam: RISBO.

Veenhoven, R. (1984). *Conditions of happiness.* Dordrecht, Netherlands: Reidel Publishing.

Vitaliano, P.P., Russo, J., Young, H.M., Becker, J., & Maiuro, R.D. (1991). The screen for caregiver burden, *The Gerontologist, 31,* 76-83.

Watson, D., & Clark, L.A. (1984). Negative affectivity: The disposition to experience aversive emotional states. *Psychological Bulletin, 96,* 465-490.

Watson, D., Clark, L.A., & Tellegen, A. (1988). Development and validation of a brief measure of positive and negative affect: The PANAS scales. *Journal of Personality and Social Psychology, 54,* 1063-1070.

Variance in Fertility due to Sex-Related Differentiation in Child - Rearing Practices

Bilge Ataca, Queen's University, Kingston, Canada
Diane Sunar & Çigdem Kagitçibasi, Bogaziçi University, Istanbul, Türkiye

The problem of population growth has drawn a great deal of attention from demographers, economists, and sociologists, who have carried out large numbers of studies concerned with the causes of fertility behavior. Typically, these studies have tried to measure the socioeconomic, cultural, and environmental variables influencing fertility, but have disregarded psychological factors underlying the behavior. However, since it is individuals, not cohorts or social strata, who have babies, attempts to develop sound policies of population control should take psychological factors into account as an essential set of variables in the analysis of determinants of fertility (Smith, 1973). If fertility is analysed according to its indirect determinants (e.g., culture, social class, education, physical environment) and direct determinants (e.g., age at marriage, contraceptive use, hygienic practices), as suggested by Bongaarts (1978), psychological factors may be regarded as constituting another group of variables operating between these two sets of determinants. Sunar (1987) has suggested that an important link between indirect and direct determinants of fertility is in the child-rearing practices of parents and their consequences for the child's personality and attitudes. Thus, she has attempted to interpose two layers of variables, child-rearing practices and their personality/attitudinal consequences, between the indirect and direct determinants of fertility. One important aspect of child-rearing is the differential socialization of male and female children which, along with its associated consequences for sex-role typing and self esteem, is predicted to have a significant impact on fertility behavior.

Sex-Role Socialization Processes

Each individual is socialized into a male or a female world from earliest childhood. Differences in the socialization experiences of boys and girls reflect the different role expectations of society, and are a major cause of sex differences (Hoffman, 1977). Children acquire the values, motives, and behaviors appropriate to either males or females in their culture by the process of sex-typing. Since sex-typed behaviors appear early and the family is both the child's first social environment and the social group with which s/he has the closest and most frequent contact, the role of the parents is particularly crucial in the development of sex-typing (Hetherington, 1975).

Many studies from various perspectives have identified different domains in which parents treat their sons and daughters differently. Adults respond differently to the same infant depending on whether they believe it is a boy or a girl (Frisch, 1977; Seavey, Katz, & Zalk, 1975; Smith & Lloyd, 1978). Close observations of parent-child interaction reveal that parents encourage sex-typed activities and interests in their children (Langlois & Downs, 1980; Lewis, 1972; Tauber, 1979). Parental reports of child-rearing emphases and young adults' perceptions of their own parents' practices have strikingly similar sex-differentiated socialization patterns in common, although collected from independent and unrelated samples (Block, 1979; Hoffman, 1977).

Sex-Role Socialization in Türkiye

The present study is an attempt to assess the usefulness of the intervening variables of child-rearing practices and their personality/attitudinal sequelae in predicting fertility in a sample of middle-class Turkish women. Rapid population growth in Türkiye over the last 50 years has made fertility an important topic in social and demographic studies of Türkiye (see Shorter, 1995); therefore a Turkish sample is an appropriate one for investigating this problem. Also, since urban Türkiye is in the process of rapid social change, including changes in marriage relationships (Imamoglu, 1994) and child-rearing practices (Sunar, 1994), the kinds of variation necessary for testing the hypotheses put forth here can be expected to exist in an urban middle class sample. A short description of sex-role socialization aims and practices in traditional Turkish culture will make clear the particular differences which will be examined.

Sex differences in socialization exist in the Turkish family even before the child's birth. As in most countries throughout the world, there is a preference for male children, and parents have different expectations of the two sexes (Kagitçibasi & Sunar, 1992). Findings of the Turkish Value of Children (VOC) study (Kagitçibasi, 1982a), conducted with a nationally representative sample and focusing on motivations for childbearing and values attributed to children by parents, reveal significant differences in parental expectations, preferences, and values attributed to sons and daughters, indicating clear sex-role differentiation. The male child has a special importance in the interdependent family system. Parents' expectations of economic help and security from sons, and the sons' privilege of carrying on the family name, both of which reflect the traditional values of the patriarchal, patrilineal and patrilocal family system, lead to preference for sons over daughters.

Since sons are depended on for financial support, families do their best to educate their sons. Professional achievement is desired much more for sons than for daughters - parents want sons to be successful and to "make something" of themselves (Kagitçibasi, 1982a). Girls, on the other hand, out of concern for proper sexual behavior and maintenance of family honor, are strictly restricted in their behavior, given less freedom, and kept under much closer adult supervision in the traditional family. Girls reared in this over-pressured and dependent context grow up to be more traditional, conservative, religious, less successful and more dependent on control by parents than boys (Kagitçibasi, 1972). Of course there is considerable variation among families, and particularly among the urban population these patterns may be moderated to some degree, notably in the direction of more education for girls.

This traditional sex-differentiated socialization is the means by which boys and girls internalize the separate values and behaviors appropriate to each sex in the culture. Hence, a high level of sex-related differentiation in child-rearing is likely to reproduce the patriarchal family structure in the next generation (Sunar, 1987). Patriarchal norms and family structure basically include male dominance, boy preference, relative female powerlessness, rigid sex-role stereotyping, early age at marriage for women, and expectations of old age security from sons (Kagitçibasi, 1982b; Sunar, 1987; Kagitçibasi & Sunar, 1992). Since these are all factors which contribute to high fertility, the reproduction of patriarchy through a high level of sex differentiation in child rearing is likely to lead to high fertility over generations. In contrast, female children of less patriarchal, more equalitarian families are more likely to be better educated, have greater autonomy, efficacy, and self sufficiency than those of more patriarchal, traditional families. They will be less dependent on their children for economic support, resulting in less boy preference and a lower rate of fertility over generations (Kagitçibasi, 1990; Sunar, 1987).

Sex Roles, Self-Esteem, and Fertility

Sex-differentiated child-rearing is crucial in the development of different sex roles for males and females. A high level of sex differentiation in child-rearing is expected to lead to high sex-role identification in the offspring. Furthermore, since the traditional family roles emphasize women's childbearing function, sex-role identification in women is likely to be positively related to fertility and negatively related to direct determinants such as age at marriage and contraceptive use.

Studies investigating the impact of sex-role orientation on self-esteem in various age groups have all agreed that, whereas masculinity and androgyny are associated with high levels of self-esteem, femininity is associated with lower levels (Gürbüz, 1988; Lamke, 1982; Puglisi & Jackson, 1981). Hence, sex-related differentiation, encouraging different sex roles for males and females, is expected to bring lower levels of self-esteem in female offspring.

Female self-esteem, in turn, may have an effect on fertility. The woman with higher self-esteem is expected to be more active in decision-making, more likely to discuss desired family size with her husband, and more likely to use contraceptives than the woman with lower self-esteem (Sunar, 1987). Since these are all factors which contribute to a lower level of fertility, women with higher self-esteem are expected to have fewer children than women with lower self-esteem.

Figure 1. The Relationship between Sex-Related Differentiation in Child-Rearing, Sex-Role Identification, Self Esteem, Relationship with Spouse, Age at Marriage, Contraceptive Use, and Fertility

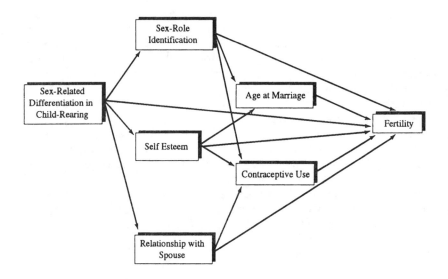

The power relation underlying the marital bond is characterized by male dominance and female subordination in the traditional Turkish family. Since sex-related differentiation in child-rearing is expected to produce feminine sex-role identification in girls, women who were exposed to these sex differentiated practices in childhood may be expected to have less power in their relationships with their spouses than women who were not. Regarding the relationship between husband-wife dominance and family size, equalitarian intra-family relations, with increased communication and role-sharing between spouses and decreased male decision- making, are associated with lower levels of fertility. Also, male decision-making and low intra-family status of women are associated with economic VOC, which in turn is associated with high fertility (Kagitçibasi, 1982a).

Theoretical Model and Hypotheses

A complex model of relationships among variables, including two layers of "intervening" variables, some "direct" determinants of fertility, and the outcome variable of fertility, is presented in Figure 1. Here, variation in sex-related differentiation in child-rearing (implicitly assumed to result from, among other things, "indirect" determinants of fertility, such as demographic and cultural factors) is hypothesized to give rise to differences in: (1) sex-role identification, (2) self esteem, and (3) relationship with spouse. This second set of intervening variables in turn, is hypothesized to affect: (1) age at marriage and (2) contraceptive use, both of which are "direct" determinants of fertility. In addition, each layer of variables is hypothesized to have a direct as well as mediated effect on the outcome variable, fertility.

Specific hypotheses are as follows: (1) Sex-related differentiation in child-rearing practices is positively related to fertility; that is, women reared in a traditional manner will have more children than those reared in an equalitarian manner. (2) Sex-related differentiation in child-rearing practices is positively related to feminine sex-role identification. (3) Sex-related differentiation in child-rearing practices is negatively related to female self-esteem. (4) Sex-related differentiation in child-rearing practices is negatively related to power in relationship with the husband. (5) Feminine sex-role identification is negatively related to age at marriage. (6) Self-esteem is positively related to age at marriage. (7) Feminine sex-role identification is negatively related to use of contraceptives. (8) Self-esteem is positively related to contraceptive use. (9) Power in relationship with the husband is positively related to contraceptive use. (10) Feminine sex-role identification is positively related to fertility (number of children). (11) Self-esteem is negatively related to fertility. (12) Power in relationship with the husband is negatively related to fertility.

Method

Respondents

Seventy five women of middle socioeconomic status living in Istanbul participated in the study. All respondents, who were recruited from among mothers of children attending school in a middle-class neighborhood in Istanbul, were married housewives, living with their husbands. Each had at least one brother of similar age with whom she had spent her childhood. The respondents were divided into two groups on the basis of their number of children. Those with one or two children ($n = 38$) were classified as the Low Fertility (LF) group, while those with three, four, or five children, ($n = 37$) were classified as the High Fertility (HF) group. Mean age of the LF group was 38.8, while that of the HF group was 39.6. Mean years of schooling for the LF group was 8.0, and that for the HF group was 7.8. These differences in age and schooling were not statistically significant.

Instruments

The Sex-Trait Stereotype Measure (STSM)
Sex-role identification was measured with the Sex-Trait Stereotype Measure (STSM) developed by Gürbüz (1988) for use in Türkiye. The validity of Bem's Sex-Role Inventory (Bem, 1974) was found to be less than satisfactory when used with Turkish respondents; thus, following Bem's method, Gürbüz developed a new scale in Turkish to measure the degree of sex-trait stereotyping in an individual's self- description.

The Self-Esteem Inventory (SEI)
Self-esteem was measured with a revised version of the Self-Esteem Inventory (SEI) designed by Coopersmith (1967). This instrument was translated into Turkish, using standard back-translation methods (Onur, 1981; Sunar, 1978), and reliability and validity of the Turkish version were reported to be satisfactory (Gürbüz, 1987; Onur, 1981).

Procedure

Each respondent was visited by appointment in her home, where the interview was carried out by the first author. At the conclusion of the interview, the STSM and SEI were completed by the respondent, following instructions by the interviewer. The whole procedure took approximately one hour.

The Interview
The interview schedule was composed of two parts, the first designed to obtain the respondent's retrospective perception of sex-related differentiation in child-rearing practices applied by her parents to herself and her brother, and the second designed to investigate the nature of the respondent's relationship with her husband. The interview was tested in a pilot application to ten women and was revised in accordance with the feedback received.

In the first part of the interview, the respondent was asked to name one brother with whom she had had the most frequent contact and the most intimate relationship in her childhood. Then she was asked to compare herself with her brother during their childhood in terms of various parental behaviors related to dependence/obedience vs. independence/autonomy, educational and occupational aspiration, toys given to children, assignment of household chores, expression of affect, amount of pocket money given, expected responsibilities, reaction to relationships with the opposite sex, and emphasis upon chastity. Thus, sex-related differentiation in child-rearing practices is defined in terms of the respondents' perceptions of the degree of difference in their parents' child-rearing practices which were applied to themselves and their brothers while they were growing up.

The second part of the interview focused on the nature of the respondent's relationship with her husband. Female power in relationship with the husband is defined as the degree of equalitarianism in the conjugal relationship, which is measured in terms of role-sharing, mutual or alternating decision-making, and marital communication as reported in this part of the interview. Other questions regarding contraceptive use, age at marriage, sex preference for children, and values attributed to children were also asked in this section.

Results

In order to test the hypotheses, an index of sex-related differentiation in child-rearing practices was derived from the interview. The items chosen for the index were those regarding sex-related differentiation between the respondent and her brother in the domains of *education, dependence/obedience* vs. *independence/autonomy, household chores,* and *chastity*. The respondent's response to each item was categorized as *traditional* (differentiation in favor of the brother), *nontraditional* (differentiation in favor of the female respondent), or *equalitarian* (no differentiation). Respondents who were scored as traditional on half or more of the index items were classified as Traditional, while those who were scored as nontraditional or equalitarian on more than half of the index items were classified as Equalitarian (since there were very few nontraditional responses, they were combined with equalitarian responses).

Power in relationship with the spouse was measured in an analogous manner, using the 7 items regarding decision-making for the power index. Respondents who were scored for husband dominance on more than half of the items were classified as Husband Dominant, while those who were scored for wife dominance or shared decision making on more than half of the items were classified as Shared Decision Making.

Tests of Hypotheses

Hypothesis 1, that sex-related differentiation in child-rearing practices is related to fertility, was supported; the difference between the mean number of children of the traditional (2.83) and equalitarian (2.18) groups was found to be significant ($t = -2.44, p = .02$).

Hypothesis 3, that sex-related differentiation in child-rearing practices is negatively related to female self-esteem, was supported, as the difference in self esteem scores between equalitarian and traditional groups was significant ($t = 1.97, p = .05$).

Hypothesis 4, that sex-related differentiation in child-rearing practices is negatively related to power in relationship with the husband, received only weak support, since close to 90% of the sample reported shared decision making. There was, however, a significant positive correlation between equalitarian upbringing and shared decision making ($r = .30, p = .01$), and a significant negative correlation between traditional child-rearing and shared decision making ($r = -.30, p = .01$).

Hypothesis 10, that feminine sex-role identification is positively related to fertility, was supported ($r = .27, p = .05$).

Hypothesis 12, that power in relationship with the husband is inversely related to fertility, was supported. The difference between the mean number of children in the husband-dominant group (3.37) and in the shared-decision group (2.43) was significant ($t = -2.17, p = 05$). Communication with the spouse, which may be considered a component of the power relationship, is also related to fertility: 78.4% of the LF group, compared with 40.5% of the HF group, reported talking with their husbands about the number of children they would have (chi square = 6.86, p < .05).

The other hypotheses were not supported.

Table 1. Intercorrelations Among Variables (n = 75)

	1	2	3	4	5	6	7	8	9	10
1. Equalitarian child rearing										
2. Traditional child rearing	-.98**									
3. Femininity	-.18	.18								
4. Masculinity	-.03	.01	-.16							
5. Self esteem	.23*	-.24*	-.27*	.26*						
6. Husband dominance	-.19	.20	.30**	-.30**	-.02					
7. Wife dominance	-.16	.15	-.22	.29*	-.05	-.37**				
8. Shared decision	.30**	-.30**	-.03	-.04	.06	-.43**	-.67**			
9. Age at marriage	.08	-.07	-.20	.04	.01	-.20	-.26*	.39**		
10. Birth control	.09	-.10	-.07	.03	.03	-.31**	-.05	.24*	.29*	
11. Number of children	-.34**	.32**	.27*	-.12	-.17	.38**	.17	-.47**	-.33**	-.02

*p < .05.   **p < .01

Table 1 presents the correlations among all the variables under analysis, along with probability levels. Significant relations from this table may be summarized as follows:

1. Number of children, or fertility, is positively correlated with scores on the traditionalism index, scores on the femininity subscale of the STSM, and scores on the husband-dominance index; it is negatively correlated with scores on the equalitarianism index, scores on the shared decision-making index, and age at marriage.
2. Equalitarian upbringing (score on the equalitarianism index) is positively correlated with self esteem scores and scores on the index of shared decision making in marriage, while it is negatively correlated with number of children.
3. Traditional sex-differentiated upbringing (score on the traditionalism index) is positively correlated with number of children and negatively correlated with self-esteem and scores on the index of shared decision making in marriage.
4. Self esteem is positively correlated with scores on the equalitarianism index and with scores on the masculinity subscale of the STSM; it is negatively correlated with scores on the traditionalism index and with scores on the femininity subscale of the STSM.
5. Scores on the femininity subscale of the STSM (feminine sex role identification) are correlated positively with decision making by the husband and number of children, and correlated negatively with self-esteem scores.

Table 2. Stepwise Regression Analysis for Variables Predicting Fertility (N = 75)

Variables in the equation	B	SE B	ß	R^2
Age at marriage	-.13	.04	-.32	.11*
Trad/equal child-rearing	.60	.25	.25	.07*
Spousal relations	.86	.40	.22	.05*

$R^2 = .23$ (p < .01)   *p < .01.

6. Scores on the shared decision making index are correlated positively with equalitarian upbringing, age at marriage, and use of contraceptives, and correlated negatively with traditional upbringing and number of children.

A stepwise multiple regression analysis was also carried out to determine the relative contributions of traditional-equalitarian child-rearing, femininity, self-esteem, power in relationship with the spouse, and age at marriage to number of children. As shown in Table 2, three variables were retained in the regression equation (age at marriage, sex differentiation in child-rearing, and power in relationship with the spouse); together these variables accounted for 23% of the variance in fertility. Self-esteem and femininity were not included in the equation.

Discussion

The model proposed in the introductory section above received a considerable amount of support, particularly in regard to the connections between child-rearing practices, their effects on personality, and fertility. On the other hand, certain features of the model were found to be in error, calling for revisions, especially with regard to the roles of self-esteem and the "direct" determinants of fertility.

There was a clear effect of sex-related differentiation in child-rearing practices on fertility, with the traditional group having significantly more children than the equalitarian group. Sex-related differentiation also showed significant effects on the intervening variables of self-esteem and power in relationship with the husband, although only the latter was related to fertility.

While feminine sex role identification as measured by the STSM was only tenuously related to child-rearing practices and to age at marriage, it nevertheless had a direct relationship with fertility.

Self-esteem, rather than functioning as an intervening variable with an effect on fertility, appeared to be an outcome of child-rearing practices and sex role identification, with equalitarian upbringing and higher masculinity scores being associated with higher self-esteem scores.

Power in relationship with the husband had the predicted effect on number of children: fewer children born to couples who share decision making, more to those in which the husband makes the decisions. It was also, as predicted, related positively to equalitarian upbringing and negatively to traditional sex-differentiated upbringing.

Of the "direct" determinants of fertility studied here, contraceptive use did not appear to be related to family size. The population represented by this sample appears to be rather fully informed regarding contraceptives and inclined to use them (see Shorter, 1995). It can be inferred that larger family sizes were desired by the couples who had them and that contraceptives were used to control the spacing of children and to avoid further pregnancies when the desired family size was reached. The distinguishing characteristics of couples who had larger numbers of children tended to be, rather than contraceptive use, husband domination in decision making and a wife who had been brought up in a traditional sex-differentiated manner and had relatively higher feminine identification.

On the other hand, age at marriage had the predicted relationship to fertility, with women who married later having fewer children. Age at marriage, however, correlated with only one of the predicted intervening variables, power in relationship with the husband, and in this case was probably a determining factor rather than an effect: a woman who marries at a later age is less likely to be dominated by her husband.

Three major conclusions may be drawn from these findings:

First, intra-family status of the wife, as measured by decision-making power (and also by communication with the husband) is an important determinant of fertility. This result was also found previously in the VOC study (Kagitçibasi, 1982a). Since training interventions have also been found to have an effect on women's intra-family status (Kagitçibasi, Sunar & Bekman, 1988), in groups where information about contraceptives is readily available, interventions focusing on empowerment of women may have a greater effect on fertility.

Second, age at marriage, while a "direct" determinant of fertility in Bongaarts' (1978) scheme, is also related to the power relationship between the spouses, and may exert some of its effect through the woman's status and communication patterns between husband and wife.

Third, traditional/equalitarian child-rearing is an important determinant of fertility, but the present study did not fully succeed in identifying the crucial intervening mechanisms through which the effect is achieved. Of the two intervening variables proposed, feminine identification was related to fertility but not to child-rearing practices, while self-esteem was related to child-rearing practices but not to fertility.

The positive correlation between feminine identification and fertility lends support to the idea that sex role identification may be an important variable intervening between socialization and fertility. However, the failure to find a significant relation between child-rearing practices and femininity needs further investigation. It is possible that this failure is due to the nature of the STSM; further investigations comparing this instrument and other instruments measuring sex-role identification will be useful. Likewise, the findings regarding self-esteem are suggestive but non-definitive; it is possible that self-esteem's effect on fertility is weakened by the mediation of other variables.

Finally, it is to be noted that the main findings of the study, the significant relations observed both between sex-related differentiation in child-rearing practices and fertility, and power in relationship with the spouse and fertility, possibly reflect important and robust links between some psychological variables and fertility outcome. This is because significant relations emerged between sex-differentiated child-rearing and spousal power relations and number of children even in this urban, middle class, relatively educated sample where there is general acceptance of family planning. In a more heterogeneous sample, with greater variability in both the predictor variables of differential socialization and spousal power differentials and the fertility outcome, the relations would be expected to be even stronger. Thus, the model of fertility with the chain of effects presented at the beginning should not be completely rejected, but the validity of the model should be investigated with a wider range of people.

References

Bem, S.L. (1974). The measurement of psychological androgyny. *Journal of Consulting and Clinical Psychology, 42*, 155-162.

Block, J.H. (1979). Another look at sex differentiation in the socialization behavior of mothers and fathers. In J. Sherman & F.L. Denmark (Eds.), *Psychology of women: Future directions of research* (pp. 31-87). New York: Psychological Dimensions.

Bongaarts, J. (1978). A framework for analysing the proximate determinants of fertility. *Population and Development Review, 4*(1), 105-132.

Coopersmith, S. (1967). *The antecedents of self esteem.* San Francisco: Freeman.

Frisch, H.L. (1977). Sex stereotypes in adult-infant play. *Child Development, 48*, 1671-1675.

Gürbüz, E. (1988). *A measurement of sex-trait stereotypes.* Unpublished master's thesis, Bogaziçi University, Istanbul, Türkiye.

Hetherington, E.M. (1975). The effects of familial variables on sex typing, on parent-child similarity, and on imitation in children. In P.H. Mussen, J.J. Conger, & J. Kagan (Eds.), *Basic and contemporary issues in developmental psychology* (pp. 259-271). New York: Harper and Row.

Hoffman, L.W. (1977). Changes in family roles, socialization, and sex differences. *American Psychologist, 32*, 644-657.

Imamoglu, O. (1994). A model of gender relations in the Turkish family. *Bogaziçi Journal: Review of Social, Economic and Administrative Studies, 8*(1-2): Symposium on Gender and Society, 165-176.

Kagitçibasi, Ç. (1972). *Sosyal degismenin psikolojik boyutlari* [Psychological dimensions of social change]. Ankara: Social Science Association Publications.

Kagitçibasi, Ç. (1982a). *The changing value of children in Turkey.* Publication No. 60-E. Honolulu: East-West Center.

Kagitçibasi, Ç. (1982b). Introduction. In C. Kagitçibasi (Ed.), *Sex roles, family and community in Turkey* (pp. 1-32). Bloomington: Indiana University Press.

Kagitçibasi, Ç. (1990). Family and socialization in cross-cultural perspective: A model of family change. In J. Berman (Ed.), *Nebraska Symposium on Motivation (Vol. 37,* pp. 135-200). Lincoln: University of Nebraska Press.

Kagitçibasi, Ç., & Sunar, D. (1992). Family and socialization in Turkey. In J.P. Roopnarine & D.B. Carter (Eds.), *Parent child relations in diverse cultural settings: Socialization for instrumental competency.* New Jersey: Ablex Publishing Corp.

Kagitçibasi, Ç., Sunar, D., & Bekman, S. (1988). *Comprehensive preschool education project: Final report.* Ottawa: IDRC. Manuscript Report 209e.

Lamke, L.K. (1982). The impact of sex-role orientation on self esteem in early adolescence. *Child Development, 53*, 1530-1535.

Langlois, J.H., & Downs, A.C. (1980). Mothers, fathers, and peers as socialization agents of sex-typed play behaviors in young children. *Child Development, 51*, 1237-1247.

Lewis, M. (1972). Parents and children: sex-role development. *School Review, 80*(2), 229-240.

Onur, E.P. (1981). *Self-esteem in children and its antecedents.* Unpublished master's thesis, Bogaziçi University, Istanbul, Türkiye.

Puglisi, J.T., & Jackson, D.W. (1981). Sex-role identity and self esteem in adulthood. *International Journal of Aging and Human Development, 12*, 129-138.
Seavey, C.A., Katz, P.A., & Zalk, S.R. (1975). Baby X: The effect of gender labels on adult responses to infants. *Sex Roles, 1,* 103-109.
Shorter, F.C. (1995). The crisis of population knowledge in Turkey. *New Perspectives on Turkey, 12,* 1-31.
Smith, M.B. (1973). A social-psychological view of fertility. In J. T.Fawcett (Ed.), *Psychological perspectives on population* (pp. 3-18). New York: Basic Books.
Smith, C., & Lloyd, B. (1978). Maternal behavior and perceived sex of infant: Revisited. *Child Development, 49,* 1263-1266.
Sunar, D. (1978). *Coopersmith's Self-Esteem Inventory: A Turkish version*. Unpublished manuscript, Middle East Technical University, Ankara.
Sunar, D. (1987). *Proposal for an investigation of intergenerational and sex differences in Turkish child rearing practices*. Unpublished manuscript, Bogaziçi University, Istanbul.
Sunar, D. (1994, August). *Changes in child rearing practices and their effect on self esteem in three generations of Turkish families*. Paper presented at the American Psychological Association Annual Convention, Los Angeles, CA.
Tauber, M.A. (1979). Sex differences in parent-child interaction styles during a free-play session. *Child Development, 50,* 981-988.

The Psychosocial Ecology of Child Maltreatment: A Cross - Cultural and Discriminant Analysis

Gonzalo Musitu Ochoa, Juan Herrero Olaizola, & Enrique Gracia Fuster
Universitat de Valencia, Valencia, Spain

Many authors consider child maltreatment to be culturally determined (Gracia, 1991). Thus, the limits within which a behavior may be labeled as "maltreatment" depends on the practices and traditions of the culture regarding the family realm, the socialization practices or the privacy of the family. In this sense, whether the definition of "abuse" should be one or another, the perception that a child is being abused in a particular society is more a question of shared beliefs and values about raising children in that society than a question of scientific asignment of labels to different behaviors or practices. Obviously, the social construction of the family is especially relevant here.

On the other hand, evidence that some parents' practices have a negative influence on the further psychosocial development of the child, e.g., increasing the number and intensity of physical and psychosocial symptoms of maladjustment in the adult life of the child, has led researchers to pursue a cross-cultural conceptualization of child abuse (Rohner, 1975, 1980). In this sense, the identification of those aspects regarding child abuse shared by different cultures has been achieved.

As a socially conditioned process, the causes of abusive practices in parents may be understood from an ecological perspective. Historically, parental behavior involving abuse was regarded as an evidence for the parents' psychopathology. However, this individual model of the etiology of child abuse has been overcome by more comprehensive models that analyze the influences of the individual, family and social contexts in the origins of abusive behaviors. Thus, the existence of parental psychopathology in a deprived social context may encourage abusive behavior that affects the child adjustment in his/her social relations. Furthermore, this circumstance may produce social isolation not only from the institutions and formal systems of social support but also from the informal sources of social support such as relatives, neighbors, and friends. The effect of this circumstance is accumulative due to the limited access to resources observed in the socially isolated people, i.e., lack of social support, symptoms of mental disorder, etc, (Lin, Dean, & Ensel, 1986). In addition, the lack of psychosocial resources may lead to a position in which the effects of the stressful situations are extreme, producing deviate behaviors such as child maltreatment.

A large body of scientific evidence has confirmed the strong relationship between social isolation and child maltreatment. The absence of social relationships and poor participation in groups and organizations, as well as attitudinal factors related to neighborhood and community, have been variables strongly associated with maltreated children (Justice & Duncan, 1976; Garbarino & Crouter, 1978; Garbarino & Sherman, 1980; Egeland, Breitenbucher, & Rosenberg, 1980; Gaudin & Pollance, 1983; Salzinger, Kaplan, & Artemyeff, 1983; Howze & Kotch, 1984; Justice, Calvert, & Justice, 1985; Straus & Kantor, 1987; Gracia & Musitu, 1990).

Garbarino (1977) has suggested that family social isolation from social and psychological resources is a central constituent in the explanation of child maltreatment. The underlying premise in this statement is that parent-child relationships are deeply embedded in the social context that surrounds the family-relatives, friends, neighboring, community, culture, etc. (Bronfenbrenner, 1977, 1979; Belsky, 1980; Garbarino, Guttmann, & Seeley 1986).

Nevertheless, social isolation is not always a direct cause in child maltreatment. It is well known in the literature that social loneliness involves, on the one hand, the absence of contact with social networks that can provide the family with behavior patterns, feedback, material and emotional support, as well as the opportunities and resources to counteract the negative effects of stress (Garbarino & Stocking, 1980; Tiejten, 1980), and on the other hand, it also implies the frustration of needs such as affiliation, belonging or social recognition (Caplan, 1974; Turner, 1981; Aneshensel & Stone. 1982). The impoverishment and social deprivation of the family can increase the risk of deterioration of the family climate and consequently generate a dangerous context in parent-child relationships that may lead to abusive behavior.

Furthermore, a negative circular relationship seems to exist between violent families and their community. In this sense, Polansky and Gandin (1983) confirmed that not only violent and neglecting parents perceive the community as a non supportive context - avoiding any contact or social relationship - but community members also show a marked tendency towards parents who maltreat their children, increasing in this way the spiral of isolation.

Beyond the influence of different contexts in the etiology of the child abuse, the question of cultural determinants in child abuse literature remains the same. Do different cultures perceive child-abusive behavior in the same way? Are the causes of the child-abusive behavior the same in different cultures? May these causes be operationalized and assessed?

A few studies have explored in detail the influences that different categories and dimensions of an ecological system have in different cultures. This study aims to compare the psychosocial ecology of child maltreatment in Spanish and Colombian cultures.

Method

Sample

A group of 68 abusive families, 34 from Spain and 34 from Colombia, all from the same sociocultural level, were selected according to information provided by the child's teacher. Previously all teachers had received from the researcher a special document with all the information required to register the kind and intensity of maltreatment. The parents had an average age of 36 in the Spanish culture and 34 in the Colombian culture. The average age of the children was 6 in both cultures.

A comparison group of non abusive families with the same sociocultural level and age was used.

Measures

Individual System

Pathological Symptomatology. Inventory of Symptoms (Derogatis et al., 1974). The SCL-90 provides a global index of parental psychopathological symptomatology. The SCL-90 shows a high correlation with a wide range of depression measurements, disfunctional attitudes, anxiety, lack of assertiveness, and emotional problems (Gotlib, 1984). The standardized *alpha* coefficient of internal consistency for the total of the scale is .99.

Child Behavioral Problems. Inventory of Child Behavior (Achenbach & Edelbrock, 1983). The CBC is a standardized instrument based on a thorough review of the clinical and research literature about behavioral problems in childhood and adolescence. This instrument aims to obtain a description of the child's behavior and the behavior of the people around him/her; one version is for parents and another for teachers. The instrument consists of two parts: (a) the child's level of social competence and school adjustment, and (b) the child's behavioral problems.

Children's Behavioral and Personality Disposition. Questionnaire of Personality Assessment (Rohner, 1978b, 1984). This self-report inventory allows us to evaluate the children's perception of his/her own personality and behavioral dispositions. The PAQ consists of seven scales aimed to evaluate the children's perceptions on seven different aspects of personality and behavior. These aspects have been associated with parental Acceptance-Rejection (Rohner 1975, 1984). The seven scales are: *Hostility/Aggression; Dependence; Self-Esteem; Self-Adequacy; Emotional Responsiveness; Stability; and World View*. The *alpha* coefficient of internal consistency obtained for this questionnaire is .96.

Family System
Parental Acceptance-Rejection. For the purposes of this study, we have considered it appropriate to use Rohner's dimensions of parental behavior (1984) taken from his theory of Parental Acceptance-Rejection. The conceptual validity of this theory has been cross-culturally established with a sample of 101 cultures (Rohner, 1975; Rohner & Rohner, 1981). Questionnaire of Parental Acceptance-Rejection (Rohner, 1978a, 1984). This self-report inventory provides an evaluation of parental behavior towards their children, and also the children's perception of their parents' behavior towards them in relation to four dimensions: *Warmth/Affection; Hostility/Aggression; Indifference/ Negligence; and Indifferent Rejection.* The standardized *alpha* coefficients obtained from this questionnaire are: .99 (parents), .99 (child in relation to father), and .99 (child in relation to mother). This study used the Spanish versions of the PAQ and PARQ instruments, which were provided by the author.

Family Climate. Scale of Family Social Climate (Moos & Moos, 1981; TEA, 1984). This scale accounts for the socio-environmental characteristics of all different types of families. It evaluates and describes the interpersonal relations between the family members, the significant developmental aspects of the family and its basic structure. Administered both to parents and children, as in this study, it records the differences between parents and children in terms of their family perception. The FES includes ten sub-scales which define three basic dimensions: (1) Relations: this dimension evaluates the level of communication and freedom of expression within the family, as well as the level of conflictive interaction. This dimension consists of three sub-scales: Cohesion, Expressivity, and Conflict. (2) Development: evaluates the importance of certain personal developmental processes within the family, which are either encouraged or discouraged in the family's everyday life. This dimension is composed of the sub-scales: Autonomy, Performance, Intellectual-Cultural, Social-Recreational, and Morality-Religiousness. (3) Stability: this dimension provides information about the family structure and organization. Moreover, it also deals with the level of control which some family members frequently have on the others ones. It consists of two sub-scales: Organization and Control. The *alpha* coefficients of internal consistency for this questionnaire are: .98 (children), and .98 (parents).

Intra-Familial Sources of Stress. Family Inventory of Vital Events and Changes (McCubbin et al., 1985). FILE provides an index of the global stress level experienced by the parents taking into account the vital events and changes which cause stress within the family environment. This inventory includes a list of 71 vital events and changes related to different aspects in the family. This section only deals with the following areas or sources of stress: Family, Conjugal Relations, Pregnancy and Childbirth, Sickness, Death, and Mobility within the family. The *alpha* coefficient of internal consistency obtained for the total scale is .96.

Social System
Extra-Familial Sources of Stress. In this section only three extra-familial areas or sources of stress of FILE have been studied: Economic Difficulties, Work Environment, and Legal Problems. The global index of stress was obtained from the sum of the various vital events and changes in these areas.
Social Support. Questionnaire of Community Social Support (Gracia & Musitu, 1990). This AC-90 evaluates the structural aspects of social support (Cohen & Wills, 1985; Cohen & Syme, 1985; Gracia et al., 1989), specifically, the external levels and stages of social relations. The AC-90 provides an index of the level of integration in the wider social network and of the sense of belonging to a community (Sarason, 1974). The AC-90 includes three scales: Integration and Satisfaction with the Community, Community Association and Participation, and Institutional and Community Resources of Social Support. The standardized *alpha* coefficient was .97.

Results

Analysis of Variance for Spain

Individual System
Differences between the Abuse and Non-Abuse groups for all the variables in the model were tested with *Univariate F* tests. In the individual system differences between Non-Maltreated and Maltreated children on all variables were significant: (a) Parental Psychopathology ($F = 8.868; p < .004$; $M = 140, M = 170.88$); (b) Child Behavior: Internal ($F = 22.43; p < .001$; $M = 13.06, M = 24.94$); Child Behavior: External ($F = 33.08; p < .001$; $M = 16.32, M = 35.61$); Child Personality ($F = 32.87; p < .001$; $M = 95.47, M = 112.85$) (see Table 1).

Family System
In the Family System significant differences were found between the two groups with almost all the variables (Table 1).

Social System
Significant differences in the Social System between both groups of children were found in Social Stress ($p < .05$), Community Integration, and Social Class ($p < .001$) (Table 1).

Analysis of Variance for Colombia

Individual System

In the Individual system almost all the variables were significant ($p < .001$) between Non-Maltreated and Maltreated children: (a) Child Behavior: Internal ($F = 12.73; p < .001; M = 16.23, M = 24.76$);(b) Child Behavior: External ($F = 11.02; p < .001; M = 17.32, M = 25.32$); Child Personality ($F = 11.087; p < .001; M = 93.82, M = 104.85$) (Table 1) .

Family System

In the Family System significant differences were found between the two groups with almost all the variables (Table 1).

Table 1. *Variables, F, and Means in the Individual, Family, and Social Systems in Abused and Non-abused Children in Spain and Colombia*

Variables	Spain			Colombia		
Individual system	F	Non-abuse	Abuse	F	Non-abuse	Abuse
Parents psychopathology	2.71	140.00	170.8··	2.71	162.35	182.3
Child behavior: internal	12.74	13.06	24.9···	12.74	16.24	24.7··
Child behavior: external	11.02	16.32	35.6···	11.02	17.32	25.3··
Child personality	11.09	95.47	112.8···	11.09	93.82	104.8··
Family system	F	Non-abuse	Abuse	F	Non-abuse	Abuse
Relationships (child perspective)	2.77	15.21	14.2	1.82	14.88	14.1
Personal growth (child perspective)	20.63	26.47	20.4···	14.57	27.21	22.3···
Structure (parental perspective)	1.02	10.68	9.8	.63	11.62	11.1
Relationships (parental perspective)	1.59	16.24	15.4	3.14	15.65	14.4
Personal growth (parental perspect.)	8.75	26.18	22.0··	10.41	27.71	22.7··
Structure (parental perspective)	1.36	10.35	9.4	2.32	12.15	10.9
Acceptance (child mother)	28.85	71.74	56.5···	12.84	72.06	61.1··
Rejection (child mother)	43.00	72.32	106.2···	24.29	74.24	100.5···
Acceptance (child father)	29.11	70.65	53.4···	16.65	72.21	60.2···
Rejection (child father)	48.79	69.24	106.7···	28.28	69.85	98.3···
Acceptance (parental perspective)	10.74	73.70	64.4··	9.22	73.50	65.7··
Rejection (parental perspective)	27.19	66.41	90.9···	21.52	72.03	94.1···
Family stress	6.30	5.38	8.8·	9.78	6.97	11.5··
Social System	F	Non-abuse	Abuse	F	Non-abuse	Abuse
Social stress	6.62	3.44	6.0	7.30	5.26	8.1··
Community integration	14.92	30.88	25.4	7.58	30.68	26.7··
Community participation	.02	4.94	4.9··	5.14	5.44	4.9·
Community resourses	1.07	8.41	8.7··	1.74	8.32	8.0
Social class	13.73	3.73	2.3··	7.44	3.55	2.6··

* p < .05; ** p < .01; *** p < .001

Social System
Significant differences in the Social System were found between both groups of children in Social Stress ($p < .009$), Community Integration ($p < .008$), Community Participation ($p < .027$), and Social Class ($p < .008$) (Table 1).

Discriminant Analysis of Abuse and Non-Abuse Groups in Spain and Colombia

Discriminant analysis was performed between the Abuse and Non-Abuse groups in the two cultures, with the statistically significant variables obtained in the ANOVA analysis. The results are shown in Table 2. The profiles are very similar. There are only two variables that differentiate between the two cultures: Community Resources and Parental Psychopathology.

Table 2. Discriminant Analysis, F, and p-levels in Spain and Colombia

Variables	Spain[1]			Colombia[2]		
	F	p	Corr.	F	p	Corr.
Parental psychopathology	8.87	.004	.31	-	-	-
Child behavior: internal	22.44	<.001	.48	12.74	.001	.50
Child behavior: external	33.09	<.001	.59	11.02	.001	.46
Child personality	32.87	<.001	.59	11.09	.001	.46
Personal growth (child perspective)	20.63	<.001	-.47	14.57	<.001	-.53
Personal growth (parental persp.)	8.75	.004	-.30	10.41	.002	-.45
Acceptance (child mother)	28.85	<.001	-.55	12.84	.001	-.45
Rejection (child mother)	43.00	<.001	.67	24.29	<.001	.69
Acceptance (child father)	29.11	<.001	-.55	16.65	<.001	-.57
Rejection (child father)	48.79	<.001	.72	28.28	<.001	.74
Acceptance (parental perspective)	10.74	.002	-.34	9.22	.003	-.42
Rejection (parental perspective)	27.19	<.001	.53	21.52	<.001	.65
Family stress	6.30	.015	.26	9.78	.003	.44
Social stress	6.62	.012	.26	7.30	.009	.38
Community integration	14.92	<.001	-.40	7.58	.008	-.38
Community resources	-	-	-	5.14	.027	-.32
Social class	13.73	<.001	-.38	7.44	.008	-.38

1. Wilk's Lambda = .409; $F = 4.599$, $d.f. = 16, 51$, $p < .01$;
   Chi-square = 51.799, $d.f. = 16$, $p < .01$, Canonical correlation = .769
2. Wilk's Lambda = .562; $F = 2.488$, $d.f. = 16, 51$, $p < .01$;
   Chi-square = 33.463, $d.f. = 16$, $p < .01$, Canonical correlation = .662

As regards the discrimination power, 86.76% of the cases were correctly assigned in the Spanish sample in the Non-Abuse group and 79.41% in the Abuse group whereas 79.41% of Non-Abused subjects and 76.47% of the Abused subjects were correctly assigned in the Colombian sample (see Table 3).

Table 3. Percent of Correct Classification of Cases of the Discriminant Analysis

Group	Spain			Colombia		
	Non-abuse	Abuse	Total	Non-abuse	Abuse	Total
Non-abuse	86.76	13.24	100.00	79.41	20.59	100.00
Abuse	23.53	76.47	100.00	23.53	76.47	100.00
Total	55.15	44.85	100.00	51.47	48.53	100.00

Discussion

Two major findings emerge from these results. First, the relative importance of each context in the etiology of child abuse. Most variables yielded a significant relationship with the existence of child maltreatment, supporting the hypothesis that the etiology of child abusive behavior is multidimensional and derives from different contexts: individual, family, and social. Second, the discriminant analysis showed that the ecological model that intends to explain the etiology of child maltreatment is useful in both cultures, supporting our conviction that it is possible to achieve a wide and comprehensive conceptualization of child abuse in different cultures.

Looking at the results, it appears that the variables in the individual, family, and social system have the same meaning in the Colombian and Spanish cultures. These results very strongly support others' findings regarding the significance of the variables considered in this study, and at the same time, represent a scientific advance in the ecological approach.

From a cross-cultural point of view, we should emphasize the similar profile detected in both cultures. In this sense, the etiology of child abuse follows a similar pattern in both samples. This pattern includes the three domains postulated by the theory: individual, family, and social or community aspects. Nevertheless, there are some differences between the Spanish and Colombian samples. Thus, the lack of community resources does not seem to have a special importance in the appearance of child maltreatment in the Spanish sample whereas parental psychopathology does not influence abusive behavior in the Colombian sample.

These differences between the Spanish and Colombian cultures probably exist due to the fact that the family concept in Spain is largely focussed on privacy, as families regard themselves as a world apart, with their own rules and authority. Therefore, the psychopathological maladjustment of some Spanish parents may influence the occurrence of child-abusive behaviors. On the other hand, the influence of community resources on the occurrence of child abuse in Colombia reflects a higher participation of the community in family socialization of the children and education, either by solving family problems or by helping and advising the parents. According to Arango (1993), institutional policy promotes the creation of formal means leading to active participation.

Informal groups and organizations also exist which promote and support processes of social participation and the ideology of collective participation, which is effective and, moreover, part of their culture. Thus, the ecological model of this study is able to account for the presence of child-abusive behaviors in different cultures. Nevertheless, some differences were found in relation to which variables are more related to abuse in each culture. Therefore, it would be of great interest to compare these results in other cultures.

Finally, in order to prevent and treat abuse, the promotion of community programs and projects is necessary. These are aimed to foster social interaction and create participative behaviors in people and families, especially when facing their own day-to-day problems and their parent-child relations. In Spain and other western countries with similar social circumstancies, it is necessary to create strategies aimed to integrate the community participation practice within the institutional and popular cultures, in order to make people more participant in a collective (rather than individual) problem-solving processes.

References

Achenbach, T., & Edelbrock, C. (1983). *Manual for the Child Behavior Checklist and Revised Child Behavior Profile*. Burlington, VT: University of Vermont Press.

Aneshensel, C.S., & Stone, J.D. (1982). Stress and Depression: A test of the buffering model of social support. *Archives of General Psychiatry, 39*, 1392-1396.

Belsky, J. (1980). Child maltreatment: An ecological integration. *American Psychologist, 35*, 320-335.

Bronfenbrenner, U. (1977). Toward an experimental ecology of human development. *American Psychologist, 32*, 513-531.

Bronfenbrenner, U. (1979). *The experimental ecology of human development*. Cambridge, Mass: Harvard University Press.

Caplan, G. (1974). Support-Systems. In G. Caplan (Ed.). *Support systems and community mental health*. New York: Basic Books.

Cohen, S., & Syme, S.L. (1985). Issues in the study and application of social support. In S. Cohen & S.L. Syme (Eds.). *Social support and health*. New York: Academic Press.

Cohen, S.Y., & Wills, T.A. (1985). Stress, social support, and the buffering hypothesis. *Psychological Bulletin, 98*, 310-57.

Egeland, B., Breitenbucher, M., & Rosenberg, D. (1980). Prospective study of the significance of life stress in the etiology of child abuse. *Journal of Consulting and Clinical Psychology, 48*, 195-205.

Garbarino, J. (1977). The human ecology of child maltreatment: A conceptual model for research. *Journal of Marriage and Family, 39*, 721-736.

Garbarino, J., & Crouter, A.C. (1978). Defining the community context of parent-child relations. *Child Development, 49*, 604-616

Garbarino, J., & Sherman, D. (1980). High-risk neighborhoods and high-risk families: The human ecology of child maltreatment. *Child Development, 51,* 188-198.

Garbarino, J., & Stocking, S.H. (1980). The social context of child maltreatment. In J. Garbarino, & S.H. Stocking (Eds.). *Protecting children from abuse and neglect.* London, Jossey-Bass.

Garbarino, J., Guttmann, D. & Seeley, J.W. (1986). *The psychologically battered child.* London: Jossey-Bass.

Gaudin, J.M., & Pollane, L.P. (1983). Social networks, stress and child abuse. *Children and Youth Services Review, 5,* 91-102.

Gotlib, I. (1984). Depression and general psychopathology in university students. *Journal of Abnormal Psychology, 93,* 19-30.

Gracia, E., & Musitu, G. (1990). Integraciá¢án y participaciá¢án en la comunidad: Una conceptualizaciá¢án empáiárica del apoyo social comunitario. In G. Musitu, E. Berjano, & J.R. Bueno (Eds.), *Psicologáiáa Comunitaria.* Valencia: Nau Llibres.

Gracia, E., Musitu, G., & Garcia, F. (1989). *El apoyo social en los programas de intervenciá¢án comunitaria: Una propuesta de evaluaciá¢án.* Lisboa: Conferencia Internacional "A Psicologia e os Psicá¢álogos Hoje".

Howze, D.C., & Kotch, J.B. (1984). Disentangling life events, stress and social support: Implications for the primary prevention of child abuse and neglect. *Child Abuse and Neglect, 8,* 401-409.

Justice, B., & Duncan, D.F. (1976). Life crisis as a precursor to child abuse. *Public Health Reports, 91,* 110-115.

Justice, B., Calvert, A., & Justice, R. (1985). Factors mediating child abuse as a response to stress. *Child Abuse and Neglect, 9,* 359-363.

Lin, N., Dean, A., & Ensel, W. (1986). *Social support, life events and depression.* New York: Academic Press.

McCubbin, H., Patterson, J., & Wilson, I. (1985). FILE: Family Inventory of Life Events and Changes. In D. Olson et al. (Eds.), *Family inventories.* St Paul, MN: University of Minnesota Press.

Moos, R., & Moos, H. (1981). *Family environment scale manual.* Palo Alto, CA: Consulting Psychologists Press.

Polansky, N.A., & Gaudin, J. (1983). *Preventing child abuse through public awareness activities.* Working paper num. 019. Chicago. National Committee for Prevention of Child Abuse.

Rohner, R (1975). *They love me, they love me not: a world wide study of the effects of parental acceptance-rejection.* New Haven, CT: HRAF.

Rohner, R (1984). *Handbook for the study of parental acceptance and rejection (Revised edition).* Storrs, Centre for the Study of Parental Acceptance and Rejection: University of Connecticut.

Rohner, R., Saavedra, J., & Granum (1978). *Development and validation of the Personality Assessment Questionnaire: Test manual.* Ann Arbor, MI: ERIC/ CAPS.

Salzinger, S., Kaplan, S., & Artemyeff, C. (1983). Mother's personal social networks and child maltreatment. *Journal of Abnormal Psychology, 92,* 68-76.

Sarason, S.B. (1974). *The psychological sense of community: Prospects for a community psychology.* San Francisco: Josey-Bass.

Straus, M.A., & Kantor, G.K. (1987). Stress and child abuse. In R.E. Helfer & R.S. Kempe (Eds.), *The battered child* (4th edition). Chicago: University of Chicago Press.

Tiejten, A.N. (1980). Integrating formal and informal support systems: The Swedish experience. In J. Garbarino & S.H. Stocking (Eds.), *Protecting children from abuse and neglect*. London: Jossey-Bass.

Turner, R.J. (1981). Social support as a contingency in psychological well-being: Theoretical possibilities. In I.G. Sarason & B. Sarason (Eds.), *Social support: Theory, research and applications*. The Hague, NL: Martinus Nijhof.

The Application of an Ecocultural Framework to the Study of Disability Programs in a Rural and Urban Indian Community

Philip H. Cook
University of Victoria, Victoria, B.C., Canada

It is now generally recognized that programs aimed at preventing disease, treating illness, and promoting well being must incorporate an understanding of culture and patterns of family and community interaction (Dasen, Berry, & Sartorius, 1988). Similarly, in the last thirty years, the field of cross-cultural psychology has advanced to the point where it has substantial theoretical and methodological armamentarium to address a variety of health issues in developing countries, and conceptual tools are now available to study many cultural phenomena and the linkages between them at both the cultural and psychological level (Kagitcibasi & Berry, 1989).

While cross-cultural psychological approaches have been applied to a variety of primary health care foci ranging from malaria to oral rehydration (Dasen, Berry, & Sartorius, 1988), a comprehensive framework has yet to be developed which examines health behavior at the individual, community, and cultural level. The present research applied such an approach to the study of disability health behavior within a community based rehabilitation (CBR) program in India.

The current prevalence of disability in the world is thought to be approximately 10 per cent of the total global population (Noble, 1981; UNICEF, 1993). It is estimated that 75 per cent of these persons with a disability live in less developed countries, and that by the year 2000 this figure will rise to 80 per cent, or approximately 600 million people (Noble, 1981; Simeonsson, 1991).

One approach that is increasingly being applied to these disability problems in less developed countries is Community Based Rehabilitation. CBR consists of programs run by Government, and Non-governmental organizations that attempt to address the divergent needs of people with a disability within the context of a particular community. CBR is generally founded on rehabilitation measures taken at the community level that build on the resources of the community, the person with a disability, and their family. In this situation local participants supposedly take "ownership" of their problems and their rehabilitation responsibilities (Helander, Mendis, & Nelson, 1979; Miles, 1985; O'Toole, 1987).

The purpose of the present research was to develop and evaluate a cross-cultural psychological framework of CBR that attempts to predict health seeking behavior of people with a physical disability in two geographically and culturally different communities in Bombay and the neighboring Thane district, Maharastrha State, India. The framework developed for this research is set within an ecological context, is based on previous health research in cross-cultural psychology, and is compared with the Fishbein, Triandis, and Health Belief models. A methodology incorporating variables representing universal processes, that are formulated in indigenous terms was used to collect and interpret the data.

A number of socio-cultural models have been developed to better describe the "ecology" or context of health behavior. In this project three tried and tested models of health behavior were applied to the situation of persons with a physical disability in two communities in Bombay, India. The first two conceptual frameworks, the Fishbein and Triandis models of reasoned action, describe the relationship between a specific behavior and a number of predictor variables, and are expressed in terms of a multiple regression equation in which either two or three components are represented as influencing a person's behavior.

In Fishbein's model an individual's behavior is assumed to be a function of intention to perform the behavior. The behavioral intention is then assumed to be a function of: (1) attitude toward performing the act, and (2) beliefs about what others think should be done weighted by motivation to comply with those others.

The Triandis model is made up of two separate multiple regression equations. In this cross-culturally tested model (Davidson, Jaccard, Triandis, Morales, & Diaz-Guerrero, 1976) an individual's behavior is a function of: (1) intention to perform that behavior, and (2) the "habit" of the individual to perform this act (i.e., frequency with which the individual has performed the behavior in the past). The behavioral intention is further a function of: (1) affect toward performing the act, (2) beliefs about consequences of performing that behavior and the evaluation of those consequences, (3) perceived appropriateness of a particular behavior for members of, (a) the specific reference groups (norms) and (b) specific positions in the social structure (roles), and (4) personal normative beliefs about what ought to be done with regard to the behavior of interest.

The HBM was developed in the early 1950's by a group of social psychologists in an attempt to better understand the relationship between health attitudes and preventive health action. Later the model was applied to patient's health seeking behavior, and compliance with prescribed medical regimens (Janz & Becker, 1984). The model contains the following elements:

1. The individual's subjective state of readiness to take action, which is determined by both the individuals perceived likelihood of "susceptibility" to the particular illness; as well as by a person's perceptions of the probable "severity" of the consequences (organic an/or social) of contracting or having the disease.

2. The individual's evaluation of the advocated health behavior in terms of it's feasibility and efficaciousness (i.e., a subjective estimate of the

actions potential "benefits" in reducing susceptibility and/or severity) weighed against perceptions of physical, financial and other costs ("barriers") involved in the action.

3. A "cue to action" must occur to trigger the appropriate health behavior, coming from either internal (e.g., symptoms) or external (e.g., interpersonal factors, mass media communications) sources (Becker & Maiman, 1983).

An Ecocultural Framework for Community Based Rehabilitation

The Ecocultural CBR Framework (See Figure 1) is a general sociocultural framework used for testing hypotheses relating to health seeking behavior among disabled persons within a community framework. The framework was developed by the author for the present research. The framework draws on previous theories and research on health behavior and community health care and is, therefore, generalizable and could be applied to any number of community health concerns in less developed and developed countries.

This research attempted to understand the specific Indian context influencing attitudes towards persons with a disability as well as their health seeking behavior. A framework was applied that incorporates pertinent variables from each of the three models, as well as scales measuring culturally appropriate attitudes and beliefs pertaining to disability. These variables are chosen for their suitability to the Indian cultural setting influencing CBR programs. As such, this framework is "ecological" in its focus on context and "cultural" in its attention to the cultural basis of disability. It is best described as an Ecocultural CBR Framework.

The Ecocultural CBR Framework draws on existing contextually based ecological frameworks developed in cross-cultural psychology, and combines the psychological level of analysis common to health psychology with the cultural level of analysis found in medical anthropology and models such as the Triandis and Fishbein theories of reasoned action. The framework has closest theoretical proximity to the general cross-cultural ecological framework described by Berry, Poortinga, Segal, and Dasen (1991). The origins of this and other ecological approaches in cross-cultural psychology trace their roots to earlier research in the field of culture and personality, and in particular to the work of Kardiner and Linton (1945), and J.W. Whiting (1974).

Another application of an ecological approach employed in cross-cultural psychology, has been developed by Super and Harkness (1986) in their conceptualization of the "developmental niche". This framework examines the ecological interface between sociocultural systems and childhood development. More recent applications of this framework have focused on weaning diarrhoea, child survival (Harkness & Super, 1987), and general dysfunctions of development (Super, 1987).

Figure 1. The Ecocultural CBR Model

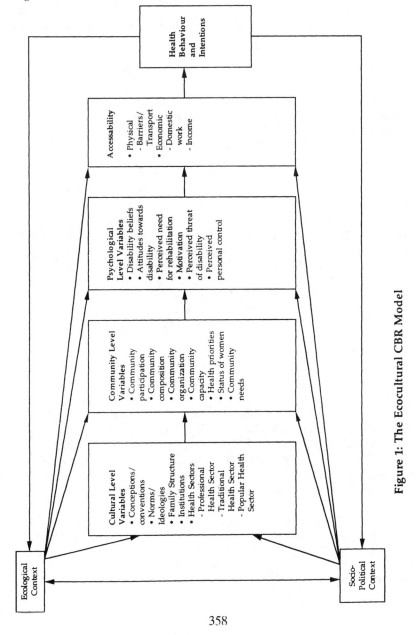

Figure 1: The Ecocultural CBR Model

Whereas, the "developmental niche" primarily addresses aspects of cultural and human development, the Ecocultural CBR Framework can be used to study the interactive process between a person with a disability and their health behavior. In addition, the Ecocultural CBR Framework can be applied to the formulation of predictive relationships among sets of psychological level, community level, and cultural level variables. In general the flow of the framework is from left to right, with the ecological and socio-political contexts both influencing cultural, community, and individual level variables, degree of accessibility to health treatment (both physical and economic), and ultimately health behavior and intentions. It is recognized, however, that the direction of influence is not necessarily unidirectional, and that behavior as well as accessibility, psychological, community, and cultural level variables all exert some influence on both background level variables.

Ecological context and Socio-Political context are background level variables. Ecological context refers to the interaction of human organisms and physical environment. The relationship is interactive rather than deterministic (Berry, 1991), and typically the central feature of the ecological context is the economic activity of a society. In the present Framework this variable would correspond to macro-level influences such as the incidence of disability due to environmental causes. The socio-political context refers to the socio-political organization of a society (e.g., democracy, dictatorship) and it's impact on disability prevention and rehabilitation. Both these variables are macro-level indices affecting behavior, and in the present research are addressed in the background information presented in the ethnography.

In applying this framework to the particular CBR setting in two Bombay communities the other four groups of predictor variables (cultural, community, psychological, and accessibility) are used to trace sociocultural influences of acceptance of persons with a disability, and perceived need for rehabilitation among both persons with a disability and without a disability. Cultural factors form the first block of the framework. This set of variables target culture-wide variables affecting health behavior. These variables are drawn from Berry's (1989b) cultural level CBR domain, the Fishbein and Triandis (Fishbein, 1967; Triandis, 1964) normative variables, and Kleinman's (1980) health sectors. Examples of cultural level variables are: social normative influences, cultural variation in family structure, and the three health sectors (Professional, Traditional, and Popular) available to disabled persons.

The next block of variables include community specific factors affecting health behavior. This set draws from the literature on community participation in health (Bichmann, Rifkin, & Shrestha, 1989) and CBR (O'Toole, 1987; Werner, 1987; Peat, 1989), and includes variables such as: degree of community participation, community composition (i.e., religious and caste groups), the status of women, and community needs.

The next block of variables address the psychological level variables predicting health behavior. This set is based on Berry's (1989b) psychological CBR variables, Health Belief variables, and a number of variables specific to the literature on health behavior (Makas et al, 1988; Cook, 1989; Groce, 1989). Included here are: disability beliefs, attitudes towards persons with a disability, motivation, and perceived control.

The degree of accessibility to health care services is seen as a key factor influencing whether a person with a disability is able to seek treatment. This block of variables focuses on both physical and economic barriers restricting access to treatment. In the present study accessibility was assessed using variables measuring family size, and number of children as economic indicators, and the Health Belief Model variable "Barriers" to assess physical barriers restricting access to treatment. Young's cross-cultural application of the Health Belief Model and the findings of Cummings and colleagues indicate that these variables account for the majority of variance in explaining health behavior in a wide range of settings.

The final block represents the criterion to be measured. In this research, self reported health behavior (reported use of treatment) was measured, however, in research such as the Fishbein and Triandis studies, behavioral intention was the measured criterion. Health behavior was assessed as the self reported degree of use of the Professional, Traditional, and Popular health sectors.

The research took place in an urban slum area and in a semi rural community adjacent to Bombay. The first site for the research was the Bombay Development and District Chawls (BDD Chawls), a predominately Hindu community with access to both biomedical and traditional institutionalized health care. The other area of investigation for the research comprised the rural community of Juchandra. The population in these villages is predominantly Agri, a "Tribal" group, with few local indigenous health resources. The Agri, also known as Agle or Kharpatil, are a sub-group of Bombay's largest indigenous people, the Kolis. Although the Agri worship Hindu Gods they are, nevertheless, perceived as a Scheduled caste and differ markedly from the Hindu population in many of their cultural traditions, norms, and beliefs. At the time of the research both research sites were participating in CBR programs through King Edward Memorial (KEM) Bombay hospital.

While the research carried out for this CBR project included interviews with persons with a disability, and persons without a disability (in order to determine the influence of community acceptance of disability), the present paper will only address the data collected with persons with a disability. This will be done in order to specifically examine the effectiveness of the Ecocultural Framework in predicting health seeking behavior in comparison with the Fishbein, Triandis, and Health Belief models.

The two primary hypotheses tested in this research were that: (1) the two cultural groups (Hindu and Agri) would differ in their health beliefs and behavior; and (2) the Ecocultural CBR Framework would predict reported behaviors better than any one of the other three models.

Method

One of the goals of the present project was the adaptation of universal health concepts to indigenous settings. A useful framework that has been developed in cross-cultural psychological research, and serves to distinguish the cultural specific from universal, is the emic-etic distinction.

In the present study, both qualitative and quantitative data were collected to develop a framework of community based rehabilitation specific to the area in India where the data was collected (emic), but whose etic constructs could possibly be used in further research on the application of this health framework. This process has been referred to as the "combined etic-emic approach" and forms an important link between construct and culturally appropriate content in cross-cultural health research. The ecological framework developed and tested in this research draws from the Fishbein and Triandis models of reasoned action and the Health Belief Model, all of which are imposed etics in the Indian context.

Sample

One hundred persons with a physical disability were contacted to partake in the study. Of these, 96 participated in this study (53 males, 43 females). Fifty of these persons were living in the BDD Chawls and 46 were living in the Juchandra area.

Information gathered as part of an epidemiological study carried out by the King Edward Memorial Hospital in 1989, was used as a "sampling frame" (Pareek and Venkateswara, 1980) for contacting physically disabled persons in both the BDD Chawls and five Juchandra communities. In the case of non-disabled persons, random samples were drawn from both communities. In the Chawls, apartments were chosen randomly from the 119 buildings, and in the five villages of Juchandra homes were selected randomly from demographic information gathered during interviews with a key informants. In order to maintain a balance in the ratio of sex of respondents, male and female subjects were interviewed whenever possible alternatively in consecutive households.

Materials

Scale Construction of Behavioral Intentions

Items representing the use of either home treatment, government rehabilitation facilities, Ayurvedic practitioner, or other form of treatment (e.g. private Physician) were generated by the present researcher to indicate past use (if any) of particular treatment, present use of treatment, as well as intention to use the particular treatment in the future. A dichotomous (yes\no) answer format was used with these items.

The Open-Ended Questionnaire
The components of all four models (Fishbein Model, Triandis Model, Health Belief Model, Ecocultural CBR Framework) are assumed to be imposed etic. Thus, the first step in applying these constructs in the present local research setting required finding their emic expression for the behavior in question. This was achieved through the administration of open-ended interviews to 20 disabled people seeking rehabilitation a KEM Hospital. Questions with these disabled persons were used to elicit culturally appropriate information pertaining to all 14 variables and demographic items.

Key informants from both BDD Chawls and Juchandra were also interviewed in the initial compilation of items so as to include culturally relevant material. All sets of qualitative data were content analyzed and the resulting data was then used to build culturally appropriate "derived etic" scales.

Fishbein (F), Triandis (T), Health Belief (H), and Ecocultural (E) Variables
Items were developed based on Fishbein (F), Triandis (T), Health Belief (H), and Ecocultural Variables (E*)*. The variables were: (1) *Referent Norms* (F): normative belief that a given other (e.g., Head of the family, sibling, coworkers) thinks s/he should or should not perform the behavior (seek treatment from either KEM, Ayurvedic, home or other source) (9 items); (2) *Personal Normative Beliefs* (T) (E): the degree to which an individual feels s/he should perform the behavior in question (11 items); (3) *Normative Beliefs* (T): perceived appropriateness of performing the behavior for a member of a reference group (e.g., spouse, older family member, friend) (11 items); (4) *Role Beliefs* (T): perceived appropriateness for a person occupying a position in the social structure (e.g., men, women, own caste, other caste) (13 items); (5) *Motivation* (F): motivation to seek help (rehabilitation) from a government hospital (10 items); (6) *Consequences* (T) (H): perceived evaluation of the consequences of seeking rehabilitation; (7) *Probability* (T): perceived probability of hospital treatment being efficacious (2 items); (8) *Affect Towards the Act* (F) (T) (E): emotional response to seeking rehabilitation (11 items); (9) *Severity* (H) perceived severity of an individuals disability (10 items); (10) *Barriers* (H): perceived barriers involved in seeking rehabilitation (e.g., having to use local transportation) (9 items); (11) *Attitudes* (E): attitudes towards disability encompassing, social (e.g., marriage), economic (e.g., financial self sufficiency), and gender (e.g., stigma for men vs women) issues (10 items); (12) *Personal Control - Causation* (E):

beliefs pertaining to degree of control exercised by a disabled person over the causes of their disability (6 items); (13) *Personal Control - Consequences* (E): beliefs pertaining to degree of control exercised by a disabled person over the consequences of their disability (9 items); (14) *Taxonomy* (E): beliefs pertaining to disability classification (13 items). The Cronbach *alphas* for these scales are available, but are not reported for lack of space.

Procedure

Each interview was conducted by the researcher with the aid of a physiotherapist and two contact persons were employed to facilitate contact with the communities. Interviews lasted approximately one and a half hours and took place in participants' homes.

While it had been hoped that disabled participants could be interviewed separately from the rest of their family, this turned out to be impractical and culturally inappropriate and other family members were therefore often present during these interviews. This may have influenced the responses of participants. However, isolating persons from their families would likely have created further discomfort for participants and having other family members present often added to the richness of the data collected on disability norms. The interviews were often carried out in an informal atmosphere and participants were encouraged at the end of the session to give any other information that they thought might aid the research.

Results

Multivariate Analyses of Variance (MANCOVA) for all Predictor Variables in the Fishbein, Triandis, Health Belief, and Ecocultural CBR Models for Persons with a Disability.

A 2 X 2 between-subjects MANCOVA, in which the independent variables were Disability Group (BDD Chawls, Juchandra) and Gender was performed on the fourteen dependent variables in the Fishbein, Triandis, Health Belief, and Ecocultural Models. Adjustment was made for the covariate Age. Analysis was carried out through SPSSPC MANCOVA, with hierarchical (default) ordering of effects to adjust for non-orthogonality. Order of entry of independent variables was groups, then gender.

With the use of the Pillis-Bartlett criterion, the combined dependent variables were significantly related to the covariate Age, approximate $F(14, 37) = 2.37, p < .01$; to Group, $F(14, 37) = 5.74, p < .001$; and Gender, $F(14, 37) = 1.91, p < .05$. The gender by group interaction was non-significant. Univariate F tests were carried out to investigate in more detail the group differences on each variable for Age, Group, and Gender.

Table 1: Multiple Stepwise Regression Analysis for the Fishbein, Triandis, Health Belief, and Ecocultural Models to predict KEM, HOME, OTHER, and AYURVEDIC.

Independent Variables	1 order	2 R	3 beta	4 Sig. of T
Eishbein Predicting KEM				
Motivation	1	.25	.25	.01
Triandis Predicting KEM				
Personal Normative Beliefs	1	.38	.38	.00
Group	2	.43	.20	.04
Triandis Predicting HOME				
Personal Normative Beliefs	1	.22	-.22	.03
Triandis Predicting AYURVEDIC				
Normative Beleifs	1	.34	-.34	.00
Health Belief Model Predicting KEM				
Barriers	1	.42	-.42	.00
Ecocultural CBR Model Predicting KEM				
Barriers	1	.42	-.32	.00
Personal Normative Beliefs	2	.52	.30	.00
Beliefs of Control over Disability Prevention	3	.56	.21	.02
Ecocultural CBR Model Predicting HOME				
Hindu	1	.27	-.25	.01
Personal Normative Beliefs	2	.35	-.20	.03
Number of Children	3	.41	-.34	.00
Number of Family Members	4	.47	.26	.01

1 = Order in which independent variable was entered in stepwise direction
2 = Multiple correlation coefficient
3 = Standardized correlation coefficient (beta weight)
4 = Significance

Multiple Regression Analyses

Multiple stepwise regression was used to predict health seeking behavior from the variables in the Fishbein, Triandis, Health Belief, and Ecocultural CBR models (See Table 1). Independent Variables were composed of all variables for each model in turn. Dependent variables were treatment criterion (KEM, home, other, Ayurvedic).

In the case of the Ecocultural CBR Framework, the multiple regression results reported included two sets of predictor variables and two corresponding sets of results. The first set was comprised of the main effects which were entered first in the equation, and the second is made up of main effects and interactions which were subsequently entered in the equation. This procedure provides a better picture of the overall steps in the equation. Analysis was first performed on both groups of persons with a disability (using "dummy coding"), then on each group independently.

Multiple Stepwise Regression: all Fishbein Variables for KEM, HOME, OTHER, and, AYURVEDIC (Both Groups of Persons With a Disability).
Independent variables in the Fishbein model were Referent Norms, Motivation, and Affect Towards the Act. The multiple R for the stepwise equation to predict use of KEM was .25 (See Table1). Only the variable Motivation (high) emerged as an independent and significant predictor of intention to use KEM. No variables from the Fishbein Model significantly predicted intention to use HOME, OTHER, or, AYURVEDIC treatment.

Multiple Stepwise Regression: all Triandis Variables for KEM,
HOME, OTHER, and AYURVEDIC (Both Groups of Persons with a Disability). Independent variables in the Triandis model are Personal Normative Beliefs, Consequences, Probability, and Affect Towards the Act. The multiple R for the stepwise equation to predict use of KEM was .43 (See Table 1). Two variables emerged as independent and significant predictors of intention to use KEM: (1) Personal Normative Beliefs (high), and (2) Group (BDD Chawls).

For intention to use HOME, the overall multiple R was .22, with Personal Normative Beliefs (low) emerging as the only significant predictor. Normative and Role Beliefs (low) was the only significant predictor of intention to use AYURVEDIC, with a multiple R of .34 (See Table 1).

Multiple Stepwise Regression all Health Belief Variables for KEM, HOME, OTHER, and AYURVEDIC (Both Groups of Person with a Disability).
Independent variables in the Health Belief Model were: Susceptibility, Consequences, and Barriers. The overall multiple R for the stepwise equation to predict intention to use KEM was .42 (See Table 1). Barriers (less) emerged as the only significant predictor of this dependent variable, and no significant predictors emerged for HOME, OTHER, or, AYURVEDIC.

Independent variables in the Ecocultural Model were: Personal Normative Beliefs, Barriers, Attitudes, Karmic and Supernatural Beliefs, Beliefs about Efficacy of Disability Prevention, Beliefs of Disability Consequences, and Taxonomy. Demographic variables pertinent to the cultural context were also entered in the equation. These variables were: Severity of Disability, Size of family, Number of Children, Gender, Age, and Religion. Interaction of group by gender and religion by gender are entered in the second equation.

The overall multiple R to predict intention to use KEM was .56 (See Table 1). Three variables emerged as independent and significant predictors of KEM: (1) Barriers (less), (2) Personal Normative Beliefs (high), and (3) Beliefs of Control over Disability Prevention (high). The inclusion of the interactions did not change the initial equation.

For intention to use HOME, the overall multiple R was .47 (see Table 1). Four variables emerged as significant predictors of this treatment option: Hindu (more likely than Buddhists), Personal Normative Beliefs (low), Number of Children (low), and Number of Family Members (high). For intention to use HOME the inclusion of the interactions does not change the equation.

None of the main effects from the Ecocultural Framework were significant predictors of intention to use OTHER. However, when the interactions were added to the equation the multiple R was .29 (See Table 1) for the variable Buddhist Females (more likely than Hindu females).

For the intention to use AYURVEDIC, Karmic and Supernatural Disability Beliefs (high) emerged as the only predictor with an overall multiple R of .30 (See Table 1). The addition of the interactions hasdno significant effect on the regression equation.

Discussion

Results from the multivariate analyses of variance revealed, as hypothesized, that significant differences exist between both groups of persons with a disability for responses on all scales in each of the four models. Univariate F-tests showed significant differences for the variables Probability, Attitudes, and Beliefs of Personal Control, thus supporting initial hypotheses that the Agri sample in Juchandra would display more negative beliefs regarding disability than their urban counterparts in BDD Chawls.

The rural sample's significantly higher negative attitudes could be due to the role of beliefs of divine retribution associated with disability (Dalal & Pande, 1988; Groce, 1989). Negative attitudes towards persons with a disability could also be more prevalent among the Agri because of the increased financial importance of children being suitable marriage partners. Traditional Indian arranged marriages place a high priority on physical purity, particularly for brides. Thus, persons with a physical disability may be seen as more of a liability for the traditional rural Indian family due to the added difficulty in finding a suitable marriage partner for the disabled child.

In comparison, the modernization effects experienced in the city have perhaps led to greater opportunities for persons with a physical disability. Mental disabilities have been found to be more widely accepted than physical disabilities in many rural communities in less developed countries. This is often due to the greater opportunities for physical employment typically available to persons with cognitive handicaps in most rural communities

(Werner, 1987; Dalal & Pande, 1988). The reverse is often the case, however, in urban areas where persons with a physical disability can find better employment opportunities than someone who is cognitively challenged.

Significant gender differences were also discovered between groups of persons with a disability. Univariate F-tests showed women having more understanding of disability (higher mean scores on the Taxonomy scale), while also feeling less personal moral obligation to seek treatment (low mean scores on PN), and a greater affinity for Karmic Beliefs about Disability. These results revealed a seemingly contradictory situation in which women with a disability have greater knowledge of biomedical classification of disability, yet are less willing to seek biomedical treatment than men, and hold more supernatural beliefs concerning disability causation. One explanation for this situation is that while Indian women, as primary care givers, have more familiarity than men with the biomedical taxonomy of disability, they still receive less formal education than men and, therefore, hold more traditional illness beliefs. Similarly, as women with a disability in India are often perceived as being less important in the family than men with a disability, they are frequently denied the health care attention accorded males.

MacCormack's (1988) research in traditional agrarian based cultures has focused on health and the social power of women. In her studies, MacCormack discusses the relationship between ecology, economy, and the value placed on girls. In cultures where women do little socially acknowledged work, dowry must add value to young women to induce other families to support them. This typifies views in most of the Indian subcontinent, and is linked to a general cultural propensity to invest (nutritionally, educationally, etc.) in boys rather than girls. For example, Elliot and Sorsby (1979), found boys in the Punjab are breastfed longer, and given more food after weaning than girls. Similarly, Chand and Soni (1983) found that 20.3% of girls in a North Indian survey were severely malnourished compared with 0.1% of boys surveyed. The evidence from this research suggests that the Indian cultural gender bias against girls may extend to disability norms.

Results from the multiple stepwise regressions provide support for the Ecocultural CBR Framework of behavioral intention. The Ecocultural Framework accounted for more of the variance in predicting intention to use KEM (government rehabilitation) services, home health care, and other forms of treatment.

It must be noted that the Ecocultural Framework is partially made up of components from the three other models, and in some cases the majority of the variance was accounted for by variables from one of the other models. For example, while the Ecocultural Framework best predicted use of KEM, the variables accounting for the greatest variance were Barriers (fewer physical barriers: more likely to seek treatment at KEM) from the Health

Belief Model, and the Triandis variable Personal Normative Beliefs (more personal normative beliefs: more likely to seek treatment at KEM).

Where the Ecocultural Framework did explain the majority of the variance it could be argued that having more predictors might increase the probability of accounting for more variance by using such a broad based approach. The results, however, show that the variables particular to this Framework were often primary or secondary predictors of health seeking behavior. This was especially true for the use of alternative health sectors (Home, Other) where the majority of the variance is explained by variables specific to the Ecocultural framework.

In summary, four particular contributions were made by the Ecocultural research. The first of these is the multi-step procedure that was used for collecting culturally appropriate scale items. The second is the significant contribution that the variables attitudes (Ecocultural), barriers (HBM), and the beliefs of personal control (Triandis) made in explaining health seeking behavior. A third contribution was the uncovering of the significant role of cultural and geographic (Rural - "Tribal" vs. Urban - Hindu) variables in explaining health behavior, and the last contribution of the Ecocultural Framework was the importance of cultural context variables in predicting the use of non-biomedical health care.

There are a number of areas where the Ecocultural Framework could be applied in primary health care programs in India and other developing countries. An example is in developing health care programs that build on strengths within the community. In the case of community programs for persons with a disability this research points to sectors of the community, such as the elderly, where positive attitudes might be used as a starting place for community intervention. Another application of the Ecocultural Framework could be made in determining and integrating alternative sources of health care currently under-utilized in primary health care programs.

In general, cross-cultural health research using an ecocultural approach offers an opportunity to create more culturally sensitive, sustainable, community health care that better fits the reality of various sectors within a community. While this research examined the application of the Framework in the Indian context, the Framework could also be applied in other developing countries and in community care or outreach programs in developed countries.

References

Berry, J.W. (1980). Introduction to methodology. In H. Triandis & J. Berry (Ed.), *Handbook of cross-cultural psychology*, (vol. 2, pp. 1-28). Allyn & Bacon: Boston.

Berry, J.W. (1989a). Imposed etics-emics-derived etics: The operationalization of a compelling idea. *International Journal of Psychology, 24,* 721-735.

Berry, J.W. (1989b). *Remarks to CBR meeting*. Proceedings of the Conference Community Based Rehabilitation International Perspectives. Queen's University.

Berry, J.W., Poortinga, Y., Segall, M., & Dasen, P. (Eds.). (1991). *Cross-cultural psychology: Research and applications*. Cambridge: Cambridge University Press.

Bichmann, W., Rifkin, S., & Mathura, S. (1989). Toward the measurement of community participation. *World Health Forum, 10*, 467-472.

Chand, A., & Soni, M. (1983). Evaluation in primary health care: A case study from India. In D. Morley (Ed.), *Practicing health for all* (pp. 156-194). Oxford: Oxford University Press.

Cook, P. (1989). *Chronic illness beliefs and health seeking behavior among Chinese immigrants, Indian immigrants, and Anglo-Canadians*. Unpublished master's thesis. Queen's University, Kingston, Ontario.

Dalal, A., & Pande, N. (1988). Psychological recovery of accident victims with temporary and permanent disability. *International Journal of Psychology, 23*, 25-40.

Dasen, P.R., Berry, J.W., & Sartorius, N. (Eds.). (1988). *Health and cross-cultural psychology: Toward applications*. Beverly Hills: Sage.

Davidson, A.R., Jaccard, J.J., Triandis, H.C., Morales, M.L., & Diaz-Guerrero, R. (1976). Cross-cultural model testing: Towards a solution of the etic-emic dilemma. *International Journal of Psychology, 11*, 1-13.

Elliot, V., & Sorsby, V. (1979). *An investigation into evaluation of projects designed to benefit women*. Washington: AID/ORT 147-79-41 Focus International, Inc.

Fishbein, M. (1967). Attitudes and the prediction of behavior. In M. Fishbein (Ed.), *Readings in attitude theory and measurement* (pp. 23-56). New York: John Wiley.

Groce, N. (1989). *Traditional folk belief systems and disability: an important factor in policy planning*. Paper presented at the conference on community based rehabilitation: International perspectives. Queen's University, Kingston, On.

Harkness, S., & Super, C.M. (1987). Fertility change, child survival, and child development: Observations on a rural Kenyan community. In N. Scheper-Hughes (Ed.), *Child survival: Anthropological perspectives on the treatment and maltreatment of children* (pp. 59-70). Boston: Reidel.

Helander, E., Mendis, P., & Nelson, G. (1979). *Training disabled people in the community*. Geneva: WHO.

Jahoda, G. (1983). The cross-cultural emperor's conceptual clothes: The emic-etic issue revisited. In J.B. Deregowski, D. Dziurawic, & R.C. Annis (Eds.), *Episcations in cross-cultural psychology*. Lisse: Swets & Zeitlinger.

Janz, N.K., & Becker, M.H. (1984). The Health Belief Model: A decade later. *Health Education Quarterly, 11*, 1-47.

Kagitcibasi, C., & Berry, J. (1989). Cross-cultural psychology: Current research and trends. *Annual Review of Psychology, 40*, 493-531.

Kardiner, A., & Linton, R. (1945). *The individual and his society*. New York: Columbia University Press.

Kleinman, A. (1980). *Patients and healers in the context of culture*. Berkeley: University of California Press.

MacCormack, C.P. (1988). Health and the social power of women. *Social Science and Medicine, 26*, 677-683.

Miles, M. (1985). *Where there is no rehab plan*. Unpublished manuscript.

Noble, J. (1981). Social inequity in the prevalence of disability, projections for the year 2000. *Assignment Children, 53*, 23-32.

O'Toole, B. (1987). Community-based rehabilitation (CBR): Problems and possibilities. *European Journal of Special Needs Education, 2*, 177-190.

Pareek, U., & Venkateswara, T. (1980). Cross-cultural surveys and interviewing. In H. Triandis & J. Berry (Eds.), *Handbook of cross-cultural psychology: Vol. 2. Methodology*. Allyn & Bacon: Boston.

Peat, M. (1989). Community based Rehabilitation a viable alternative. *Synergy, 2*, 9.

Pike, R. (1966). *Language in relation to a unified theory of the structure of human behavior*. The Hague: Mouton.

Simeonsson, R.J. (1991). Early prevention of childhood disability in developing countries. *International Journal of Rehabilitation Research, 14*, 1-12.

Super, C.M., & Harkness, S. (1986). The developmental niche: A conceptualization at the interface of child and nature. *International Journal of Behavioral Development, 4*, 545-569.

Super, C.M. (1987). The role of culture in developmental disorder. In C.M. Super (Ed.), *The role of culture in developmental dysfunction* (pp. 96-121). New York: Academic Press.

Triandis, H.C. (1964). Exploratory factor analyses of the behavioral component of social attitudes. *Journal of Abnormal and Social Psychology, 68*, 420-430.

UNICEF (1993). *One in Ten, 12*, 1-7.

Whiting, J.W. (1974). *A model for psychocultural research*. Annual Report, American Anthropology Association, Washington, D.C.

Werner, D. (1987). *Disabled village children*. Palo Alto, California: The Hesperian Foundation.

World Health Organization and UNICEF (1981). *Primary health care*. Alma Alta Conference. Geneva and New York: WHO and Unicef.

AIDS: Barriers to Behavior Change in Malawi

Eilish Mc Auliffe
Institute of Public Administration, Dublin, Ireland

Malawi, which has a population of over nine million people, has one of the highest incidences of reported HIV positive cases in the world. Approximately 12 percent of the total population is estimated to be HIV positive. In Malawi, as in Africa as a whole, AIDS is mainly transmitted heterosexually and, to a lesser extent, from mothers to their unborn and newborn children (Barnett & Blackie, 1993). Although, during the mid 1980s the disease was thought to be predominantly an urban problem, evidence now exists that it has spread to rural areas. The prevalence differs greatly in the rural and urban areas, and among different sub-groups of the population. Between 10 and 11 percent of the rural population is estimated to be HIV positive. Approximately 25 percent of healthy male urban blood donors, and 31 percent of women attending urban-natal clinics have been found to be HIV positive. Furthermore 80 to 95 percent of "bargirls" (commercial sex workers) are infected; it is estimated that in 1993, 25 people in Malawi contracted HIV and five people died from AIDS every hour; the number of people dying from AIDS every hour is projected to rise to 12 by 1988 (Liomba, 1994).

In the context of the above, and the apparent persistence of high risk behaviours, UNICEF Malawi identified the need to investigate barriers to behavior change. The resulting date provides an opportunity to apply the AIDS Risk Reduction Model to the Malawian context.

Introduction

Much of the literature produced for AIDS campaigns fails to identify the context in which HIV is transmitted; high levels of preventable disease, inadequate health resources, and a background of poverty. Even correct information, presented in a manner sensitive to the cultural context of the community, may not have sufficient weight to bring about behavior change if it is not tied to the factors which influence and reinforce such change. Reports from Uganda (Sewankambo, Musgrave et al., 1991) and Zambia (Hira, 1990) suggest that providing information alone does not control the spread of HIV infection in the groups studied.

If AIDS is dependent on context, then so is the outcome of public education designed to control its spread. The problem is that the effectiveness of an AIDS/HIV education campaign is not guaranteed when people are simply exposed to or even understand the words on a poster/leaflet. The person receiving the information, even if s/he understands it, may decide that the message is not addressed to him or her because s/he considers that AIDS is not a threat to a "person like me". Condom promotion posters may be considered irrelevant because condoms are too expensive.

Health education programs are most effective when they are informed by actual health beliefs and practices. People may be more motivated to make behavioral changes if they believe that their personal and cultural views are understood (Wallman, 1988). Many existing AIDS prevention campaigns appear to underplay the complexities in turning attitude change into behavior change.

The Health Belief Model (Beaker, 1974) was among the first to propose that health behavior can be predicted from the individual's appraisal of the risks and benefits of compliance. Across a range of health behaviors there has been a wealth of evidence that people's evaluation of their susceptibility to the disease/condition, the severity of the disease/condition, the efficacy of the recommended behavior, and the barriers to compliance determine whether or not they will comply.

In the case of HIV infection, the Health Belief Model proposes that people who believe they are personally susceptible to HIV exposure, the consequences of exposure are severe, protective measures are effective in preventing transmission, and perceive few barriers to the adoption of condom use may be more likely to adopt such behavior to avoid HIV transmission. According to the Health Belief Model, people engage in a rational cost-benefit analysis when trying to decide whether to adopt preventive behavior. Although numerous studies have shown this to be the case (Jan & Becker, 1984), more recent studies have shown that risk behavior in relation to HIV is influenced by issues other than an individual's beliefs about AIDS. For example, the Hingson et al. (1990) study of adolescents reported condom usage revealed that condom use is less among heavy drinkers and frequent drug users, suggesting the limits of rational cost-benefit beliefs in determining condom use in a context where people are intoxicated. They also suggest that other factors not directly related to health beliefs about condoms and HIV may influence adolescent risk-reduction behaviors, among them factors such as ability to discuss contraception with sexual partners, influence of peers, and cultural and religious beliefs.

The Theory of Reasoned Action (Ajzen & Fishbein, 1980) which has similar components to the Health Belief Model does take account of some of these factors in that it incorporates the individual's perception of the social pressures to perform the behavior. Cochran et al. (1992) provide evidence for the utility of the Theory of Reasoned Action in explaining attitude-behaviour relationships concerning the practice of "safer sex" in gay men. Results of their study indicate that a positive attitude towards "safer sex" and a belief that important, influential referents encourage such behavior covaried with intention to practice risk reduction behaviors. Intention positively predicted enactment of risk reduction behaviors and lower levels of sexual exposure. However the Theory of Reasoned Action does not predict what will happen when immediate pressure from a sexual partner conflicts with normative peer pressure or the beliefs of influential referents.

The AIDS Risk Reduction Model (ARRM) (Catania et al., 1990) describes the *determinants* of behavioral risk reduction specifically for HIV/AIDS. This model proposes three stages of risk reduction: (1) recognition and labelling of one's sexual behaviors as high risk for contracting HIV; (2) making a commitment to reduce high risk sexual contacts and increase low risk activities; and (3) seeking and enacting strategies to obtain these goals (see Figure 1).

The ARRM is based on the assumption that in order to avoid HIV infection, people engaging in high risk activities must first perceive that their sexual behavior places them at risk for HIV infection (and is therefore problematic). However, labelling one's behavior as problematic may not in itself lead to behavior change. The model postulates that for change to occur a person must make a commitment to changing his/her behavior. This commitment process involves weighing the benefits against the costs of change. Finally, enacting change involves one's ability to help oneself or to seek help from others, through informal social support or professional help.

Figure 1. The AIDS Risk Reduction Model

AIDS Risk Reduction Model

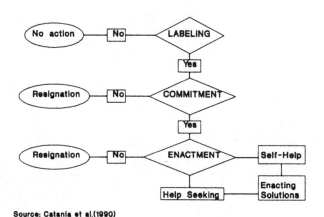

Source: Catania et al.(1990)

This study aimed to examine the variables that influence the three stages of risk reduction, as identified by the AIDS Risk Reduction Model (Catania et al.,1990), by gaining an understanding of the different conditions that influence the outcomes at each stage of the change process.

Method

This study is qualitative in nature. The methodology used is the focus group discussion (FGD). This was considered the most appropriate methodology because it provides rich qualitative data (Morgan et al., 1984; Stewart & Shamdasani, 1989), uses minimal resources, and because the concept of a group discussion is a culturally appropriate means of obtaining information from Malawian communities. Each group discussion was conducted with 6 to 10 participants.

Sample

The FGDs attempted to gain group-specific information. Each focus group therefore consisted of a single-sexed "homogenous" group of people. Table 1 shows the diverse categories of participants, which were chosen because they are distinctive demographic groups with potentially different levels of risk for contracting HIV. For each category three FGDs were conducted, one in each of the three administrative regions of Malawi. As is common with the use of focus group methodology, the sample selected was a convenience sample. Two research assistants screened households in each village until the required number of participants was recruited. One person from every third household was selected.

Table 1. Category of FGD Participant by Number of Participants and Age Range.

Category of Group	Total Number of Participants (3 FGDS)	Age Range
Out-of-school youth	27	10 - 18
Primary school youth	27	11 - 18
Secondary school youth	27	13 - 19
Urban married men	27	23 - 48
Rural married men	24	23 - 53
Polygamously married men	25	25 - 66
Urban married women	25	20 - 45
Rural married women	28	20 - 45
Polygamously married women	30	20 - 56
Total	240	10 - 66

Procedure

For each FGD, between six and ten participants gathered at a quiet meeting place, away from the distractions of their everyday tasks. The participants were seated in a circle, either on chairs or on mats, with the FGD facilitators forming part of the circle. Facilitator were social science students from the University of Malawi who had undergone a five-day training course in the use of focus group methodology. A recorder (also a social science student with training in FGDs) sat outside of the group circle, facing the facilitator and took detailed notes as the discussion proceeded.

Immediately after the discussion finished the facilitator and recorder wrote a one-page summary of their observations and impressions of the discussion. They then wrote a detailed report in English on the content of the discussion using the recorder's notes. The discussions were conducted in the vernacular language (Chitumbuku in the northern region; Chichewa in the central and southern regions). The duration of the discussions was 60 - 90 minutes. The following topics formed the agenda for discussion in all of the FGDs.

(1) *AIDS knowledge and practices*: knowledge about AIDS; sources from which AIDS knowledge was obtained; methods/media of AIDS-related communication having greatest impact; sources from which one would like to receive further information on AIDS/HIV; knowledge of methods of prevention; knowledge of routes of transmission; other perceptions (misinformation) about the nature of HIV infection or AIDS; perceived risk HIV infection; personal risk reduction behaviors; and barriers to risk reduction behaviors.

(2) *Cultural rules regarding sex and fidelity*: marital and extramarital sex; attitudes towards fidelity and infidelity; and social norms in sexual behavior.

(3) *Individual attitudes towards sex and fidelity*: marital and extramarital sex; attitudes towards fidelity and infidelity; and sexual behavior (including age at first intercourse, sexual practices etc.).

(4) *Condom knowledge and use*: attitudes and practices regarding condoms; and condom availability.

The FGDs were content analyzed, key concepts were identified within each discussion topic; and the most frequently mentioned concepts were presented graphically in matrix form.

Results

Results for the categories were combined and presented in three categories: youth (10 - 19 years old); men; and women. Where notable differences were found in subcategories these are reported.

Barriers to Labelling Behavior as Problematic

All participants were very knowledgeable about the routes of HIV transmission. Thus they were able to identify the behaviors which place them at high risk of contracting HIV. Table 2 shows the identified routes of HIV transmission and the categories of participants who identified each of these routes. Although all groups identified the most common routes of transmission, in each of the groups there were participants who mentioned mythical routes of transmission (Table 2). Belief in these AIDS myths suggests that people are labelling much of their everyday behavior as problematic.

Barriers to the Commitment to Behavior Change

Out-of-school males were of the opinion that AIDS cannot be prevented because there are many ways of contracting it. Discussion about behavior change indicates that those youth who have not changed have not done so because, "there are many ways of getting AIDS apart from through sexual intercourse", an example being "if someone defecates in the upper part of a stream and someone drinks below the point he can get AIDS." These concerns indicate that youth do not fully understand the mechanism by which the HIV virus is transmitted, nor do they seem to understand that certain behaviors are more high risk than others.

Table 2. Routes of HIV Transmission Identified by Participants

Routes Of HIV Transmission	Youth	Men	Women
Sexual intercourse	***	***	***
Sharing razor blades	***	***	***
Unsterilized injection	***	***	***
Receiving infected blood	***	***	***
Sharing toothbrushes	***		***
Sharing sewing needles/pins/scissors	**		
HIV positive mother breastfeeding	*		
Borrowing clothes	*	*	**
Promiscuity	**		
Kissing	*		
Touching wound of AIDS victim	*		
Mosquitoes		*	*
Drinking from same cup or beer carton	*	*	

* No. of groups who mentioned this route

Women in polygamous marriages were concerned because their husbands' other wives may get AIDS from other men and pass it on to them through the husband. Many of these women felt it is difficult for polygamously married women to change their behavior because they lack things like soap (the implication being that they need to exchange sex for such commodities).

Some urban women held the belief that everyone has the AIDS virus, but it is inert and can only be activated through sexual intercourse. By implication if one wants to prevent AIDS one should totally avoid sexual intercourse.

In general, the concerns of rural married men seemed to focus on contracting the virus from needles and syringes, razor blades, one's wife, and from flies that might have landed on the wound of an AIDS victim. The participants from one group said there is a danger of contracting AIDS if one touches the fluid from the woman's vagina when removing the condom. Many said that it is difficult to prevent AIDS in the rural areas because people always touch the corpse of an AIDS victim and can get AIDS in this way. Some men also felt that condoms did not offer protection from HIV as "it only protects the penis, leaving the rest of the body at risk". One participant argue that it is difficult to prevent AIDS in a family because, "...your sperms just end up in the condom and the wife will be sick because she is not satisfied, and as a result she will be going out with other men which she will be satisfied with and from there she will get AIDS".

Urban men also felt that it is difficult to prevent AIDS within the family. They argued that if one uses a condom he "can be regarded as barren". They argued that even if you stick to one partner one is still at risk because "the wife can get AIDS through injections at the hospital". Many of these men felt that AIDS cannot be prevented because there are so many ways of contracting it. "One can use a condom, even three at one act, but one can still get AIDS in other ways." It is evident from the discussions that these men not only feel there are many ways of contracting AIDS but are also confused about how the virus is transmitted. This is illustrated by the following: "AIDS is caused by a virus. At school we learned that viruses only survive on living tissue. How do they survive on razor blades?"

Barriers to the Enactment of Behavior Change

The main route of HIV transmission in Malawi is heterosexual intercourse. Therefore the main behavior change promoted by the National AIDS Control Programme is condom usage. However very few people mentioned condom use when asked about changes in their behavior. Also the recent Demographic and Health Survey in Malawi (1992) indicates that only 7.2 percent of Malawian men are currently using condoms. This study therefore focused on the reasons for low levels of condom usage, which would constitute a major form of enactment.

Attitudes towards condom use among the youth groups were generally negative. One person expressed a concern that, "when removing the condom after use you can catch it and then don't wash your hands. Then you can eat food using these hands so you can contract it". Many raised concerns about condoms bursting or penetrating into the vagina and becoming lodged there. Several myths abound concerning condom use, for example that the lubricant causes tuberculosis and cancer. A male pupil said, "the fluid which is in the condom can be destructive to the body cells".

Communication about condom use is also poor among these groups, as mentioning condom use in a relationship is thought to raise suspicions from both partners. Table 3 summarizes, according to the different categories of male and female FGDs, opinions on who should suggest condom use, and why.

The majority of youth believe that it is the female's responsibility to suggest condom use, while it is the male's responsibility to provide condoms for use in a sexual relationship.

Table 3. Youth's Reasons why Males/Females should Suggest Condom Use

Male Groups	Female Groups
Female	Female
- she is the one who can be impregnated	- men are usually carefree
- if she gets AIDS she will have problems	- if man refuses woman should force him
- she knows her menstruation and fertile periods	- she can prevent herself from STDs/AIDS
- she knows when she has a disease	- it is woman who is afraid of STDs
- she knows the conditions to use it	- boys are not trustworthy
- fear of pregnancy and AIDS	- can have many sexual partners and get diseases
- men fear condom will remain in her womb	
Male	Male
- man has responsibilities in family	- because woman would be ashamed to suggest
- has the power to command	
- he is the one who demands sex	- he is the one who buys them
- he is the head of the family	- he is afraid if transmitting diseases to women
- he is the leader	
- he is the one who makes things happen	- he knows more about condoms
- a woman can't command a man	- he is the one who doesn't trust the woman
- men look for many sexual partners	
- a man has the power to command	
Both	Both
- they both have to prevent AIDS	- if one forgets the other can remind him/her
	- because either cannot be trustworthy
	- none of them would like to get AIDS

Tables 3 and 4 also portray the perceived power imbalance in male-female relationships. This imbalance may be a major barrier, not only to condom use but to male-female communication within sexual relationships. Also there appears to be much suspicion and mistrust between the sexes: "boys are not trustworthy"; "he is the one who doesn't trust the woman".

Several of the women had not seen a condom prior to this discussion (research assistants unwrapped a condom during the discussion about condom use). Communication about condom usage is very poor in the marriages of these women. Some have suggested to their husbands that they use condoms for child spacing, but the majority of husbands refuse. The most commonly cited reasons why people may not want to use condoms are summarised in Table 5.

Women were also asked their opinions on who should suggest and provide condoms for use in a relationship. Their responses are summarized in Table 6. The majority of polygamously married women and rural married women believe that it is the female's responsibility to both suggest and provide condoms for use in a relationship whereas the majority of urban married women believe it is the male's responsibility.

Table 4 Youth's Reasons why Males/Females should Provide Condoms

Male Groups	Female Groups
Female	Female
- she knows if conception is possible at that time	- many men don't think about using condoms
- she wants child spacing	- she cannot know what a man has in his body
- she should give it to man so that he can wear it	- men don't like using condoms
- if she knows she has an STD she should provide	
- she knows when she can become pregnant	
Male	Male
- he is the one who wears it	- people may suspect female if she buys condoms
- difficult for woman to buy	- she may be considered a prostitute if buying condoms
- the man commands something to take place	- he has money to buy them
- a man forces a woman to have sex	- he puts it on
- when the two people are not married	- he asks for sex
- he is supposed to prevent AIDS	- he is the head of the family
- he is the head of the family	
Both	Both
- in case either one forgets	- both may contract diseases

All men who participated in the group discussions had heard about condoms, although some of the rural men had not previously seen a condom. Many reasons for not using condoms were forthcoming from these men, the most common being that sex with a condom is not "sweet" or pleasurable (see Table 7). Discussion about condom use indicates that men are not aware of exactly how condoms prevent HIV infection.

Attitudes towards condoms were rather negative, and condoms were strongly associated with promiscuity or child spacing (see Table 8). Consequently communication about condom use within relationships is poor. Many fear that suggesting condom use within their marital relationships would raise suspicions of unfaithfulness for both partners. Some of the civil service employees indicated that they use condoms with their girlfriends, but not with their wives.

Table 5. Barriers to Condom Use as Identified by Women

Category	Barriers To Condom Use
Polygamously married women	fear condom will slip into vagina men don't like condoms men sometimes hole the condoms women want children men afraid of developing sores men will not use because penis too big men don't want to discard sperms
Urban married women	faithful to spouse takes time to put condom on do not like sperm remaining in condom not pleasurable
Rural married women	fear of condom bursting fear condom will lodge in vagina sex is not nice with condom some believe doctors put medicine in condoms to make people infertile men refuse to use condoms ignorance of benefits of condom

Table 6. Women's Reasons why Males/Females should Suggest and Provide Condoms

Who Should Suggest	Who Should Provide
Female - men are unfaithful - when she is nursing a child - she can get diseases - men forget - knows when to use it - for childspacing	Female - for childspacing - men forget
Male - he asks for sex - he puts it on - he is the head of the family	Male - he is head of family - he asks for sex - he puts it on - he buys condoms

Discussion

It is encouraging that the majority of participants who took part in the focus group discussions are knowledgeable about AIDS and were able to identify many of their own behaviors as being high risk for HIV transmission. However it is evident from the discussions that participants, by contrast, possess little understanding of the relative risk associated with different routes of transmission. People expressed the futility of using condoms, when there are so many alternative ways of contracting HIV. Most of the groups seemed equally, if not more concerned about contracting HIV through razor blades as through sexual intercourse.

Table 7. Barriers to Condom Use as Identified by Men

Category	Barriers To Condom Use
Polygamously married men	sex with condom is not sweet
people want children
don't know how to use condoms
slips into woman's vagina
condoms burst
some are ignorant about condoms
Civil service men	"you don't eat sweets in a wrapper"
sex is not pleasurable with condom
sores develop on penis if condom is expired
condoms are unhealthy
condoms smell
condoms encourage promiscuity
Rural married men	fear of condom bursting
fear condom will lodge in vagina
knowledge of other methods of child-spacing
condoms give false sense of security
sex should be "skin to skin"
condoms are a waste of money
man should not wear a condom with his wife

Many of the youth indicated that they have changed their behavior since hearing about AIDS. Unfortunately, those who have not changed seem to have a sense of hopelessness about the AIDS situation, because they believe there are so many ways of contracting AIDS. Again, it seems that although youth are aware of routes of transmission, they possess little understanding of the relative risk associated with different routes of transmission. The participants do not realize for instance that having protected sex will reduce risk considerably more than using your own toothbrush or razor blade.

The women who participated in the study labelled their behaviors and the behaviors of other women in the community as high-risk in terms of contracting HIV. However, many of these women, particularly those in polygamous marriages seem to have difficulty making the commitment to change. If these women are engaging in high-risk behavior in order to meet basic subsistence needs, it may be difficult for them to perceive the benefits of risk-reduction behaviors as outweighing the costs. They may rightly perceive the benefits of risk-reduction behavior as long-term survival, whilst at the same time perceive that the cost jeopardizes their (and their childrens') short-term survival.

Table 8. Men's Reasons why Males/Females should Suggest and Provide Condoms

Who Should Suggest	Who Should Provide
Female	Female
- for childspacing	- for childspacing
- men are unfaithful	
- when she is nursing a child	
- knows when to use it	
- to avoid pregnancy	
- when nursing child	
- when menstruating	
Male	Male
- he asks for sex	- he asks for sex
- he is the head of the family	- he is head of family
- he puts it on	- he buys condom
- men fear AIDS/STDs	- he puts it on

The men who participated in this study seemed to be focused on the idea that there are many ways of contracting HIV and therefore there is little point in using condoms. This indicates that these men are having difficulty estimating the relative costs and benefits of different behavior changes, because they lack information on the relative risks associated with the various routes of transmission. They recognize that their behaviors are problematic or high-risk. However, they lack the information necessary to decide which particular behaviors place them at highest risk. It is feasible therefore that their perception may be that any behavior change they instigate is not likely to completely protect them against HIV infection. If the benefits of behavior change are not fully appreciated, a strong commitment to change may be difficult to achieve with these men.

The ARRM model discussed in the Introduction section of this report proposes three stages of risk reduction: (1) recognition and labelling of one's sexual behaviors as high risk for contracting HIV; (2) making a commitment to reduce high risk sexual contacts and increase low risk activities; and (3) seeking and enacting strategies to obtain these goals (see Figure 1). In terms of the ARRM it seems that people have progressed through stage one of the model in that they recognize and label their behaviors as being problematic. However some people appear to have become stuck at this stage, not being able to make the commitment to behavior change that is required to move through stage two of the model. In order to make a commitment to change it is necessary to undertake a cost-benefit analysis of various options or behavior changes. This creates a difficulty for people, as they cannot estimate the costs and benefits of specific behaviors in the absence of knowledge about relative risks of the various transmission routes. The consequence is that some people become exasperated, decide it is too difficult to make all the necessary changes to fully protect oneself, and therefore perceive little benefit in making one change because of the perceived magnitude of risk from other sources. It is strongly recommended that in order to overcome this problem, education materials that emphasize the relative risk (as far as it is known) associated with each of the transmission routes should be produced and disseminated.

Another factor which may be influencing people's difficulty in making a commitment to change is the "commonsense epidemiology of the layperson" phenomenon, as described by Jemmott et al. (1988). In a study of 110 undergraduates they found that subjects who rated a particular disease as more prevalent also rated that disease as less life-threatening. Thus it may be that in Malawi, where HIV prevalence is high and is perceived to be high, it is commonly perceived as less life-threatening than other diseases.

Youth's opinions on who should suggest and provide condoms in a sexual relationship portray the perceived power imbalance in male-female relationships. This imbalance may be a major barrier, not only to condom use but to male-female communication within sexual relationships (psychological barrier). There appears to be much suspicion and mistrust between the sexes.

Many of the married women who took part in this study feel powerless to suggest condoms to husbands whom they strongly suspect of unfaithfulness. Men do not use condoms with their wives because they believe condoms are for extra-marital affairs. Throughout the discussions with women several references were made to the exchange of sex for money to meet basic needs. This practice of occasionally exchanging sex for money to make ends meet (economic barrier) creates further power imbalance between the sexes.

Even if some of these women can make the commitment to behavior change, many of them lack the context necessary to effect the change. A further factor influencing these women's decision and ability to change is their perceived norms regarding sexual relationships within their societies. Many indicated that sexual exchange is common within their communities. This lends an element of legitimacy to the practice and in so doing adds to the difficulty of behavior change (cultural barriers).

The ARRM suggests that one of the factors which influences the decision to change is having the skills and self-efficacy to bring about the behavior change. The findings of this study, particularly those from the FGDs with women, illustrate that perceived self-efficacy may be strongly influenced by norms and cultural values. This is particularly true in the context of sexual relationships. Women believe that the male, as head of the family, is the decision maker, and therefore if condoms are to be introduced in the relationship, the male has to make this decision (psychological barrier). In cultures, such as the Malawian culture, where there is an obvious power imbalance between the sexes, enactment of behavior change proves extremely difficult. Effective health education may produce a commitment to behavior change. However if this commitment is to lead to enactment the powerless position of women needs to be directly addressed.

Community interventions have met with some success in increasing safe sex practices amongst gay men (Kelly et al., 1989, 1990, 1992, 1993). In particular, the use of trained opinion leaders by Kelly et al. (1993) appears to have resulted in significant reduction in the percentage of men who engaged in unprotected anal intercourse. It is possible that similar interventions could have an impact on the adoption of safer sexual practices in Malawi, particularly as opinion leaders are prevalent in the Malawian culture. Small group counselling appears to have had similar success amongst the gay community (Kelly et al., 1989). This could also be employed as a method to increase self-efficacy, particularly amongst Malawian women. However, it is important that any such intervention involves both sexes in order to address the underlying problem of the power imbalance between the sexes.

The AIDS Relative Risk Evaluation Model outlines the variables which influence behavior change in the Malawian context. The model highlights the steps to behavior change: identification of risk behavior; assessment of relative risk; desire to reduce high risk behavior; and overcoming the barriers to change. AIDS prevention programs in Malawi need to produce AIDS education materials that assist people to assess the relative risk from each of the HIV transmission routes. Failure to do so will result in frustration or fatalism, attitudes which are not conducive to change. If the person can accurately assess relative risk it is suggested that s/he will then engage in a cost-benefit analysis of behaviour change. If the costs of behavior change outweigh the benefits, the person may decide not to change. If the benefits outweigh the costs, this leads to a desire to reduce high risk behavior. In order to move forward from this point in the model the individual becomes dependent on others. Overcoming the social barriers to behavior change requires a contextual change rather than an individual one. The individual may therefore be rendered powerless at this stage of the process. Malawi AIDS prevention programs need to find ways of reducing these social barriers in order to create an enabling environment which allows the individual to achieve behavior change.

References

Ajzen, I., & Fishbein, M. (1980). *Understanding attitudes and predicting social behavior*. Englewood Cliffs, N. J: Prentice-Hall.

Becker, M.H. (1974). The Health Belief Model and sick role behavior. *Health Education Monographs*, 2, 409-419.

Barnett, T., & Blackie, P. (1993). Simple methods for monitoring the socio-economic impact of AIDS: Lessons from Sub-Saharan Africa. In S. Cross & A. Whiteside, (Eds) *Facing up to AIDS: The Socio-economic impact in Southern Africa*. New York: St. Martin's Press.

Catania, J.A., Kegeles, S.M., & Coates, T.J. (1990). Towards an understanding of risk behavior: An AIDS risk reduction model (ARRM). *Health Education Quarterly*, 17, 53-72.

Cochran, S.D., Mays, V.M., Ciaretta, J., Caruso, C., & Mallon, D. (1992). Efficacy of the theory of reasoned action in predicting AIDS-related sexual risk reduction among gay men. *Journal of Applied Social Psychology*, 22, 1482-1501.

Hingson, R.W., Strunin, L., Berlin, B.M., & Heeren, T. (1990). Beliefs about AIDS, use of alcohol and drugs, and unprotected sex among Massachusetts adolescents. *Journal of Public Health*, 8, 295-299.

Hira, S.K. (1990). Epidemiology of human immunodeficiency. Virus in families in Lusaka, Zambia. *Journal of AIDS*, 3, 383-386.

Jans, N.K., & Beeker, M.H. (1984). The Health belief model. A decade later. *Health education Quarterly*, 11, 1-47.

Jemmort, J.B., Croyle, R.T., & Ditto P.H. (1988). Commonsense epidemiology: Self-based judgements from laypersons and physicians. *Health Psychology*, 7, 55-73.

Kelly, J.A., St. Lawrence, J.S., Hood, H.V., & Brasfield, T.L. (1989). Behavioural Interventions to reduce AIDS risk activities. *Journal of Consulting Psychology*, 60-67.

Kelly, J.A., St. Lawrence, J.S., Betts, R., Brasfield, T.L., & Hood, H.V. (1990). Skills-training group intervention model to assist persons in reducing risk behaviours for HIV infection. *AIDS and Education Prevention*, 2, 24-35.

Kelly, J.A., St. Lawrence, J.S., Stevenson, L.Y. (1992). Community AIDS/HIV risk reduction: the effects of endorsements by popular people in three cities. *American Journal of Public Health*, 82, 1483-1489.

Kelly, J.A., Winett, R.A., & Roffman, R.A. (1993). *Social diffusion models can produce population-level HIV risk-behaviour reduction: Field trial results and mechanisms underlying change*. IX International Conference on AIDS/HIV STD World Congress, Berlin.

Liomba, G. (1994). *National AIDS Control Programme. 1993 Statistics*. Aids Secretariat, Ministry of Health, Malawi.

Malawi demographic and Health Survey (1992). National Statistics Office, Zomba, Malawi & Marco International Inc. Calverton, Maryland, USA.

Morgan, D.L., Spanish, M.T. (1984). Focus Groups: A new tool for qualitative research. *Qualitative Psychology* 7, 253-270.

Sewankambo, N.K., Musgrave, S., Wawer, M.J., & Serwadda, D. (1991). *Preliminary HIV-1 incidence rates in Rakai district, Uganda*. VI International Conference on AIDS in Africa, Dakar, Senegal.
Stewart, D.W., Shamdasani, P.N. (1989). *Applied Social research Methods Series: Vol 20. Focus Groups Therapy and Practice*. London: Sage Publications.
Wallman, S. (1988). Sex and death: The AIDS crisis in social and cultural context. *Journal Acquired Immune Deficieny Syndrome, I,* 571-578.

Acknowledgement

This research was commissioned and funded by UNICEF, Malawi.

Note

Mc Auliffe, E. (1994) AIDS: The Barriers to Behaviour Change. Report commissioned by UNICEF, Malawi.

Previous books in the series:

Selected Papers from the International Conference of the International Association for Cross-Cultural Psychology (IACCP)

Diversity and Unity in Cross-Cultural Psychology
5th International Conference, Bhubaneswar India, 1981
Editors: R. Rath, H.S. Asthana, J.B.H. Sinha and D. Sinha
1982, 380 pages, ISBN 90 265 0431 4

Expiscations in Cross-Cultural Psychology
6th International Conference, Aberdeen Scotland, 1982
Editors: J.B. Deregowski, S. Dziurawiec and R.C. Annis
1983, 460 pages, ISBN 90 265 0450 0

From a Different Perspective
7th International Conference, Acapulco Mexico, 1984
Editors: I. Reyes Lagunes and Y. Poortinga
1985, 396 pages, ISBN 90 265 0672 4

Growth and Progress in Cross-Cultural Psychology
8th International Conference, Istanbul Turkey, 1986
Editor: Ç. Kağitçibaşi
1987, 418 pages, ISBN 90 265 0852 2

Heterogeneity in Cross-Cultural Psychology
9th International Conference, Newcastle Australia., 1988
Editors: D.M. Keats, D. Munro and L. Mann
1989, 592 pages, ISBN 90 265 1018 7

Innovations in Cross-Cultural Psychology: Selected Papers
10th International Conference, Nara Japan, 1990
Editors: S. Iwawaki, Y. Kashima and K. Leung
1992, 492 pages, ISBN 90 265 1232 5

Journeys into Cross-Cultural Psychology
11th International Conference, Liège Belgium, 1992
Editors: A.M. Bouvy, F.J.R. van de Vijver, P. Boski and P. Schmitz
1994, 410 pages, ISBN 90 265 1403 4